CAMBRIDGE MONOGRAPHS ON
APPLIED AND COMPUTATIONAL
MATHEMATICS

21 Spectral Methods for Time-Dependent Problems

The *Cambridge Monographs on Applied and Computational Mathematics* reflects the crucial role of mathematical and computational techniques in contemporary science. The series publishes expositions on all aspects of applicable and numerical mathematics, with an emphasis on new developments in this fast-moving area of research.

State-of-the-art methods and algorithms as well as modern mathematical descriptions of physical and mechanical ideas are presented in a manner suited to graduate research students and professionals alike. Sound pedagogical presentation is a prerequisite. It is intended that books in the series will serve to inform a new generation of researchers.

Also in this series:

1. A Practical Guide to Pseudospectral Methods, *Bengt Fornberg*
2. Dynamical Systems and Numerical Analysis, *A. M. Stuart and A. R. Humphries*
3. Level Set Methods and Fast Marching Methods, *J. A. Sethian*
4. The Numerical Solution of Integral Equations of the Second Kind, *Kendall E. Atkinson*
5. Orthogonal Rational Functions, *Adhemar Bultheel, Pablo González-Vera, Erik Hendiksen, and Olav Njåstad*
6. The Theory of Composites, *Graeme W. Milton*
7. Geometry and Topology for Mesh Generation, *Herbert Edelsbrunner*
8. Schwarz-Christoffel Mapping, *Tofin A. Driscoll and Lloyd N. Trefethen*
9. High-Order Methods for Incompressible Fluid Flow, *M. O. Deville, P. F. Fischer, and E. H. Mund*
10. Practical Extrapolation Methods, *Avram Sidi*
11. Generalized Riemann Problems in Computational Fluid Dynamics, *Matania Ben-Artzi and Joseph Falcovitz*
12. Radial Basis Functions: Theory and Implementations, *Martin D. Buhmann*
13. Iterative Krylov Methods for Large Linear Systems, *Henk A. van der Vorst*
14. Simulating Hamiltonian Dynamics, *Ben Leimkuhler and Sebastian Reich*
15. Collocation Methods for Volterra Integral and Related Functional Equations, *Hermann Brunner*
16. Topology for computing, *Afra J. Zomordia*
17. Scattered Data Approximation, *Holger Wendland*
19. Matrix Preconditioning Techniques and Applications, *Ke Chen*
22. The Mathematical Foundations of Mixing, *Rob Sturman, Julio M. Ottino and Stephen Wiggins*

Spectral Methods for Time-Dependent Problems

JAN S. HESTHAVEN

Brown University

SIGAL GOTTLIEB

University of Massachusetts, Dartmouth

DAVID GOTTLIEB

Brown University

CAMBRIDGE
UNIVERSITY PRESS

CAMBRIDGE UNIVERSITY PRESS
Cambridge, New York, Melbourne, Madrid, Cape Town,
Singapore, São Paulo, Delhi, Tokyo, Mexico City

Cambridge University Press
The Edinburgh Building, Cambridge CB2 8RU, UK

Published in the United States of America by Cambridge University Press, New York

www.cambridge.org
Information on this title: www.cambridge.org/9780521792110

First published 2007

A catalogue record for this publication is available from the British Library

ISBN 978-0-521-79211-0 Hardback

For our children and grandchildren

Contents

	Introduction	1
1	**From local to global approximation**	5
1.1	Comparisons of finite difference schemes	9
1.2	The Fourier spectral method: first glance	16
2	**Trigonometric polynomial approximation**	19
2.1	Trigonometric polynomial expansions	19
2.2	Discrete trigonometric polynomials	24
2.3	Approximation theory for smooth functions	34
3	**Fourier spectral methods**	43
3.1	Fourier–Galerkin methods	43
3.2	Fourier–collocation methods	48
3.3	Stability of the Fourier–Galerkin method	52
3.4	Stability of the Fourier–collocation method for hyperbolic problems I	54
3.5	Stability of the Fourier–collocation method for hyperbolic problems II	58
3.6	Stability for parabolic equations	62
3.7	Stability for nonlinear equations	64
3.8	Further reading	65
4	**Orthogonal polynomials**	66
4.1	The general Sturm–Liouville problem	67
4.2	Jacobi polynomials	69
5	**Polynomial expansions**	79
5.1	The continuous expansion	79
5.2	Gauss quadrature for ultraspherical polynomials	83

5.3 Discrete inner products and norms 88
5.4 The discrete expansion 89

6 Polynomial approximation theory for smooth functions 109
6.1 The continuous expansion 109
6.2 The discrete expansion 114

7 Polynomial spectral methods 117
7.1 Galerkin methods 117
7.2 Tau methods 123
7.3 Collocation methods 129
7.4 Penalty method boundary conditions 133

8 Stability of polynomial spectral methods 135
8.1 The Galerkin approach 135
8.2 The collocation approach 142
8.3 Stability of penalty methods 145
8.4 Stability theory for nonlinear equations 150
8.5 Further reading 152

9 Spectral methods for nonsmooth problems 153
9.1 The Gibbs phenomenon 154
9.2 Filters 160
9.3 The resolution of the Gibbs phenomenon 174
9.4 Linear equations with discontinuous solutions 182
9.5 Further reading 186

10 Discrete stability and time integration 187
10.1 Stability of linear operators 188
10.2 Standard time integration schemes 192
10.3 Strong stability preserving methods 197
10.4 Further reading 202

11 Computational aspects 204
11.1 Fast computation of interpolation and differentiation 204
11.2 Computation of Gaussian quadrature points and weights 210
11.3 Finite precision effects 214
11.4 On the use of mappings 225

12 Spectral methods on general grids 235
12.1 Representing solutions and operators on general grids 236
12.2 Penalty methods 238

12.3 Discontinuous Galerkin methods 246
12.4 References and further reading 248

Appendix A Elements of convergence theory 249

Appendix B A zoo of polynomials 252
B.1 Legendre polynomials 252
B.2 Chebyshev polynomials 255

Bibliography 260
Index 272

Introduction

The purpose of this book is to collect, in one volume, all the ingredients necessary for the understanding of spectral methods for time-dependent problems, and, in particular, hyperbolic partial differential equations. It is intended as a graduate-level text, covering not only the basic concepts in spectral methods, but some of the modern developments as well. There are already several excellent books on spectral methods by authors who are well-known and active researchers in this field. This book is distinguished by the exclusive treatment of time-dependent problems, and so the derivation of spectral methods is influenced primarily by the research on finite-difference schemes, and less so by the finite-element methodology. Furthermore, this book is unique in its focus on the stability analysis of spectral methods, both for the semi-discrete and fully discrete cases. In the book we address advanced topics such as spectral methods for discontinuous problems and spectral methods on arbitrary grids, which are necessary for the implementation of pseudo-spectral methods on complex multi-dimensional domains.

In Chapter 1, we demonstrate the benefits of high order methods using phase error analysis. Typical finite difference methods use a local stencil to compute the derivative at a given point; higher order methods are then obtained by using a wider stencil, i.e., more points. The Fourier spectral method is obtained by using all the points in the domain. In Chapter 2, we discuss the trigonometric polynomial approximations to smooth functions, and the associated approximation theory for both the continuous and the discrete case. In Chapter 3, we present Fourier spectral methods, using both the Galerkin and collocation approaches, and discuss their stability for both hyperbolic and parabolic equations. We also present ways of stabilizing these methods, through super viscosity or filtering.

Chapter 4 features a discussion of families of orthogonal polynomials which are eigensolutions of a Sturm–Liouville problem. We focus on the Legendre and Chebyshev polynomials, which are suitable for representing functions on finite domains. In this chapter, we present the properties of Jacobi polynomials, and their associated recursion relations. Many useful formulas can be found in this chapter. In Chapter 5, we discuss the continuous and discrete polynomial expansions based on Jacobi polynomials; in particular, the Legendre and Chebyshev polynomials. We present the Gauss-type quadrature formulas, and the different points on which each is accurate. Finally, we discuss the connections between Lagrange interpolation and electrostatics. Chapter 6 presents the approximation theory for polynomial expansions of smooth functions using the ultraspherical polynomials. Both the continuous and discrete expansions are discussed. This discussion sets the stage for Chapter 7, in which we introduce polynomial spectral methods, useful for problems with non-periodic boundary conditions. We present the Galerkin, tau, and collocation approaches and give examples of the formulation of Chebyshev and Legendre spectral methods for a variety of problems. We also introduce the penalty method approach for dealing with boundary conditions. In Chapter 8 we analyze the stability properties of the methods discussed in Chapter 7.

In the final chapters, we introduce some more advanced topics. In Chapter 9 we discuss the spectral approximations of non-smooth problems. We address the Gibbs phenomenon and its effect on the convergence rate of these approximations, and present methods which can, partially or completely, overcome the Gibbs phenomenon. We present a variety of filters, both for Fourier and polynomial methods, and an approximation theory for filters. Finally, we discuss the resolution of the Gibbs phenomenon using spectral reprojection methods. In Chapter 10, we turn to the issues of time discretization and fully discrete stability. We discuss the eigenvalue spectrum of each of the spectral spatial discretizations, which provides a necessary, but not sufficient, condition for stability. We proceed to the fully discrete analysis of the stability of the forward Euler time discretization for the Legendre collocation method. We then present some of the standard time integration methods, especially the Runge–Kutta methods. At the end of the chapter, we introduce the class of strong stability preserving methods and present some of the optimal schemes. In Chapter 11, we turn to the computational issues which arise when using spectral methods, such as the use of the fast Fourier transform for interpolation and differentiation, the efficient computation of the Gauss quadrature points and weights, and the effect of round-off errors on spectral methods. Finally, we address the use of mappings for treatment of non-standard intervals and for improving accuracy in the computation of higher order derivatives. In Chapter 12, we talk about the implementation of spectral methods on general grids. We discuss how the penalty method formulation enables the use of spectral methods on general grids in one dimension, and in complex domains in multiple dimensions, and illustrate this using both the Galerkin and collocation approaches. We also show how penalty methods allow us to easily generalize to complicated boundary conditions and on triangular meshes. The discontinuous Galerkin method is an alternative way of deriving these schemes, and penalty methods can thus be used to construct methods based on multiple spectral elements.

Chapters 1, 2, 3, 5, 6, 7, 8 of the book comprise a complete first course in spectral methods, covering the motivation, derivation, approximation theory and stability analysis of both Fourier and polynomial spectral methods. Chapters 1, 2, and 3 can be used to introduce Fourier methods within a course on numerical solution of partial differential equations. Chapters 9, 10, 11, and 12 address advanced topics and are thus suitable for an advanced course in spectral methods. However, depending on the focus of the course, many other combinations are possible.

A good resource for use with this book is PseudoPack. PseudoPack Rio and PseudoPack 2000 are software libraries in Fortran 77 and Fortran 90 (respectively) for numerical differentiation by pseudospectral methods, created by Wai Sun Don and Bruno Costa. More information can be found at http://www.labma.ufrj.br/bcosta/pseudopack/main.html and http://www.labma.ufrj.br/bcosta/pseudopack2000/main.html.

As the oldest author of this book, I (David Gottlieb) would like to take a paragraph or so to tell you my personal story of involvement in spectral methods. This is a personal narrative, and therefore may not be an accurate history of spectral methods. In 1973 I was an instructor at MIT, where I met Steve Orszag, who presented me with the problem of stability of polynomial methods for hyperbolic equations. Working on this, I became aware of the pioneering work of Orszag and his co-authors and of Kreiss and his co-authors on Fourier spectral methods. The work on polynomial spectral methods led to the book *Numerical Analysis of Spectral Methods: Theory and Applications* by Steve Orszag and myself, published by SIAM in 1977. At this stage, spectral methods enjoyed popularity among the practitioners, particularly in the meteorology and turbulence community. However, there were few theoretical results on these methods. The situation changed after the summer course I gave in 1979 in France. P. A. Raviart was a participant in this course, and his interest in spectral methods was sparked. When he returned to Paris he introduced his postdoctoral researchers, Claudio Canuto and Alfio Quarteroni, and his students, Yvon Maday and Christine Bernardi, to these methods. The work of this European group led to an explosion of spectral methods, both in theory and applications. After this point, the field became too extensive to further review it. Nowadays, I particularly enjoy the experience of receiving a paper on spectral methods which I do not understand. This is an indication of the maturity of the field.

The following excellent books can be used to deepen one's understanding of many aspects of spectral methods. For a treatment of spectral methods for incompressible flows, the interested reader is referred to the classical book by C. Canuto, M. Y. Hussaini, A. Quarteroni and T. A. Zang, *Spectral Methods: Fundamentals in single domains* (2006), the more recent *Spectral Methods for Incompressible Viscous Flow* (2002) by R. Peyret and the modern text *High-Order Methods in Incompressible Fluid Flows* (2002) by M. Deville, P. F. Fischer, and E. Mund (2002). The book *Spectral/hp Methods in Computational Fluid Dynamics*, by G. E. Karniadakis and S. J. Sherwin (2005), deals with many important practical aspects of spectral methods computations for large scale fluid dynamics application. A comprehensive discussion of approximation theory may be found in *Approximations Spectrales De Problemes Aux Limites Elliptiques* (1992) by C. Bernardi and Y. Maday and in *Polynomial Approximation of Differential Equations* (1992) by D. Funaro. Many interesting results can be found in the book by B. -Y. Guo, *Spectral Methods and their Applications* (1998). For those wishing to implement spectral methods in Matlab, a good supplement to this book is *Spectral Methods in Matlab* (2000), by L. N. Trefethen.

For the treatment of spectral methods as a limit of high order finite difference methods, see *A Practical Guide to Pseudospectral Methods* (1996) by B. Fornberg. For a discussion of spectral methods to solve boundary value and eigenvalue problems, as well as Hermite, Laguerre, rational Chebyshev, sinc, and spherical harmonic functions, see *Chebyshev and Fourier Spectral Methods* (2000) by J. P. Boyd.

This text has as its foundation the work of many researchers who make up the vibrant spectral methods community. A complete bibliography of spectral methods is a book in and of itself. In our list of references we present only a partial list of those papers which have direct relevance to the text. This necessary process of selection meant that many excellent papers and books were excluded. For this, we apologize.

1

From local to global approximation

Spectral methods are global methods, where the computation at any given point depends not only on information at neighboring points, but on information from the entire domain. To understand the idea of a global method, we begin by considering local methods, and present the global Fourier method as a limit of local finite difference approximations of increasing orders of accuracy. We will introduce phase error analysis, and using this tool we will show the merits of high-order methods, and in particular, their limit: the Fourier method. The phase error analysis leads to the conclusion that high-order methods are beneficial for problems requiring well resolved fine details of the solution or long time integrations.

Finite difference methods are obtained by approximating a function $u(x)$ by a local polynomial interpolant. The derivatives of $u(x)$ are then approximated by differentiating this local polynomial. In this context, *local* refers to the use of nearby grid points to approximate the function or its derivative at a given point.

For slowly varying functions, the use of local polynomial interpolants based on a small number of interpolating grid points is very reasonable. Indeed, it seems to make little sense to include function values far away from the point of interest in approximating the derivative. However, using low-degree local polynomials to approximate solutions containing very significant spatial or temporal variation requires a very fine grid in order to accurately resolve the function. Clearly, the use of fine grids requires significant computational resources in simulations of interest to science and engineering. In the face of such limitations we seek alternative schemes that will allow coarser grids, and therefore fewer computational resources. Spectral methods are such methods; they use *all* available function values to construct the necessary approximations. Thus, they are *global* methods.

Example 1.1 Consider the wave equation

$$\frac{\partial u}{\partial t} = -2\pi \frac{\partial u}{\partial x} \quad 0 \leq x \leq 2\pi, \tag{1.1}$$

$$u(x, 0) = e^{\sin(x)},$$

with periodic boundary conditions.

The exact solution to Equation (1.1) is a right-moving wave of the form

$$u(x, t) = e^{\sin(x - 2\pi t)},$$

i.e., the initial condition is propagating with a speed 2π.

In the following, we compare three schemes, each of different order of accuracy, for the solution of Equation (1.1) using the uniform grid

$$x_j = j\Delta x = \frac{2\pi j}{N+1}, \quad j \in [0, \ldots, N]$$

(where N is an even integer).

Second-order finite difference scheme A quadratic local polynomial interpolant to $u(x)$ in the neighborhood of x_j is given by

$$u(x) = \frac{1}{2\Delta x^2}(x - x_j)(x - x_{j+1})u_{j-1} - \frac{1}{\Delta x^2}(x - x_{j-1})(x - x_{j+1})u_j + \frac{1}{2\Delta x^2}(x - x_{j-1})(x - x_j)u_{j+1}. \tag{1.2}$$

Differentiating this formula yields a second-order centered-difference approximation to the derivative du/dx at the grid point x_j:

$$\left.\frac{du}{dx}\right|_{x_j} = \frac{u_{j+1} - u_{j-1}}{2\Delta x}.$$

High-order finite difference scheme Similarly, differentiating the interpolant based on the points $\{x_{j-2}, x_{j-1}, x_j, x_{j+1}, x_{j+2}\}$ yields the fourth-order centered-difference scheme

$$\left.\frac{du}{dx}\right|_{x_j} = \frac{1}{12\Delta x}(u_{j-2} - 8u_{j-1} + 8u_{j+1} - u_{j+2}).$$

Global scheme Using all the available grid points, we obtain a global scheme. For each point x_j we use the interpolating polynomial based on the points $\{x_{j-k}, \ldots, x_{j+k}\}$ where $k = N/2$. The periodicity of the problem furnishes us with the needed information at any grid point. The derivative at the grid points is calculated using a matrix-vector product

$$\left.\frac{du}{dx}\right|_{x_j} = \sum_{i=0}^{N} \tilde{\mathrm{D}}_{ji} u_i,$$

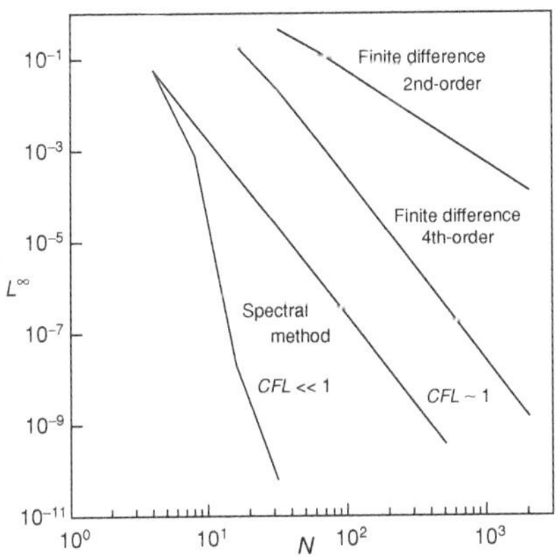

Figure 1.1 The maximum pointwise (L^∞) error of the numerical solution of Example 1.1, measured at $t = \pi$, obtained using second-order, fourth-order and global spectral schemes as a function of the total number of points, N. Here the Courant–Friedrichs–Lewy coefficient, CFL $= \Delta t/\Delta x$.

where the entries of the matrix is $\tilde{\mathrm{D}}$ are

$$\tilde{\mathrm{D}}_{ji} = \begin{cases} \frac{(-1)^{j+i}}{2}\left[\sin\left(\frac{(j-i)\pi}{N+1}\right)\right]^{-1} & i \neq j, \\ 0 & i = j. \end{cases}$$

The formal proof of this will be given in Section 1.1.3. This approach is equivalent to an infinite-order finite difference method as well as a Fourier spectral method.

To advance Equation (1.1) in time, we use the classical fourth-order Runge–Kutta method with a sufficiently small time-step, Δt, to ensure stability

Now, let's consider the dependence of the maximum pointwise error (the L^∞-error) on the number of grid points N. In Figure 1.1 we plot the L^∞-error at $t = \pi$ for an increasing number of grid points. It is clear that the higher the order of the method used for approximating the spatial derivative, the more accurate it is. Indeed, the error obtained with $N = 2048$ using the second-order finite difference scheme is the same as that computed using the fourth-order method with $N = 128$, or the global method with only $N = 12$. It is also evident that by lowering Δt for the global method one can obtain even more accurate results, i.e., the error in the global scheme is dominated by time-stepping errors rather than errors in the spatial discretization.

Figure 1.2 shows a comparison between the local second-order finite difference scheme and the global method following a long time integration. Again, we clearly observe that the global scheme is superior in accuracy to the local scheme, even though the latter scheme employs 20 times as many grid points and is significantly slower.

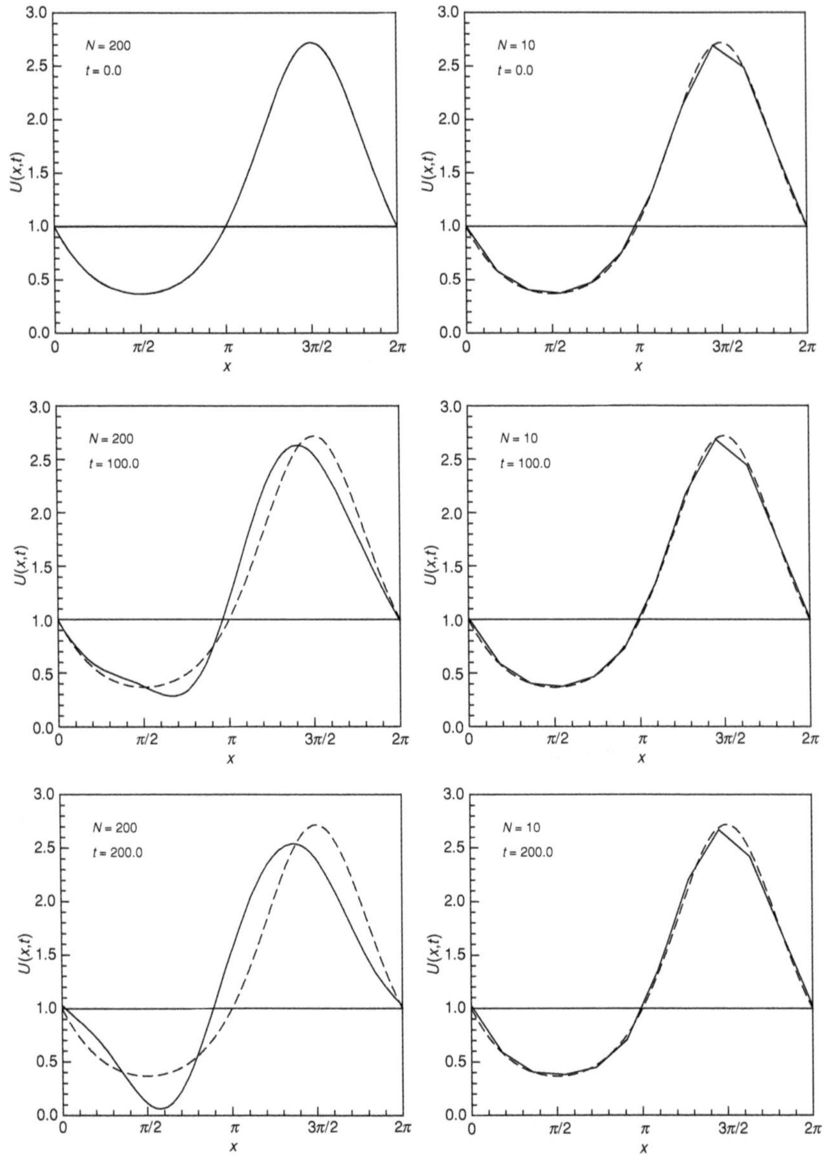

Figure 1.2 An illustration of the impact of using a global method for problems requiring long time integration. On the left we show the solution of Equation (1.1) computed using a second-order centered-difference scheme. On the right we show the same problem solved using a global method. The full line represents the computed solution, while the dashed line represents the exact solution.

1.1 Comparisons of finite difference schemes

The previous example illustrates that global methods are superior in performance to local methods, not only when very high spatial resolution is required but also when long time integration is important. In this section, we shall introduce the concept of phase error analysis in an attempt to clarify the observations made in the previous section. The analysis confirms that high-order and/or global methods are a better choice when very accurate solutions or long time integrations on coarse grids are required. It is clear that the computing needs of the future require both.

1.1.1 Phase error analysis

To analyze the phase error associated with a particular spatial approximation scheme, let's consider, once again, the linear wave problem

$$\begin{aligned} \frac{\partial u}{\partial t} &= -c\frac{\partial u}{\partial x} \quad 0 \leq x \leq 2\pi, \\ u(x,0) &= e^{ikx}, \end{aligned} \tag{1.3}$$

with periodic boundary conditions, where $i = \sqrt{-1}$ and k is the wave number. The solution to Equation (1.3) is a travelling wave

$$u(x,t) = e^{ik(x-ct)}, \tag{1.4}$$

with phase speed c.

Once again, we use the equidistant grid

$$x_j = j\Delta x = \frac{2\pi j}{N+1}, \quad j \in [0, \ldots, N].$$

The $2m$-order approximation of the derivative of a function $f(x)$ is

$$\left.\frac{df}{dx}\right|_{x_j} = \sum_{n=1}^{m} \alpha_n^m \mathcal{D}_n f(x_j), \tag{1.5}$$

where

$$\mathcal{D}_n f(x_j) = \frac{f(x_j + n\Delta x) - f(x_j - n\Delta x)}{2n\Delta x} = \frac{f_{j+n} - f_{j-n}}{2n\Delta x}, \tag{1.6}$$

and the weights, α_n^m, are

$$\alpha_n^m = -2(-1)^n \frac{(m!)^2}{(m-n)!(m+n)!}. \tag{1.7}$$

In the semi-discrete version of Equation (1.3) we seek a vector $\boldsymbol{v} = (v_0(t), \ldots, v_N(t))$ which satisfies

$$\frac{dv_j}{dt} = -c \sum_{n=1}^{m} \alpha_n^m \mathcal{D}_n v_j, \tag{1.8}$$

$$v_j(0) = e^{ikx_j}.$$

We may interpret the grid vector, $\boldsymbol{v}$, as a vector of grid point values of a trigonometric polynomial, $v(x, t)$, with $v(x_j, t) = v_j(t)$, such that

$$\frac{\partial v}{\partial t} = -c \sum_{n=1}^{m} \alpha_n^m \mathcal{D}_n v(x, t), \tag{1.9}$$

$$v(x, 0) = e^{ikx}.$$

If $v(x, t)$ satisfies Equation (1.9), the solution to Equation (1.8) is given by $v(x_j, t)$. The solution to Equation (1.9) is

$$v(x, t) = e^{ik(x - c_m(k)t)}, \tag{1.10}$$

where $c_m(k)$ is the numerical wave speed. The dependence of c_m on the wave number k is known as the dispersion relation.

The phase error $e_m(k)$, is defined as the leading term in the relative error between the actual solution $u(x, t)$ and the approximate solution $v(x, t)$:

$$\left| \frac{u(x, t) - v(x, t)}{u(x, t)} \right| = \left| 1 - e^{ik(c - c_m(k))t} \right| \simeq |k(c - c_m(k))t| = e_m(k).$$

As there is no difference in the amplitude of the two solutions, the phase error is the dominant error, as is clearly seen in Figure 1.2.

In the next section we will compare the phase errors of the schemes in Example 1.1. In particular, this analysis allows us to identify the most efficient scheme satisfying the phase accuracy requirement over a specified period of time.

1.1.2 Finite-order finite difference schemes

Applying phase error analysis to the second-order finite difference scheme introduced in Example 1.1, i.e.,

$$\frac{\partial v(x, t)}{\partial t} = -c \frac{v(x + \Delta x, t) - v(x - \Delta x, t)}{2\Delta x},$$

$$v(x, 0) = e^{ikx},$$

we obtain the numerical phase speed

$$c_1(k) = c\frac{\sin(k\Delta x)}{k\Delta x}.$$

For $\Delta x \ll 1$,

$$c_1(k) = c\left(1 - \frac{(k\Delta x)^2}{6} + \mathcal{O}((k\Delta x)^4)\right),$$

confirming the second-order accuracy of the scheme.

For the fourth-order scheme considered in Example 1.1,

$$\frac{\partial v(x,t)}{\partial t} = -\frac{c}{12\Delta x}(v(x-2\Delta x,t) - 8v(x-\Delta x,t) + 8v(x+\Delta x,t) - v(x+2\Delta x,t)),$$

we obtain

$$c_2(k) = c\frac{8\sin(k\Delta x) - \sin(2k\Delta x)}{6k\Delta x}.$$

Again, for $\Delta x \ll 1$ we recover the approximation

$$c_2(k) = c\left(1 - \frac{(k\Delta x)^4}{30} + \mathcal{O}((k\Delta x)^6)\right),$$

illustrating the expected fourth-order accuracy.

Denoting $e_1(k,t)$ as the phase error of the second-order scheme and $e_2(k,t)$ as the phase error of the fourth-order scheme, with the corresponding numerical wave speeds $c_1(k)$ and $c_2(k)$, we obtain

$$e_1(k,t) = kct\left|1 - \frac{\sin(k\Delta x)}{k\Delta x}\right|, \tag{1.11}$$

$$e_2(k,t) = kct\left|1 - \frac{8\sin(k\Delta x) - \sin(2k\Delta x)}{6k\Delta x}\right|.$$

When considering wave phenomena, the critical issue is the number p of grid points needed to resolve a wave. Since the solution of Equation (1.3) has k waves in the domain $(0, 2\pi)$, the number of grid points is given by

$$p = \frac{N+1}{k} = \frac{2\pi}{k\Delta x}.$$

Note that it takes a minimum of two points per wavelength to uniquely specify a wave, so p has a theoretical minimum of 2.

It is evident that the phase error is also a function of time. In fact, the important quantity is not the time elapsed, but rather the number of times the solution returns to itself under the assumption of periodicity. We denote the number of periods of the phenomenon by $\nu = kct/2\pi$.

Rewriting the phase error in terms of p and ν yields

$$e_1(p,\nu) = 2\pi\nu\left|1 - \frac{\sin(2\pi p^{-1})}{2\pi p^{-1}}\right|, \tag{1.12}$$
$$e_2(p,\nu) = 2\pi\nu\left|1 - \frac{8\sin(2\pi p^{-1}) - \sin(4\pi p^{-1})}{12\pi p^{-1}}\right|.$$

The leading order approximation to Equation (1.12) is

$$e_1(p,\nu) \simeq \frac{\pi\nu}{3}\left(\frac{2\pi}{p}\right)^2, \tag{1.13}$$
$$e_2(p,\nu) \simeq \frac{\pi\nu}{15}\left(\frac{2\pi}{p}\right)^4,$$

from which we immediately observe that the phase error is directly proportional to the number of periods ν i.e., the error grows linearly in time.

We arrive at a more straightforward measure of the error of the scheme by introducing $p_m(\varepsilon_p,\nu)$ as a measure of the number of points per wavelength required to guarantee a phase error, $e_p \le \varepsilon_p$, after ν periods for a $2m$-order scheme. Indeed, from Equation (1.13) we directly obtain the lower bounds

$$p_1(\varepsilon,\nu) \ge 2\pi\sqrt{\frac{\nu\pi}{3\varepsilon_p}}, \tag{1.14}$$
$$p_2(\varepsilon,\nu) \ge 2\pi\sqrt[4]{\frac{\pi\nu}{15\varepsilon_p}},$$

on p_m, required to ensure a specific error ε_p.

It is immediately apparent that for long time integrations (large ν), $p_2 \ll p_1$, justifying the use of high-order schemes. In the following examples, we will examine the required number of points per wavelength as a function of the desired accuracy.

Example 1.2

$\varepsilon_p = \mathbf{0.1}$ Consider the case in which the desired phase error is $\le 10\%$. For this relatively large error,

$$p_1 \ge 20\sqrt{\nu}, \quad p_2 \ge 7\sqrt[4]{\nu}.$$

We recall that the fourth-order scheme is twice as expensive as the second-order scheme, so not much is gained for short time integration. However, as ν increases the fourth-order scheme clearly becomes more attractive.

$\varepsilon_p = \mathbf{0.01}$ When the desired phase error is within 1%, we have

$$p_1 \ge 64\sqrt{\nu}, \quad p_2 \ge 13\sqrt[4]{\nu}.$$

Here we observe a significant advantage in using the fourth-order scheme, even for short time integration.

$\varepsilon_p = \mathbf{10^{-5}}$ This approximately corresponds to the minimum error displayed in Figure 1.1. We obtain

$$p_1 \geq 643\sqrt{\nu}, \quad p_2 \geq 43\sqrt[4]{\nu},$$

as is observed in Figure 1.1, which confirms that high-order methods are superior when high accuracy is required.

Sixth-order method As an illustration of the general trend in the behavior of the phase error, we give the bound on $p_3(\varepsilon_p, \nu)$ for the sixth-order centered-difference scheme as

$$p_3(\varepsilon_p, \nu) > 2\pi \sqrt[6]{\frac{\pi \nu}{70\varepsilon_p}},$$

for which the above special cases become

$$p_3(0.1, \nu) = 5\sqrt[6]{\nu}, \quad p_3(0.01, \nu) = 8\sqrt[6]{\nu}, \quad p_3(10^{-5}, \nu) = 26\sqrt[6]{\nu},$$

confirming that when high accuracy is required, a high-order method is the optimal choice. Indeed, sixth-order schemes are the methods of choice for many modern turbulence calculations.

While the number of points per wavelength gives us an indication of the merits of high-order schemes, the true measure of efficiency is the work needed to obtain a predefined bound on the phase error. We thus define a measure of work per wavelength when integrating to time t,

$$W_m = 2m \times p_m \times \frac{t}{\Delta t}.$$

This measure is a product of the number of function evaluations $2m$, points per wavelength $p_m = \frac{2\pi}{k\Delta x}$, and the total number of time steps $\frac{t}{\Delta t}$. Using $\nu = \frac{kct}{2\pi}$ we obtain

$$\begin{aligned} W_m &= 2mp_m \frac{t}{\Delta t} \\ &= 2mp_m \frac{2\pi\nu}{kc\Delta t} \\ &= 2mp_m \frac{2\pi\nu}{kc\frac{\Delta t}{\Delta x}\Delta x} \\ &= 2mp_m \frac{2\pi\nu}{kCFL_m\Delta x} \\ &= 2mp_m \frac{\nu p_m}{CFL_m}, \end{aligned}$$

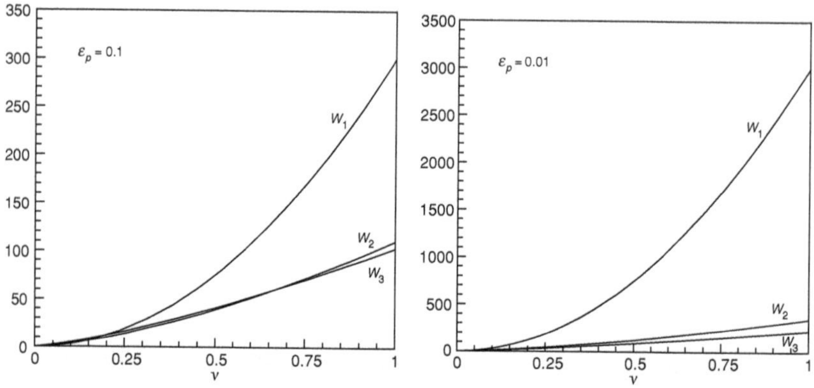

Figure 1.3 The growth of the work function, W_m, for various finite difference schemes is given as a function of time, ν, in terms of periods. On the left we show the growth for a required phase error of $\varepsilon_p = 0.1$, while the right shows the result of a similar computation with $\varepsilon_p = 0.01$, i.e., a maximum phase error of less than 1%.

where $CFL_m = c\frac{\Delta t}{\Delta x}$ refers to the CFL bound for stability. We assume that the fourth-order Runge–Kutta method will be used for time discretization. For this method it can be shown that $CFL_1 = 2.8$, $CFL_2 = 2.1$, and $CFL_3 = 1.75$. Thus, the estimated work for second, fourth, and sixth-order schemes is

$$W_1 \simeq 30\nu\frac{\nu}{\varepsilon_p}, \quad W_2 \simeq 35\nu\sqrt{\frac{\nu}{\varepsilon_p}}, \quad W_3 \simeq 48\nu\sqrt[3]{\frac{\nu}{\varepsilon_p}}. \tag{1.15}$$

In Figure 1.3 we illustrate the approximate work associated with the different schemes as a function of required accuracy and time. It is clear that even for short time integrations, high-order methods are the most appropriate choice when accuracy is the primary consideration. Moreover, it is evident that for problems exhibiting unsteady behavior and thus needing long time integrations, high-order methods are needed to minimize the work required for solving the problem.

1.1.3 Infinite-order finite difference schemes

In the previous section, we showed the merits of high-order finite difference methods for time-dependent problems. The natural question is, what happens as we take the order higher and higher? How can we construct an infinite-order scheme, and how does it perform?

In the following we will show that the limit of finite difference schemes is the global method presented in Example 1.1. In analogy to Equation (1.5), the infinite-order method is given by

$$\left.\frac{du}{dx}\right|_{x_j} = \sum_{n=1}^{\infty} \alpha_n^{\infty} \frac{u_{j+n} - u_{j-n}}{2n\Delta x}.$$

To determine the values of α_n^{∞}, we consider the function e^{ilx}. The approximation formula should be exact for all such trigonometric polynomials. Thus, α_n^{∞} should satisfy

$$\begin{aligned} ile^{ilx} &= \sum_{n=1}^{\infty} \alpha_n^{\infty} \frac{e^{i(x+n\Delta x)l} - e^{i(x-n\Delta x)l}}{2n\Delta x} \\ &= \sum_{n=1}^{\infty} \alpha_n^{\infty} \frac{e^{in\Delta xl} - e^{-in\Delta xl}}{2n\Delta x} e^{ilx} \\ &= \sum_{n=1}^{\infty} \alpha_n^{\infty} \frac{2i \sin(nl\Delta x)}{2n\Delta x} e^{ilx}, \end{aligned}$$

so

$$l = \sum_{n=1}^{\infty} \alpha_n^{\infty} \frac{\sin(nl\Delta x)}{n\Delta x}.$$

We denote $l\Delta x = \xi$, to emphasize that α_n^{∞}/n are the coefficients of the Fourier sine expansion of ξ,

$$\xi = \sum_{n=1}^{\infty} \frac{\alpha_n^{\infty}}{n} \sin(n\xi),$$

and are therefore given by $\alpha_n^{\infty} = 2(-1)^{n+1},\ n \geq 1$. Extending this definition over the integers, we get

$$\alpha_n^{\infty} = \begin{cases} 2(-1)^{n+1} & n \neq 0 \\ 0 & n = 0. \end{cases}$$

Substituting $\Delta x = \frac{2\pi}{N+1}$, we obtain

$$\left.\frac{du}{dx}\right|_{x_j} = \frac{N+1}{4\pi} \sum_{n=-\infty}^{\infty} \frac{\alpha_n^{\infty}}{n} u_{j+n}.$$

As we assume that the function $u(x, t)$ is 2π-periodic, we have the identity

$$u_{j+n} = u_{j+n+p(N+1)}, \quad p = 0, \pm 1, \pm 2 \ldots$$

Rearranging the summation in the approximation yields

$$\begin{aligned}
\left.\frac{du}{dx}\right|_{x_j} &= \frac{N+1}{4\pi}\sum_{n=-j}^{N-j}\left(\sum_{p=-\infty}^{\infty}\frac{\alpha^{\infty}_{n+p(N+1)}}{n+p(N+1)}\right)u_{j+n} \\
&= \frac{1}{2\pi}\sum_{n=-j}^{N-j}-(-1)^n\left(\sum_{p=-\infty}^{\infty}\frac{(-1)^{p(N+1)}}{p+n/(N+1)}\right)u_{j+n} \\
&= \frac{1}{2\pi}\sum_{n=-j}^{N-j}-(-1)^n\left(\sum_{p=-\infty}^{\infty}\frac{(-1)^{p}}{p+n/(N+1)}\right)u_{j+n}.
\end{aligned}$$

Using the identity $\sum_{k=-\infty}^{\infty}\frac{(-1)^k}{k+x}=\frac{\pi}{\sin(\pi x)}$, and the substitution $i=j+n$,

$$\begin{aligned}
\left.\frac{du}{dx}\right|_{x_j} &= \frac{1}{2\pi}\sum_{n=-j}^{N-j}-(-1)^n\frac{\pi}{\sin(\pi n/(N+1))}u_{j+n} \\
&= \sum_{i=0}^{N}\frac{1}{2}(-1)^{j+i}\left[\sin\left(\frac{\pi}{N+1}(j-i)\right)\right]^{-1}u_i.
\end{aligned}$$

Hence, we obtain the remarkable result that the infinite-order finite difference approximation of the spatial derivative of a periodic function can be exactly implemented through the use of the differentiation matrix, $\tilde{\mathrm{D}}$, as we saw in Example 1.1. As we shall see in the next section, the exact same formulation arises from the application of Fourier spectral methods. As we shall also see, the number of points per wavelength for the global scheme attains the minimum of $p_\infty = 2$ for a well-resolved wave.

1.2 The Fourier spectral method: first glance

An alternative way of obtaining the global method is to use trigonometric polynomials to interpolate the function $f(x)$ at the points x_l,

$$f_N(x)=\sum_{l=0}^{N}f(x_l)h_l(x),$$

where the Lagrange trigonometric polynomials are

$$h_l(x)=\frac{1}{N+1}\frac{\sin\left(\frac{N+1}{2}(x-x_l)\right)}{\sin\left(\frac{1}{2}(x-x_l)\right)}=\sum_{k=-\frac{N}{2}}^{\frac{N}{2}}e^{ik(x-x_l)}.$$

The interpolating polynomial is thus exact for all trigonometric polynomials of degree $N/2$.

In a Fourier spectral method applied to the partial differential equation

$$\frac{\partial u}{\partial t} = -c\frac{\partial u}{\partial x} \quad 0 \le x \le 2\pi,$$
$$u(x,0) = e^{ikx}$$

with periodic boundary conditions, we seek a trigonometric polynomial of the form

$$v(x,t) = \sum_{l=0}^{N} v(x_l,t)h_l(x),$$

such that

$$\frac{\partial v}{\partial t} = -c\frac{\partial v}{\partial x}, \tag{1.16}$$

at the points $x = x_j, \quad j = 0, \ldots, N.$

The derivative of v at the points x_j is given by

$$\left.\frac{\partial v}{\partial x}\right|_{x_j} = \sum_{l=0}^{N} v(x_l,t)h_l'(x_j),$$

where

$$h_l'(x_j) = \frac{(-1)^{j+l}}{2}\left[\sin\left(\frac{\pi}{N+1}(j-l)\right)\right]^{-1}.$$

Thus,

$$\left.\frac{dv}{dt}\right|_{x_j} = -c\sum_{l=0}^{N}\frac{(-1)^{j+l}}{2}\left[\sin\left(\frac{\pi}{N+1}(j-l)\right)\right]^{-1} v(x_l,t).$$

The initial condition $v(x,0)$ is the interpolant of the function e^{ikx},

$$v(x,0) = \sum_{l=0}^{N} e^{ikx_l}h_l(x).$$

This is a Fourier–collocation method, which is identical to the infinite-order finite difference scheme derived above.

It is important to note that since both sides of Equation (1.16) are trigonometric polynomials of degree $N/2$, and agree at $N+1$ points, they must be identically equal, i.e.,

$$\frac{\partial v}{\partial t} = -c\frac{\partial v}{\partial x} \quad \text{for all } x.$$

This is a fundamental argument in spectral methods, and will be seen frequently in the analysis.

Note that if $k \leq N/2$, the same argument implies that $v(x, 0) = e^{ikx}$, and therefore

$$v(x, t) = e^{ik(x-ct)}.$$

Thus, we require two points per wavelength ($N/k = p_\infty \geq 2$) to resolve the initial conditions, and thus to resolve the wave. The spectral method requires only the theoretical minimum number of points to resolve a wave. Let's think about how spectral methods behave when an insufficient number of points is given: in the case, $N = 2k - 2$, the spectral approximation to the initial condition e^{ikx} is uniformly zero. This example gives the unique flavor of spectral methods: when the number of points is insufficient to resolve the wave, the error does not decay. However, as soon as a sufficient number of points are used, we see infinite-order convergence.

Note also that, in contrast to finite difference schemes, the spatial discretization does not cause deterioration in terms of the phase error as time progresses. The only source contributing to phase error is the temporal discretization.

1.3 Further reading

The phase error analysis first appears in a paper by Kreiss and Oliger (1972), in which the limiting Fourier case was discussed as well. These topics have been further explored in the texts by Gustafsson, Kreiss, and Oliger (1995), and by Fornberg (1996).

2

Trigonometric polynomial approximation

The first spectral methods computations were simulations of homogeneous turbulence on periodic domains. For that type of problem, the natural choice of basis functions is the family of (periodic) trigonometric polynomials. In this chapter, we will discuss the behavior of these trigonometric polynomials when used to approximate smooth functions. We will consider the properties of both the continuous and discrete Fourier series, and come to an understanding of the factors determining the behavior of the approximating series.

We begin by discussing the classical approximation theory for the continuous case, and continue with the more modern theory for the discrete Fourier approximation.

For the sake of simplicity, we will consider functions of only one variable, $u(x)$, defined on $x \in [0, 2\pi]$. Also, we restrict ourselves in this chapter to functions having a continuous periodic extension, i.e., $u(x) \in C_p^0[0, 2\pi]$. In Chapter 9, we will discuss the trigonometric series approximation for functions which are non-periodic, or discontinuous but piecewise smooth. We shall see that although trigonometric series approximations of piecewise smooth functions converge very slowly, the approximations contain high-order information which is recoverable through postprocessing.

2.1 Trigonometric polynomial expansions

The classical continuous series of trigonometric polynomials, the Fourier series $F[u]$ of a function, $u(x) \in L^2[0, 2\pi]$, is given as

$$F[u] = \hat{a}_0 + \sum_{n=1}^{\infty} \hat{a}_n \cos(nx) + \sum_{n=1}^{\infty} \hat{b}_n \sin(nx), \tag{2.1}$$

where the expansion coefficients are

$$\hat{a}_n = \frac{1}{c_n \pi} \int_0^{2\pi} u(x) \cos(nx)\, dx,$$

with the values

$$c_n = \begin{cases} 2 & n = 0, \\ 1 & n > 0, \end{cases}$$

and

$$\hat{b}_n = \frac{1}{\pi} \int_0^{2\pi} u(x) \sin(nx)\, dx, \quad n > 0.$$

Alternatively, the Fourier series can be expressed in complex form

$$F[u] = \sum_{|n| \leq \infty} \hat{u}_n e^{inx}, \tag{2.2}$$

with expansion coefficients

$$\hat{u}_n = \frac{1}{2\pi} \int_0^{2\pi} u(x) e^{-inx}\, dx = \begin{cases} \hat{a}_0 & n = 0, \\ (\hat{a}_n - i\hat{b}_n)/2 & n > 0, \\ (\hat{a}_{-n} + i\hat{b}_{-n})/2 & n < 0. \end{cases} \tag{2.3}$$

Remark The following special cases are of interest.

1. If $u(x)$ is a real function, the coefficients $\hat{a}_n$ and $\hat{b}_n$ are real numbers and, consequently, $\hat{u}_{-n} = \overline{\hat{u}_n}$. Thus, only half the coefficients are needed to describe the function.
2. If $u(x)$ is real and even, i.e., $u(x) = u(-x)$, then $\hat{b}_n = 0$ for all values of n, so the Fourier series becomes a cosine series.
3. If $u(x)$ is real and odd, i.e., $u(x) = -u(-x)$, then $\hat{a}_n = 0$ for all values of n, and the series reduces to a sine series.

For our purposes, the relevant question is how well the *truncated* Fourier series approximates the function. The truncated Fourier series

$$\mathcal{P}_N u(x) = \sum_{|n| \leq N/2} \hat{u}_n e^{inx}, \tag{2.4}$$

is a projection to the finite dimensional space

$$\hat{\mathsf{B}}_N = \text{span}\{e^{inx} \mid |n| \leq N/2\}, \quad \dim(\hat{\mathsf{B}}_N) = N + 1.$$

The approximation theory results for this series are classical.

Theorem 2.1 *If the sum of squares of the Fourier coefficients is bounded*

$$\sum_{|n|\leq\infty} |\hat{u}_n|^2 < \infty \tag{2.5}$$

then the truncated series converges in the L^2 norm

$$\|u - \mathcal{P}_N u\|_{L^2[0,2\pi]} \to 0 \quad as \quad N \to \infty.$$

If, moreover, the sum of the absolute values of the Fourier coefficients is bounded

$$\sum_{|n|<\infty} |\hat{u}_n| < \infty,$$

then the truncated series converges uniformly

$$\|u - \mathcal{P}_N u\|_{L^\infty[0,2\pi]} \to 0 \quad as \quad N \to \infty.$$

The fact that the truncated sum converges implies that the error is dominated by the tail of the series, i.e.,

$$\|u - \mathcal{P}_N u\|^2_{L^2[0,2\pi]} = 2\pi \sum_{|n|>N/2} |\hat{u}_n|^2,$$

and

$$\|u - \mathcal{P}_N u\|_{L^\infty[0,2\pi]} \leq \sum_{|n|>N/2} |\hat{u}_n|.$$

Thus, the error committed by replacing $u(x)$ with its Nth-order Fourier series depends solely on how fast the expansion coefficients of $u(x)$ decay. This, in turn, depends on the regularity of $u(x)$ in $[0, 2\pi]$ and the periodicity of the function and its derivatives.

To appreciate this, let's consider a continous function $u(x)$, with derivative $u'(x) \in L^2[0, 2\pi]$. Then, for $n \neq 0$,

$$\begin{aligned} 2\pi\, \hat{u}_n &= \int_0^{2\pi} u(x) e^{-inx}\, dx \\ &= \frac{-1}{in}(u(2\pi) - u(0)) + \frac{1}{in}\int_0^{2\pi} u'(x) e^{-inx}\, dx. \end{aligned}$$

Clearly, then,

$$|\hat{u}_n| \propto \frac{1}{n}.$$

Repeating this line of argument we have the following result for periodic smooth functions.

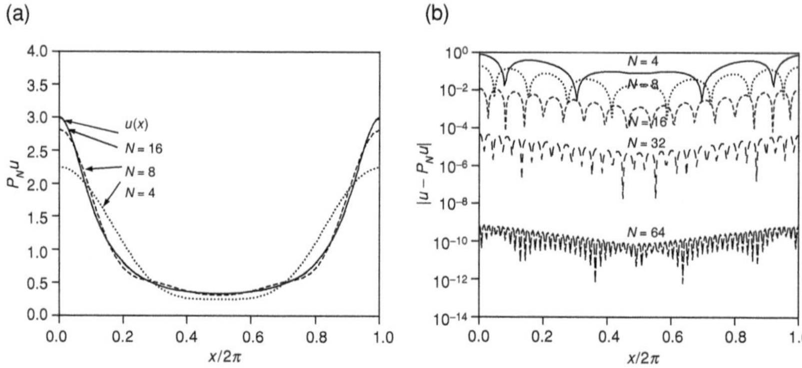

Figure 2.1 (a) Continuous Fourier series approximation of Example 2.3 for increasing resolution. (b) Pointwise error of the approximation for increasing resolution.

Theorem 2.2 *If a function $u(x)$, its first $(m-1)$ derivatives, and their periodic extensions are all continuous and if the mth derivative $u^{(m)}(x) \in L^2[0, 2\pi]$, then $\forall n \neq 0$ the Fourier coefficients, $\hat{u}_n$, of $u(x)$ decay as*

$$|\hat{u}_n| \propto \left(\frac{1}{n}\right)^m.$$

What happens if $u(x) \in C_p^\infty[0, 2\pi]$? In this case $\hat{u}_n$ decays faster than any negative power of n. This property is known as *spectral convergence*. It follows that the smoother the function, the faster the truncated series converges. Of course, this statement is asymptotic; as we showed in Chapter 1, we need at least two points per wavelength to reach the asymptotic range of convergence.

Let us consider a few examples.

Example 2.3 Consider the $C_p^\infty[0, 2\pi]$ function

$$u(x) = \frac{3}{5 - 4\cos(x)}.$$

Its expansion coefficients are

$$\hat{u}_n = 2^{-|n|}.$$

As expected, the expansion coefficients decay faster than any algebraic order of n. In Figure 2.1 we plot the continuous Fourier series approximation of $u(x)$ and the pointwise error for increasing N.

This example clearly illustrates the fast convergence of the Fourier series and also that the convergence of the approximation is almost uniform. Note that we only observe the very fast convergence for $N > N_0 \sim 16$.

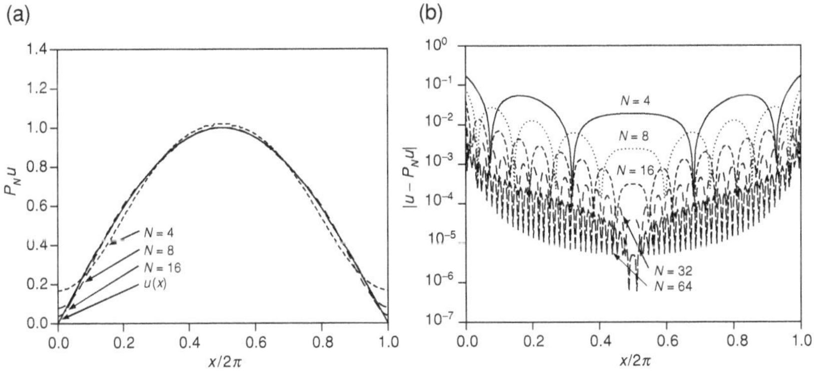

Figure 2.2 (a) Continuous Fourier series approximation of Example 2.4 for increasing resolution. (b) Pointwise error of approximation for increasing resolution.

Example 2.4 The expansion coefficients of the function

$$u(x) = \sin\left(\frac{x}{2}\right).$$

are given by

$$\hat{u}_n = \frac{2}{\pi}\frac{1}{(1-4n^2)}.$$

Note that the derivative of $u(x)$ is not periodic, and integrating by parts twice we obtain quadratic decay in n. In Figure 2.2 we plot the continuous Fourier series approximation and the pointwise error for increasing N. As expected, we find quadratic convergence except near the endpoints where it is only linear, indicating non-uniform convergence. The loss of order of convergence at the discontinuity points of the periodic extension, as well as the global reduction of order, is typical of Fourier series (and other global expansion) approximations of functions that are not sufficiently smooth.

2.1.1 Differentiation of the continuous expansion

When approximating a function $u(x)$ by the finite Fourier series $\mathcal{P}_N u$, we can easily obtain the derivatives of $\mathcal{P}_N u$ by simply differentiating the basis functions. The question is, are the derivatives of $\mathcal{P}_N u$ good approximations to the derivatives of u?

If u is a sufficiently smooth function, then one can differentiate the sum

$$\mathcal{P}_N u(x) = \sum_{|n|\le\frac{N}{2}} \hat{u}_n e^{inx},$$

term by term, to obtain

$$\frac{d^q}{dx^q}\mathcal{P}_N u(x) = \sum_{|n|\le\frac{N}{2}} \hat{u}_n \frac{d^q}{dx^q} e^{inx} = \sum_{|n|\le\frac{N}{2}} (in)^q \hat{u}_n e^{inx}.$$

It follows that the projection and differentiation operators commute

$$\mathcal{P}_N \frac{d^q}{dx^q} u = \frac{d^q}{dx^q}\mathcal{P}_N u.$$

This property implies that for any constant coefficient differention operator $\mathcal{L}$,

$$\mathcal{P}_N \mathcal{L}(\mathrm{I} - \mathcal{P}_N) u,$$

known as the *truncation error*, vanishes. Thus, the Fourier approximation to the equation $u_t = \mathcal{L}u$ is exactly the projection of the analytic solution.

2.2 Discrete trigonometric polynomials

The continuous Fourier series method requires the evaluation of the coefficients

$$\hat{u}_n = \frac{1}{2\pi}\int_0^{2\pi} u(x) e^{-inx}\, dx. \tag{2.6}$$

In general, these integrals cannot be computed analytically, and one resorts to the approximation of the Fourier integrals by using quadrature formulas, yielding the *discrete Fourier coefficients*. Quadrature formulas differ based on the exact position of the grid points, and the choice of an even or odd number of grid points results in slightly different schemes.

2.2.1 The even expansion

Define an equidistant grid, consisting of an even number N of gridpoints $x_j \in [0, 2\pi)$, defined by

$$x_j = \frac{2\pi j}{N} \quad j \in [0, \ldots, N-1].$$

The trapezoidal rule yields the discrete Fourier coefficients $\tilde{u}_n$, which approximate the continuous Fourier coefficients $\hat{u}_n$,

$$\tilde{u}_n = \frac{1}{N}\sum_{j=0}^{N-1} u(x_j) e^{-inx_j}. \tag{2.7}$$

As the following theorem shows, the trapezoidal quadrature rule is a very natural approximation when trigonometric polynomials are involved.

Theorem 2.5 *The quadrature formula*

$$\frac{1}{2\pi}\int_0^{2\pi} f(x)\,dx = \frac{1}{N}\sum_{j=0}^{N-1} f(x_j)$$

is exact for any trigonometric polynomial $f(x) = e^{inx}, \quad |n| < N$.

Proof: Given a function $f(x) = e^{inx}$,

$$\frac{1}{2\pi}\int_0^{2\pi} f(x)\,dx = \begin{cases} 1 & \text{if } n = 0, \\ 0 & \text{otherwise.} \end{cases}$$

On the other hand,

$$\begin{aligned} \frac{1}{N}\sum_{j=0}^{N-1} f(x_j) &= \frac{1}{N}\sum_{j=0}^{N-1} e^{in(2\pi j/N)} \\ &= \frac{1}{N}\sum_{j=0}^{N-1} q^j \end{aligned}$$

where $q = e^{i\frac{2\pi n}{N}}$. If n is an integer multiple of N, i.e., $n = mN$, then $\frac{1}{N}\sum_{j=0}^{N-1} f(x_j) = 1$. Otherwise, $\frac{1}{N}\sum_{j=0}^{N-1} f(x_j) = \frac{q^N - 1}{q-1} = 0$. Thus, the quadrature formula is exact for any function of the form $f(x) = e^{inx}, \; |n| < N$.
QED

The quadrature formula is exact for $f(x) \in \hat{\mathsf{B}}_{2N-2}$ where $\hat{\mathsf{B}}_N$ is the space of trigonometric polynomials of order N,

$$\hat{\mathsf{B}}_N = \text{span}\{e^{inx} \mid \; |n| \le N/2\}.$$

Note that the quadrature formula remains valid also for

$$f(x) = \sin(Nx),$$

because $\sin(Nx_j) = 0$, but it is not valid for $f(x) = \cos(Nx)$.

Using the trapezoid rule, the discrete Fourier coefficients become

$$\tilde{u}_n = \frac{1}{N\tilde{c}_n}\sum_{j=0}^{N-1} u(x_j)e^{-inx_j}, \tag{2.8}$$

where we introduce the coefficients

$$\tilde{c}_n = \begin{cases} 2 & |n| = N/2 \\ 1 & |n| < N/2 \end{cases}$$

for ease of notation. These relations define a new projection of u

$$\mathcal{I}_N u(x) = \sum_{|n| \le N/2} \tilde{u}_n e^{inx}. \tag{2.9}$$

This is the complex discrete Fourier transform, based on an even number of quadrature points. Note that

$$\tilde{u}_{-N/2} = \tilde{u}_{N/2},$$

so that we have exactly N independent Fourier coefficients, corresponding to the N quadrature points. As a consequence, $\mathcal{I}_N \sin(\frac{N}{2}x) = 0$ so that the function $\sin(\frac{N}{2}x)$ is not represented in the expansion, Equation (2.9).

Onto which finite dimensional space does $\mathcal{I}_N$ project? Certainly, the space does not include $\sin(\frac{N}{2}x)$, so it is not $\hat{\mathsf{B}}_N$. The correct space is

$$\tilde{\mathsf{B}}_N = \text{span}\{(\cos(nx), 0 \leq n \leq N/2) \cup (\sin(nx), 1 \leq n \leq N/2 - 1)\},$$

which has dimension $\dim(\tilde{\mathsf{B}}_N) = N$.

The particular definition of the discrete expansion coefficients introduced in Equation (2.8) has the intriguing consequence that the trigonometric polynomial $\mathcal{I}_N u$ interpolates the function, $u(x)$, at the quadrature nodes of the trapezoidal formula. Thus, $\mathcal{I}_N$ is the interpolation operator, where the quadrature nodes are the *interpolation points*.

Theorem 2.6 *Let the discrete Fourier transform be defined by Equations (2.8)–(2.9). For any periodic function, $u(x) \in C_p^0[0, 2\pi]$, we have*

$$\mathcal{I}_N u(x_j) = u(x_j), \quad \forall x_j = \frac{2\pi j}{N} \quad j = 0, \ldots, N-1.$$

Proof: Substituting Equation (2.8) into Equation (2.9) we obtain

$$\mathcal{I}_N u(x) = \sum_{|n| \leq N/2} \left(\frac{1}{N\tilde{c}_n} \sum_{j=0}^{N-1} u(x_j) e^{-inx_j} \right) e^{inx}.$$

Exchanging the order of summation yields

$$\mathcal{I}_N u(x) = \sum_{j=0}^{N-1} u(x_j) g_j(x), \tag{2.10}$$

where

$$\begin{aligned} g_j(x) &= \sum_{|n| \leq N/2} \frac{1}{N\tilde{c}_n} e^{in(x-x_j)} \\ &= \frac{1}{N} \sin\left[N\frac{x-x_j}{2}\right] \cot\left[\frac{x-x_j}{2}\right], \end{aligned} \tag{2.11}$$

by summing as a geometric series. It is easily verified that $g_j(x_i) = \delta_{ij}$ as is also evident from the examples of $g_j(x)$ for $N = 8$ shown in Figure 2.3.

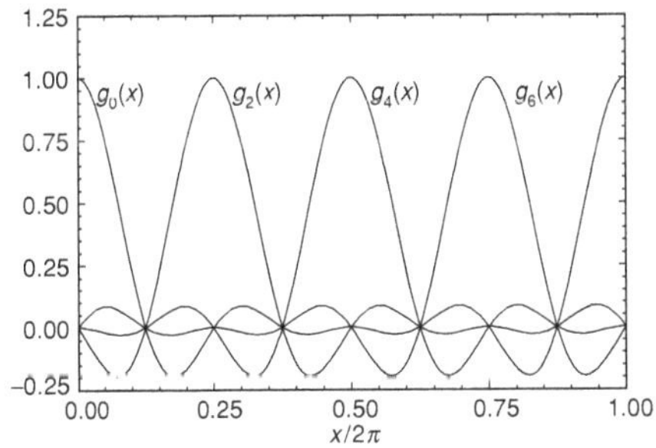

Figure 2.3 The interpolation polynomial, $g_j(x)$, for $N = 8$ for various values of j.

We still need to show that $g_j(x) \in \tilde{\mathrm{B}}_N$. Clearly, $g_j(x) \in \hat{\mathrm{B}}_N$ as $g_j(x)$ is a polynomial of degree $\leq N/2$. However, since

$$\frac{1}{2}e^{-i\frac{N}{2}x_j} = \frac{1}{2}e^{i\frac{N}{2}x_j} = \frac{(-1)^j}{2},$$

and, by convention $\tilde{u}_{-N/2} = \tilde{u}_{N/2}$, we do not get any contribution from the term $\sin(\frac{N}{2}x)$, hence $g_j(x) \in \tilde{\mathrm{B}}_N$.

QED

The discrete Fourier series of a function has convergence properties very similar to those of the continuous Fourier series approximation. In particular, the discrete approximation is pointwise convergent for $C_p^1[0, 2\pi]$ functions and is convergent in the mean provided only that $u(x) \in L^2[0, 2\pi]$. Moreover, the continuous and discrete approximations share the same asymptotic behavior, in particular having a convergence rate faster than any algebraic order of N^{-1} if $u(x) \in C_p^\infty[0, 2\pi]$. We shall return to the proof of these results in Section 2.3.2.

Let us at this point illustrate the behavior of the discrete Fourier series by applying it to the examples considered previously.

Example 2.7 Consider the $C_p^\infty[0, 2\pi]$ function

$$u(x) = \frac{3}{5 - 4\cos(x)}.$$

In Figure 2.4 we plot the discrete Fourier series approximation of u and the pointwise error for increasing N. This example confirms the spectral convergence of the discrete Fourier series. We note in particular that the approximation error is of the same order as observed for the continuous Fourier series in Example 2.3. The appearance of "spikes" in the pointwise error approaching zero in Figure 2.4 illustrates the interpolating nature of $\mathcal{I}_N u(x)$, i.e., $\mathcal{I}_N u(x_j) = u(x_j)$ as expected.

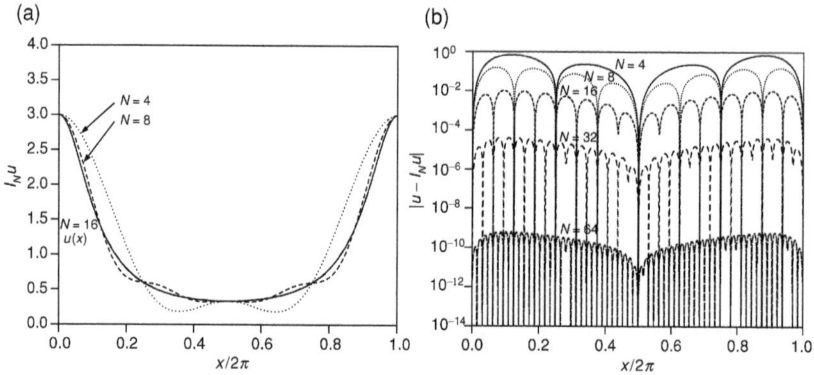

Figure 2.4 (a) Discrete Fourier series approximation of Example 2.7 for increasing resolution. (b) Pointwise error of approximation for increasing resolution.

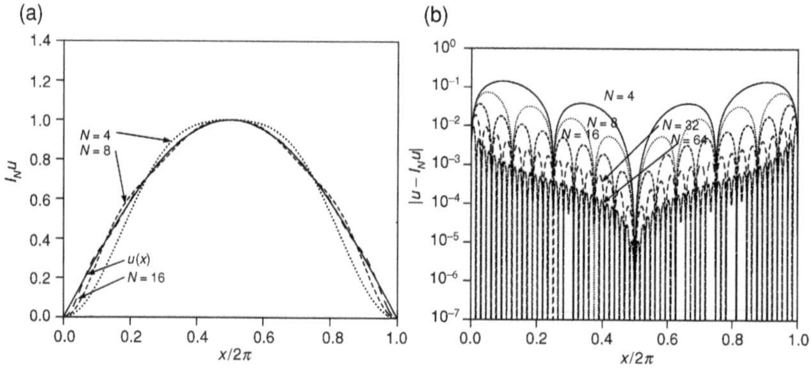

Figure 2.5 (a) Discrete Fourier series approximation of Example 2.8 for increasing resolution. (b) Pointwise error of approximation for increasing resolution.

Example 2.8 Consider again the function

$$u(x) = \sin\left(\frac{x}{2}\right),$$

and recall that $u(x) \in C_p^0[0, 2\pi]$. In Figure 2.5 we show the discrete Fourier series approximation and the pointwise error for increasing N. As for the continuous Fourier series approximation we recover a quadratic convergence rate away from the boundary points at which it is only linear.

2.2.2 The odd expansion

How can this type of interpolation operator be defined for the space $\hat{\mathsf{B}}_N$ containing an odd number of basis functions? To do so, we define a grid with an

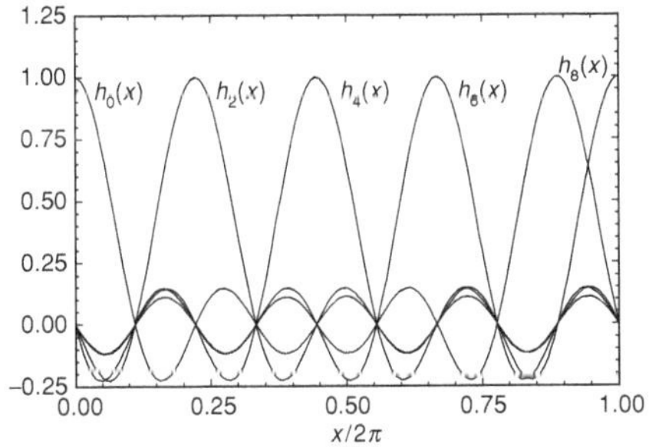

Figure 2.6 The interpolation polynomial, $h_j(x)$, for $N = 8$ for various values of j.

odd number of grid points,

$$x_j = \frac{2\pi}{N+1} j, \quad j \in [0, \ldots, N], \tag{2.12}$$

and use the trapezoidal rule

$$\tilde{u}_n = \frac{1}{N+1} \sum_{j=0}^{N} u(x_j) e^{-inx_j}, \tag{2.13}$$

to obtain the interpolation operator

$$\mathcal{J}_N u(x) = \sum_{|n| \le N/2} \tilde{u}_n e^{inx}.$$

Again, the quadrature formula is highly accurate:

Theorem 2.9 *The quadrature formula*

$$\frac{1}{2\pi} \int_0^{2\pi} f(x)\, dx - \frac{1}{N+1} \sum_{j=0}^{N} f(x_j),$$

is exact for any $f(x) = e^{inx}, \quad |n| \le N$, *i.e., for all* $f(x) \in \hat{\mathrm{B}}_{2N}$.

The scheme may also be expressed through the use of a Lagrange interpolation polynomial,

$$\mathcal{J}_N u(x) = \sum_{j=0}^{N} u(x_j) h_j(x),$$

where

$$h_j(x) = \frac{1}{N+1} \frac{\sin\left(\frac{N+1}{2}(x - x_j)\right)}{\sin\left(\frac{x - x_j}{2}\right)}. \tag{2.14}$$

One easily shows that $h_j(x_l) = \delta_{jl}$ and that $h_j(x) \in \hat{\mathrm{B}}_N$. Examples of $h_j(x)$ are shown in Figure 2.6 for $N = 8$.

Historically, the early availability of the fast fourier transform (FFT), which is highly efficient for 2^p points, has motivated the use of the even number of points approach. However, fast methods are now available for an odd as well as an even number of grid points.

2.2.3 A first look at the aliasing error

Let's consider the connection between the continuous Fourier series and the discrete Fourier series based on an even number of grid points. The conclusions of this discussion are equally valid for the case of an odd number of points.

Note that the discrete Fourier modes are based on the points x_j, for which the $(n + Nm)$th mode is indistinguishable from the nth mode,

$$e^{i(n+Nm)x_j} = e^{inx_j} e^{i2\pi mj} = e^{inx_j}.$$

This phenomenon is known as aliasing.

If the Fourier series converges pointwise, e.g., $u(x) \in C_p^1[0, 2\pi]$, the aliasing phenomenon implies that the relation between the two sets of expansion coefficients is

$$\tilde{c}_n \tilde{u}_n = \hat{u}_n + \sum_{\substack{|m| \leq \infty \\ m \neq 0}} \hat{u}_{n+Nm}, \tag{2.15}$$

In Figure 2.7 we illustrate this phenomenon for $N = 8$ and we observe that the $n = -10$ wave as well as the $n = 6$ and the $n = -2$ wave are all the same at the grid points.

In Section 2.3.2 we will show that the aliasing error

$$\|\mathcal{A}_N u\|_{L^2[0,2\pi]} = \left\| \sum_{|n| \leq N/2} \left(\sum_{\substack{m=-\infty \\ m \neq 0}}^{m=\infty} \hat{u}_{n+Nm} \right) e^{inx} \right\|_{L^2[0,2\pi]}, \tag{2.16}$$

is of the same order as the error, $\|u - \mathcal{P}_N u\|_{L^2[0,2\pi]}$, for smooth functions u. If the function is well approximated the aliasing error is generally negligible and the continuous Fourier series and the discrete Fourier series share similar approximation properties. However, for poorly resolved or nonsmooth problems, the situation is much more delicate.

2.2.4 Differentiation of the discrete expansions

To implement the Fourier–collocation method, we require the derivatives of the discrete approximation. Once again, we consider the case of an even number of grid points. The two mathematically equivalent methods given in Equations (2.8)–(2.9) and Equation (2.10) for expressing the interpolant yield two

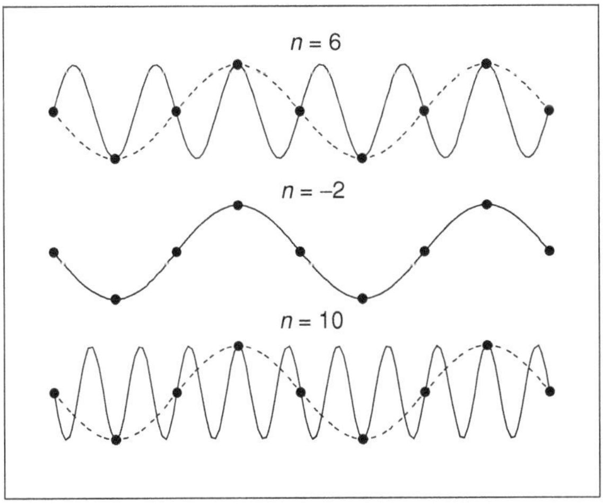

Figure 2.7 Illustration of aliasing. The three waves, $n = 6$, $n = -2$ and $n = -10$ are all interpreted as a $n = -2$ wave on an 8-point grid. Consequently, the $n = -2$ appears as more energetic after the discrete Fourier transform than in the original signal.

computationally different ways to approximate the derivative of a function. In the following subsections, we assume that our function u and all its derivatives are continuous and periodic on $[0, 2\pi]$.

Using expansion coefficients Given the values of the function $u(x)$ at the points x_j, differentiating the basis functions in the interpolant yields

$$\frac{d}{dx}\mathcal{I}_N u(x) = \sum_{|n|\le N/2} in\tilde{u}_n e^{inx}, \tag{2.17}$$

where

$$\tilde{u}_n = \frac{1}{N\tilde{c}_n}\sum_{j=0}^{N-1} u(x_j)e^{-inx_j},$$

are the coefficients of the interpolant $\mathcal{I}_N u(x)$ given in Equations (2.8)–(2.9). Higher order derivatives can be obtained simply by further differentiating the basis functions.

Note that, unlike in the case of the continuous approximation, the derivative of the interpolant is not the interpolant of the derivative, i.e.,

$$\mathcal{I}_N \frac{du}{dx} \neq \mathcal{I}_N \frac{d}{dx}\mathcal{I}_N u, \tag{2.18}$$

unless $u(x) \in \tilde{\mathrm{B}}_N$.

For example, if

$$u(x) = \sin\left(\frac{N}{2}x\right),$$

(i.e., $u(x)$ does not belong to $\tilde{\mathrm{B}}_N$), then $\mathcal{I}_N u \equiv 0$ and so $d(\mathcal{I}_N u)/dx = 0$. On the other hand, $u'(x) = N/2\cos(Nx/2)$ (which is in $\tilde{\mathrm{B}}_N$), and therefore $\mathcal{I}_N u'(x) = N/2\cos(Nx/2) \neq \mathcal{I}_N d(\mathcal{I}_N u)/dx$. If $u(x) \in \tilde{\mathrm{B}}_N$, then $\mathcal{I}_N u = u$, and therefore

$$\mathcal{I}_N \frac{du}{dx} = \mathcal{I}_N \frac{d}{dx}\mathcal{I}_N u.$$

Likewise, if we consider the projection based on an odd number of points, we have

$$\mathcal{J}_N \frac{du}{dx} \neq \mathcal{J}_N \frac{d}{dx}\mathcal{J}_N u,$$

except if $u \in \hat{\mathrm{B}}_N$.

The procedure for differentiating using expansion coefficients can be described as follows: first, we transform the point values $u(x_j)$ in physical space into the coefficients $\tilde{u}_n$ in mode space. We then differentiate in mode space by multiplying $\tilde{u}_n$ by in, and return to physical space. Computationally, the cost of the method is the cost of two transformations, which can be done by a fast Fourier transform (FFT). The typical cost of an FFT is $\mathcal{O}(N \log N)$. Notice that this procedure is a transformation from a finite dimensional space to a finite dimensional space, which indicates a matrix multiplication. In the next section, we give the explicit form of this matrix.

The matrix method The use of the Lagrange interpolation polynomials yields

$$\mathcal{I}_N u(x) = \sum_{j=0}^{N-1} u(x_j) g_j(x),$$

where

$$g_j(x) = \frac{1}{N}\sin\left[N\frac{x - x_j}{2}\right]\cot\left[\frac{x - x_j}{2}\right].$$

An approximation to the derivative of $u(x)$ at the points, x_i, is then obtained by differentiating the interpolation directly,

$$\left.\frac{d}{dx}\mathcal{I}_N u(x)\right|_{x_l} = \sum_{j=0}^{N-1} u(x_j) \left.\frac{d}{dx} g_j(x)\right|_{x_l} = \sum_{j=0}^{N-1} \mathrm{D}_{lj} u(x_j).$$

The entries of the differentiation matrix are given by

$$\mathrm{D}_{ij} = \left.\frac{d}{dx} g_j(x)\right|_{x_i} = \begin{cases} \frac{(-1)^{i+j}}{2} \cot\left[\frac{x_i - x_j}{2}\right] & i \neq j \\ 0 & i = j. \end{cases} \tag{2.19}$$

It is readily verified that D is circulant and skew-symmetric.

The approximation of higher derivatives follows the exact same route as taken for the first order derivative. The entries of the second order differentiation matrix $\mathrm{D}^{(2)}$, based on an even number of grid points, are

$$\left.\frac{d^2}{dx^2} g_j(x)\right|_{x_i} = \mathrm{D}^{(2)}_{ij} = \begin{cases} -\frac{(-1)^{i+j}}{2} \left[\sin\left(\frac{x_i - x_j}{2}\right)\right]^{-2} & i \neq j \\ -\frac{N^2+2}{12} & i = j. \end{cases} \tag{2.20}$$

It is interesting to note that, in the case of even number of points, $\mathrm{D}^{(2)}$ is not equal to the square of D. i.e.,

$$\mathcal{I}_N \frac{d^2}{dx^2} \mathcal{I}_N \neq \left(\mathcal{I}_N \frac{d}{dx}\right)^2 \mathcal{I}_N$$

To illustrate this, consider the function $\cos(\frac{N}{2}x)$,

$$\mathcal{I}_N \frac{d^2}{dx^2} \mathcal{I}_N \cos\left(\frac{N}{2}x\right) = \mathcal{I}_N \left[-\left(\frac{N}{2}\right)^2 \cos\left(\frac{N}{2}x\right)\right] = -\left(\frac{N}{2}\right)^2 \cos\left(\frac{N}{2}x\right),$$

on the other hand, since $\mathcal{I}_N \frac{d}{dx} \mathcal{I}_N \cos(\frac{N}{2}x) = 0$, the two are not the same.

The reason for this discrepancy is that the differentiation operator takes elements of $\tilde{\mathsf{B}}_N$ out of $\tilde{\mathsf{B}}_N$. In the above example, $\cos(\frac{N}{2}x)$ is in the space $\tilde{\mathsf{B}}_N$, but its derivative is not. However, elements of $\hat{\mathsf{B}}_N$, when differentiated, remain in $\hat{\mathsf{B}}_N$ and thus,

$$\mathcal{J}_N \frac{d^2}{dx^2} \mathcal{J}_N = \left(\mathcal{J}_N \frac{d}{dx}\right)^2 \mathcal{J}_N,$$

for the interpolation based on an odd number of grid points.

For the sake of completeness, we list the entries of the differentiation matrix $\tilde{\mathrm{D}}$ for the interpolation based on an odd number of points,

$$\tilde{\mathrm{D}}_{ij} = \begin{cases} \frac{(-1)^{i+j}}{2} \left[\sin\left(\frac{x_i - x_j}{2}\right)\right]^{-1} & i \neq j \\ 0 & i = j, \end{cases}$$

which is the limit of the finite difference schemes as the order increases. Once again $\tilde{\mathrm{D}}$ is a circulant, skew-symmetric matrix. As mentioned above, in this case we have

$$\tilde{\mathrm{D}}^{(q)} = \mathcal{J}_N \frac{d^q}{dx^q} \mathcal{J}_N = \tilde{\mathrm{D}}^q,$$

for all values of q.

In this method, we do not go through the mode space at all. The differentiation matrix takes us from physical space to physical space, and the act of differentiation is hidden in the matrix itself. The computational cost of the matrix method is the cost of a matrix-vector product, which is an $\mathcal{O}(N^2)$ operation, rather than the cost of $\mathcal{O}(Nlog(N))$ in the method using expansion coefficients. However, the efficiency of the FFT is machine dependent and for small values of N it may be faster to perform the matrix-vector product. Also, since the differentiation matrices are all circulant one need only store one column of the operator, thereby reducing the memory usage to that of the FFT.

2.3 Approximation theory for smooth functions

We will now rigorously justify the behavior of the continuous and discrete Fourier approximations of the function u and its derivatives. When using the Fourier approximation to discretize the spatial part of the equation

$$u_t = \mathcal{L}u,$$

where $\mathcal{L}$ is a differential operator (e.g., the hyperbolic equation $u_t = a(x)u_x$), it is important that our approximation, both to u and to $\mathcal{L}u$, be accurate. To establish consistency we need to consider not only the difference between u and $\mathcal{P}_N u$, but also the distance between $\mathcal{L}u$ and $\mathcal{L}\mathcal{P}_N u$, measured in an appropriate norm. This is critical, because the actual rate of convergence of a stable scheme is determined by the truncation error

$$\mathcal{P}_N \mathcal{L} (\mathrm{I} - \mathcal{P}_N) u.$$

The truncation error is thus determined by the behavior of the Fourier approximations not only of the function, but of its derivatives as well.

It is natural, therefore, to use the Sobolev q-norm denoted by $H_p^q[0, 2\pi]$, which measures the smoothness of the derivatives as well as the function,

$$\|u\|^2_{H_p^q[0,2\pi]} = \sum_{m=0}^{q} \int_0^{2\pi} \left|u^{(m)}(x)\right|^2 dx.$$

The subscript p indicates the fact that all our functions are periodic, for which the Sobolev norm can be written in mode space as

$$\|u\|^2_{H^q_p[0,2\pi]} = 2\pi \sum_{m=0}^{q} \sum_{|n|\leq\infty} |n|^{2m} |\hat{u}_n|^2 = 2\pi \sum_{|n|\leq\infty} \left(\sum_{m=0}^{q} |n|^{2m} \right) |\hat{u}_n|^2,$$

where the interchange of the summation is allowed provided $u(x)$ has sufficient smoothness, e.g., $u(x) \in C^q_p[0, 2\pi]$, $q > \frac{1}{2}$.

Since for $n \neq 0$,

$$(1 + n^{2q}) \leq \sum_{m=0}^{q} n^{2m} \leq (q+1)(1 + n^{2q}),$$

the norm $\|\cdot\|_{W^q_p[0,2\pi]}$ defined by

$$\|u\|_{W^q_p[0,2\pi]} = \left(\sum_{|n|\leq\infty} (1 + n^{2q}) |\hat{u}_n|^2 \right)^{1/2},$$

is equivalent to $\|\cdot\|_{H^q_p[0,2\pi]}$. It is interesting to note that one can easily define a norm $W^q_p[0, 2\pi]$ with noninteger values of q.

2.3.1 Results for the continuous expansion

Consider, first, the continuous Fourier series

$$\mathcal{P}_{2N} u(x) = \sum_{|n|\leq N} \hat{u}_n e^{inx}.$$

We start with an L^2 estimate for the distance between u and its trigonometric approximation $\mathcal{P}_{2N} u$.

Theorem 2.10 *For any $u(x) \in H^r_p[0, 2\pi]$, there exists a positive constant C, independent of N, such that*

$$\|u - \mathcal{P}_{2N} u\|_{L^2[0,2\pi]} \leq C N^{-q} \left\| u^{(q)} \right\|_{L^2[0,2\pi]},$$

provided $0 \leq q \leq r$.

Proof: By Parseval's identity,

$$\|u - \mathcal{P}_{2N} u\|^2_{L^2[0,2\pi]} = 2\pi \sum_{|n|>N} |\hat{u}_n|^2.$$

We rewrite this summation

$$\begin{aligned}
\sum_{|n|>N} |\hat{u}_n|^2 &= \sum_{|n|>N} \frac{1}{n^{2q}} n^{2q} |\hat{u}_n|^2 \\
&\leq N^{-2q} \sum_{|n|>N} n^{2q} |\hat{u}_n|^2 \\
&\leq N^{-2q} \sum_{|n|\geq 0} n^{2q} |\hat{u}_n|^2 \\
&= \frac{1}{2\pi} N^{-2q} \left\| u^{(q)} \right\|^2_{L^2[0,2\pi]}.
\end{aligned}$$

Putting this all together and taking the square root, we obtain our result.

QED

Note that the smoother the function, the larger the value of q, and therefore, the better the approximation. This is in contrast to finite difference or finite element approximations, where the rate of convergence is fixed, regardless of the smoothness of the function. This rate of convergence is referred to in the literature as spectral convergence.

If $u(x)$ is analytic then

$$\left\| u^{(q)} \right\|_{L^2[0,2\pi]} \leq Cq! \, \|u\|_{L^2[0,2\pi]},$$

and so

$$\|u - \mathcal{P}_{2N} u\|_{L^2[0,2\pi]} \leq CN^{-q} \left\| u^{(q)} \right\|_{L^2[0,2\pi]} \leq C \frac{q!}{N^q} \|u\|_{L^2[0,2\pi]}.$$

Using Stirling's formula, $q! \sim q^q e^{-q}$, and assuming that $q \propto N$, we obtain

$$\|u - \mathcal{P}_{2N} u\|_{L^2[0,2\pi]} \leq\sim C \left(\frac{q}{N}\right)^q e^{-q} \|u\|_{L^2[0,2\pi]} \sim K e^{-cN} \|u\|_{L^2[0,2\pi]}.$$

Thus, for an analytic function, spectral convergence is, in fact, exponential convergence.

Since the Fourier method is used for computation of derivatives, we are particularly interested in estimating the convergence of both the function and its derivative. For this purpose, the Sobolev norm is appropriate.

Theorem 2.11 *For any real r and any real q where $0 \leq q \leq r$, if $u(x) \in W^r_p[0, 2\pi]$, then there exists a positive constant C, independent of N, such that*

$$\|u - \mathcal{P}_{2N} u\|_{W^q_p[0,2\pi]} \leq \frac{C}{N^{r-q}} \|u\|_{W^r_p[0,2\pi]}.$$

Proof: Parseval's identity yields

$$\|u - \mathcal{P}_{2N} u\|^2_{W^q_p[0,2\pi]} = 2\pi \sum_{|n|>N} (1 + |n|^{2q}) |\hat{u}_n|^2.$$

Since $|n| + 1 \geq N$, we obtain

$$(1 + |n|^{2q}) \leq (1 + |n|)^{2q} = \frac{(1 + |n|)^{2r}}{(1 + |n|)^{2(r-q)}} \leq \frac{(1 + |n|)^{2r}}{N^{2(r-q)}} \leq (1 + r)\frac{(1 + n^{2r})}{N^{2(r-q)}},$$

for any $0 \leq q \leq r$.

This immediately yields

$$\|u - \mathcal{P}_{2N}u\|^2_{W_p^q[0,2\pi]} \leq C \sum_{|n|>N} \frac{(1 + n^{2r})}{N^{2(r-q)}} |\hat{u}_n|^2 \leq C \frac{\|u\|^2_{W_p^r[0,2\pi]}}{N^{2(r-q)}}.$$

QED

A stricter measure of convergence may be obtained by looking at the pointwise error in the maximum norm.

Theorem 2.12 *For any $q > 1/2$ and $u(x) \in C_p^q[0, 2\pi]$, there exists a positive constant C, independent of N, such that*

$$\|u - \mathcal{P}_{2N}u\|_{L^\infty} \leq C \frac{1}{N^{q-\frac{1}{2}}} \left\| u^{(q)} \right\|_{L^2[0,2\pi]}.$$

Proof: Provided $u(x) \in C_p^q[0, 2\pi]$, $q > 1/2$, we have for any $x \in [0, 2\pi]$,

$$\begin{aligned} |u - \mathcal{P}_{2N}u| &= \left| \sum_{|n|>N} \hat{u}_n e^{inx} \right| \\ &= \left| \sum_{|n|>N} n^q \hat{u}_n \frac{e^{inx}}{n^q} \right| \\ &\leq \left(\sum_{|n|>N} \frac{1}{n^{2q}} \right)^{1/2} \left(\sum_{|n|>N} n^{2q} |\hat{u}_n|^2 \right)^{1/2}, \end{aligned}$$

using the Cauchy–Schwartz inequality. The second term in the product above is bounded by the norm. The first term is the tail of a power series which converges for $2q > 1$. Thus, for $q > \frac{1}{2}$, we can bound the tail, and so we obtain

$$|u - \mathcal{P}_{2N}u| \leq \frac{C}{N^{q-1/2}} \left\| u^{(q)} \right\|_{L^2[0,2\pi]}.$$

QED

Again, we notice spectral accuracy in the maximum norm, with exponential accuracy for analytic functions. Finally, we are ready to use these results to bound the truncation error for the case of a constant coefficient differential operator.

Theorem 2.13 *Let $\mathcal{L}$ be a constant coefficient differential operator*

$$\mathcal{L}u = \sum_{j=1}^{s} a_j \frac{d^j u}{dx^j}.$$

For any real r and any real q where $0 \le q + s \le r$, if $u(x) \in W_p^r[0, 2\pi]$, then there exists a positive constant C, independent of N, such that

$$\|\mathcal{L}u - \mathcal{L}\mathcal{P}_N u\|_{W_p^q[0,2\pi]} \le C N^{-(r-q-s)} \|u\|_{W_p^r[0,2\pi]}.$$

Proof: Using the definition of $\mathcal{L}$,

$$\begin{aligned}
\|\mathcal{L}u - \mathcal{L}\mathcal{P}_N u\|_{W_p^q[0,2\pi]} &= \left\| \sum_{j=1}^{s} a_j \frac{d^j u}{dx^j} - \sum_{j=1}^{s} a_j \frac{d^j \mathcal{P}_N u}{dx^j} \right\|_{W_p^q[0,2\pi]} \\
&\le \max_{0 \le j \le s} |a_j| \left\| \sum_{j=1}^{s} \frac{d^j}{dx^j} (u - \mathcal{P}_N u) \right\|_{W_p^q[0,2\pi]} \\
&\le \max_{0 \le j \le s} |a_j| \sum_{j=1}^{s} \|u - \mathcal{P}_N u\|_{W_p^{q+s}[0,2\pi]} \\
&\le C \|u - \mathcal{P}_N u\|_{W_p^{q+s}[0,2\pi]}
\end{aligned}$$

This last term is bounded in Theorem 2.7, and the result immediately follows.
QED

2.3.2 Results for the discrete expansion

The approximation theory for the discrete expansion yields essentially the same results as for the continuous expansion, though with more effort. The proofs for the discrete expansion are based on the convergence results for the continuous approximation, and as the fact that the Fourier coefficients of the discrete approximation are close to those of the continuous approximation.

Recall that the interpolation operator associated with an even number of grid points is given by

$$\mathcal{I}_{2N} u = \sum_{|n| < N} \tilde{u}_n e^{inx},$$

with expansion coefficients

$$\tilde{u}_n = \frac{1}{2N\tilde{c}_n} \sum_{j=0}^{2N-1} u(x_j) e^{-inx_j}, \quad x_j = \frac{2\pi}{2N} j.$$

Rather than deriving the estimates of the approximation error directly, we shall use the results obtained in the previous section and then estimate the difference between the two different expansions, which we recognize as the aliasing error.

The relationship between the discrete expansion coefficients $\tilde{u}_n$, and the continuous expansion coefficients $\hat{u}_n$, is given in the following lemma.

Lemma 2.14 *Consider $u(x) \in W_p^q[0, 2\pi]$, where $q > 1/2$. For $|n| \leq N$ we have*

$$\tilde{c}_n \tilde{u}_n = \hat{u}_n + \sum_{\substack{|m| \leq \infty \\ m \neq 0}} \hat{u}_{n+2Nm}.$$

Proof: Substituting the continuous Fourier expansion into the discrete expansion yields

$$\tilde{c}_n \tilde{u}_n = \frac{1}{2N} \sum_{j=0}^{2N-1} \sum_{|l| \leq \infty} \hat{u}_l e^{i(l-n)x_j}.$$

To interchange the two summations we must ensure uniform convergence, i.e., $\sum_{|l| \leq \infty} |\hat{u}_l| < \infty$. This is satisfied, since

$$\begin{aligned} \sum_{|l| \leq \infty} |\hat{u}_l| &= \sum_{|l| \leq \infty} (1 + |l|)^q \frac{|\hat{u}_l|}{(1 + |l|)^q} \\ &\leq \left(2q \sum_{|l| \leq \infty} (1 + l^{2q}) |\hat{u}_l|^2 \right)^{1/2} \left(\sum_{|l| \leq \infty} (1 + |l|)^{-2q} \right)^{1/2}, \end{aligned}$$

where the last expression follows from the Cauchy–Schwarz inequality. As $u(x) \in W_p^q[0, 2\pi]$ the first part is clearly bounded. Furthermore, the second term converges provided $q > 1/2$, hence ensuring boundedness.

Interchanging the order of summation and using orthogonality of the exponential function at the grid yields the desired result.

QED

As before, we first consider the behavior of the approximation in the L^2-norm. We will first show that the bound on the aliasing error, $\mathcal{A}_N$, in Equation (2.16) is of the same order as the truncation error. The error caused by truncating the continuous expansion is essentially the same as the error produced by using the discrete coefficients rather than the continuous coefficients.

Lemma 2.15 *For any $u(x) \in W_p^r[0, 2\pi]$, where $r > 1/2$, the aliasing error*

$$\|\mathcal{A}_N u\|_{L^2[0,2\pi]} = \left(\sum_{|n| \leq \infty} |\tilde{c}_n \tilde{u}_n - \hat{u}_n|^2 \right)^{1/2} \leq C N^{-r} \left\| u^{(r)} \right\|_{L^2[0,2\pi]}.$$

Proof: From Lemma 2.14 we have

$$|\tilde{c}_n\tilde{u}_n - \hat{u}_n|^2 = \left| \sum_{\substack{|m|\leq\infty \\ m\neq 0}} \hat{u}_{n+2Nm} \right|^2 .$$

To estimate this, we first note that

$$\begin{aligned}
\left| \sum_{\substack{|m|\leq\infty \\ m\neq 0}} \hat{u}_{n+2Nm} \right|^2 &= \left| \sum_{\substack{|m|\leq\infty \\ m\neq 0}} |n+2Nm|^r \hat{u}_{n+2Nm} \frac{1}{|n+2Nm|^r} \right|^2 \\
&\leq \left(\sum_{\substack{|m|\leq\infty \\ m\neq 0}} |n+2Nm|^{2r} |\hat{u}_{n+2Nm}|^2 \right) \left(\sum_{\substack{|m|\leq\infty \\ m\neq 0}} \frac{1}{|n+2Nm|^{2r}} \right),
\end{aligned}$$

using the Cauchy–Schwartz inequality.

Since $|n| \leq N$, bounding of the second term is ensured by

$$\sum_{\substack{|m|\leq\infty \\ m\neq 0}} \frac{1}{|n+2Nm|^{2r}} \leq \frac{2}{N^{2r}} \sum_{m=1}^{\infty} \frac{1}{(2m-1)^{2r}} = C_1 N^{-2r},$$

provided $r > 1/2$. Here, the constant C_1 is a consequence of the fact that the power series converges, and it is independent of N.

Summing over n, we have

$$\begin{aligned}
\sum_{|n|\leq N} \left| \sum_{\substack{|m|\leq\infty \\ m\neq 0}} \hat{u}_{n+2Nm} \right|^2 &\leq \sum_{|n|\leq N} C_1 N^{-2r} \sum_{\substack{|m|\leq\infty \\ m\neq 0}} |n+2mN|^{2r} |\hat{u}_{n+2Nm}|^2 \\
&\leq C_2 N^{-2r} \left\| u^{(r)} \right\|^2_{L^2[0,2\pi]}.
\end{aligned}$$

QED

We are now in a position to state the error estimate for the discrete approximation.

Theorem 2.16 *For any $u(x) \in W_p^r[0, 2\pi]$ with $r > 1/2$, there exists a positive constant C, independent of N, such that*

$$\|u - \mathcal{I}_{2N}u\|_{L^2[0,2\pi]} \leq CN^{-r} \left\| u^{(r)} \right\|_{L^2[0,2\pi]}.$$

Proof: Let's write the difference between the function and its discrete approximation

$$\|u - \mathcal{I}_{2N}u\|^2_{L^2[0,2\pi]} = \|(\mathcal{P}_{2N} - \mathcal{I}_{2N})u + u - \mathcal{P}_{2N}u\|^2_{L^2[0,2\pi]}$$
$$\leq \|(\mathcal{P}_{2N} - \mathcal{I}_{2N})u\|^2_{L^2[0,2\pi]} + \|u - \mathcal{P}_{2N}u\|^2_{L^2[0,2\pi]}.$$

Thus, the error has two components. The first one, which is the difference between the continuous and discrete expansion coefficients, is the aliasing error, which is bounded in Lemma 2.15. The second, which is the tail of the series, is the truncation error, which is bounded by the result of Theorem 2.10. The desired result follows from these error bounds.

QED

Theorem 2.16 confirms that the approximation errors of the continuous expansion and the discrete expansion are of the same order, as long as $u(x)$ has at least half a derivative. Furthermore, the rate of convergence depends, in both cases, only on the smoothness of the function being approximated.

The above results are in the L^2 norm. We can obtain essentially the same information about the derivatives, using the Sobolev norms. First, we need to obtain a Sobolev norm bound on the aliasing error.

Lemma 2.17 *Let $u(x) \in W_p^r[0, 2\pi]$, where $r > 1/2$. For any real q, for which $0 \leq q \leq r$, the aliasing error*

$$\|\mathcal{A}_N u\|_{W_p^q[0,2\pi]} = \left(\sum_{n=-\infty}^{\infty} |\tilde{c}_n \tilde{u}_n - \hat{u}_n|^2 \right)^{1/2} \leq C N^{-(r-q)} \|u\|_{W_p^r[0,2\pi]}.$$

Proof:

$$\left| \sum_{\substack{|m|\leq\infty \\ m\neq 0}} \hat{u}_{n+2Nm} \right|^2 = \left| \sum_{\substack{|m|\leq\infty \\ m\neq 0}} (1 + |n + 2Nm|)^r \hat{u}_{n+2Nm} \frac{1}{(1 + |n + 2Nm|)^r} \right|^2,$$

such that

$$\left| \sum_{\substack{|m|\leq\infty \\ m\neq 0}} \hat{u}_{n+2Nm} \right|^2 \leq \left(\sum_{\substack{|m|\leq\infty \\ m\neq 0}} (1 + |n + 2Nm|)^{2r} |\hat{u}_{n+2Nm}|^2 \right) \times \left(\sum_{\substack{|m|\leq\infty \\ m\neq 0}} \frac{1}{(1 + |n + 2Nm|)^{2r}} \right).$$

The second factor is, as before, bounded by

$$\sum_{\substack{|m|\leq\infty \\ m\neq 0}} \frac{1}{(1+|n+2Nm|)^{2r}} \leq \frac{2}{N^{2r}} \sum_{m=1}^{\infty} \frac{1}{(2m-1)^{2r}} = C_1 N^{-2r},$$

provided $r > 1/2$ and $|n| \leq N$.

Also, since $(1+|n|)^{2q} \leq C_2 N^{2q}$ for $|n| \leq N$ we recover

$$\sum_{|n|\leq N} (1+|n|)^{2q} \left| \sum_{\substack{|m|\leq\infty \\ m\neq 0}} \hat{u}_{n+2Nm} \right|^2$$

$$\leq \sum_{|n|\leq N} C_1 C_2 N^{-2(r-q)} \sum_{\substack{|m|\leq\infty \\ m\neq 0}} (1+|n+2mN|)^{2r} |\hat{u}_{n+2Nm}|^2$$

$$\leq C_3 N^{-2(r-q)} \|u\|^2_{W_p^r[0,2\pi]}.$$

QED

With this bound on the aliasing error, and the truncation error bounded by Theorem 2.11, we are now prepared to state.

Theorem 2.18 *Let $u(x) \in W_p^r[0, 2\pi]$, where $r > 1/2$. Then for any real q for which $0 \leq q \leq r$, there exists a positive constant, C, independent of N, such that*

$$\|u - \mathcal{I}_{2N} u\|_{W_p^q[0,2\pi]} \leq C N^{-(r-q)} \|u\|_{W_p^r[0,2\pi]}.$$

The proof follows closely that of Theorem 2.16. As in the case of the continuous expansion, we use this result to bound the truncation error.

Theorem 2.19 *Let $\mathcal{L}$ be a constant coefficient differential operator*

$$\mathcal{L}u = \sum_{j=1}^{s} a_j \frac{d^j u}{dx^j}$$

then there exists a positive constant, C, independent of N such that

$$\|\mathcal{L}u - \mathcal{L}\mathcal{I}_N u\|_{W_p^q[0,2\pi]} \leq C N^{-(r-q-s)} \|u\|_{W_p^r[0,2\pi]}.$$

2.4 Further reading

The approximation theory for continuous Fourier expansions is classical and can be found in many sources, e.g. the text by Canuto et al (1988). Many of the results on the discrete expansions and the aliasing errors are discussed by Orszag (1972), Gottlieb et al (1983), and Tadmor (1986), while the first instance of realizing the connection between the discrete expansions and the Lagrange form appears to be in Gottlieb and Turkel (1983).

3
Fourier spectral methods

We are now ready to formally present and analyze Fourier spectral methods for the solution of partial differential equations. As in the previous chapter we restrict ourselves to problems defined on $[0, 2\pi]$ and assume that the solutions, $u(x)$, can be periodically extended. Furthermore, we assume that $u(x)$ and its derivatives are smooth enough to allow for any Fourier expansions which may become required. The first two sections of this chapter feature the Fourier–Galerkin and Fourier–collocation methods. The final section discusses the stability of these methods.

3.1 Fourier–Galerkin methods

Consider the problem

$$\begin{aligned} \frac{\partial u(x,t)}{\partial t} &= \mathcal{L}u(x,t), \quad && x \in [0, 2\pi], \quad t \geq 0, \\ u(x,0) &= g(x), && x \in [0, 2\pi], \quad t = 0, \end{aligned}$$

In the Fourier–Galerkin method, we seek solutions $u_N(x,t)$ from the space $\hat{\mathsf{B}}_N \in \text{span}\{e^{inx}\}_{|n|\leq N/2}$, i.e.,

$$u_N(x,t) = \sum_{|n|\leq N/2} a_n(t) e^{inx}.$$

Note that $a_n(t)$ are unknown coefficients which will be determined by the method. In general, the coefficients $a_n(t)$ of the approximation are not equal to the Fourier coefficients $\hat{u}_n$; only if we obtain the exact solution of the problem will they be equal. In the Fourier–Galerkin method, the coefficients $a_n(t)$ are determined by the requirement that the residual

$$R_N(x,t) = \frac{\partial u_N(x,t)}{\partial t} - \mathcal{L}u_N(x,t),$$

is orthogonal to $\hat{\mathsf{B}}_N$.

If we express the residual in terms of the Fourier series,

$$R(x,t) = \sum_{|n|\leq\infty} \hat{R}_n(t)e^{inx},$$

the orthogonality requirement yields

$$\hat{R}_n(t) = \frac{1}{2\pi}\int_0^{2\pi} R_N(x,t)e^{-inx}\,dx = 0 \quad \forall\,|n| \leq \frac{N}{2}.$$

These are $(N+1)$ ordinary differential equations to determine the $(N+1)$ unknowns, $a_n(t)$, and the corresponding initial conditions are

$$u_N(x,0) = \sum_{|n|\leq N/2} a_n(0)e^{inx},$$

$$a_n(0) = \frac{1}{2\pi}\int_{-1}^{1} g(x)e^{-inx}\,dx.$$

The method is defined by the requirement that the orthogonal projection of the residual onto the space $\hat{\mathsf{B}}_N$ is zero. If the residual is smooth enough, this requirement implies that the residual itself is small. In particular, if the residual itself lives in the space $\hat{\mathsf{B}}_N$, the orthogonal complement must be zero. This is a very special case which only occurs in a small number of cases, among them linear constant coefficient problems.

Example 3.1 Consider the linear constant coefficient problem

$$\frac{\partial u(x,t)}{\partial t} = c\frac{\partial u(x,t)}{\partial x} + \epsilon\frac{\partial^2 u(x,t)}{\partial x^2},$$

with the assumption that $u(x,0) \in C_p^\infty[0,2\pi]$, and c and ϵ are constants.

We seek a trigonometric polynomial,

$$u_N(x,t) = \sum_{|n|\leq N/2} a_n(t)e^{inx},$$

such that the residual

$$R_N(x,t) = \frac{\partial u_N(x,t)}{\partial t} - c\frac{\partial u_N(x,t)}{\partial x} - \epsilon\frac{\partial^2 u_N(x,t)}{\partial x^2},$$

is orthogonal to $\hat{\mathsf{B}}_N$.

Recall that

$$\frac{\partial}{\partial x} u_N(x,t) = \sum_{|n| \le N/2} (in) a_n(t) e^{inx},$$

so that the residual is

$$R_N(x,t) = \sum_{|n| \le N/2} \left(\frac{da_n(t)}{dt} - cin\, a_n(t) + \epsilon n^2 a_n(t) \right) e^{inx}.$$

Projecting the residual onto $\hat{\mathsf{B}}_N$ and setting to zero, we obtain

$$\frac{da_n(t)}{dt} = (cin - \epsilon n^2) a_n(t) \quad \forall\, |n| \le \frac{N}{2}.$$

Observe that in this case, $R_N(x,t) \in \hat{\mathsf{B}}_N$, and therefore setting its projection onto $\hat{\mathsf{B}}_N$ to zero, is equivalent to setting the residual itself equal to zero. The constant coefficient operator $\mathcal{L}$ has the property $\mathcal{P}_N \mathcal{L} = \mathcal{L} \mathcal{P}_N$, and so the truncation error

$$\mathcal{P}_N \mathcal{L} (\mathrm{I} - \mathcal{P}_N) u = 0.$$

In this case, the approximation coefficients $a_n(t)$ are, in fact, equal to the Fourier coefficients $\hat{u}_n$ and so the approximate solution is the projection of the true solution, $\mathcal{P}_N u(x,t) = u_N(x,t)$.

Example 3.2 Next, consider the linear, variable coefficient problem

$$\frac{\partial u(x,t)}{\partial t} = \sin(x) \frac{\partial u(x,t)}{\partial x},$$

with smooth initial conditions.

We seek solutions in the form of a trigonometric polynomial

$$u_N(x,t) = \sum_{|n| \le N/2} a_n(t) e^{inx}, \tag{3.1}$$

and require that the residual

$$R_N(x,t) = \frac{\partial u_N(x,t)}{\partial t} - \sin(x) \frac{\partial u_N(x,t)}{\partial x},$$

is orthogonal to $\hat{\mathsf{B}}_N$.

The residual is

$$R_N(x,t) = \sum_{|n| \le N/2} \left(\frac{da_n(t)}{dt} - \frac{e^{ix} - e^{-ix}}{2i} (in) a_n(t) \right) e^{inx}.$$

To simplify the expression of $R_N(x,t)$ we define $a_{-(N/2+1)}(t) = a_{N/2+1}(t) = 0$, and the residual becomes

$$
\begin{aligned}
R_N(x,t) &= \sum_{|n|\le N/2} \frac{da_n(t)}{dt} e^{inx} - \frac{1}{2}\sum_{|n|\le N/2} n e^{i(n+1)x} a_n(t) \\
&\quad + \frac{1}{2}\sum_{|n|\le N/2} n e^{i(n-1)x} a_n(t) \\
&= \sum_{|n|\le N/2} \frac{da_n(t)}{dt} e^{inx} - \frac{1}{2}\sum_{|n|\le N/2} (n-1) e^{inx} a_{n-1}(t) \\
&\quad + \frac{1}{2}\sum_{|n|\le N/2} (n+1) e^{inx} a_{n+1}(t) \\
&\quad - \frac{N}{4}\left(e^{i\frac{N+2}{2}x} a_{N/2}(t) + e^{-i\frac{N+2}{2}x} a_{-N/2}(t)\right) \\
&= \sum_{|n|\le N/2} \left(\frac{da_n(t)}{dt} - \frac{n-1}{2} a_{n-1}(t) + \frac{n+1}{2} a_{n+1}(t)\right) e^{inx} \\
&\quad - \frac{N}{4}\left(e^{i\frac{N+2}{2}x} a_{N/2}(t) + e^{-i\frac{N+2}{2}x} a_{-N/2}(t)\right).
\end{aligned}
$$

The last two terms in this expression are not in the space $\hat{\mathsf{B}}_N$, and so the residual $R_N(x,t)$ is not solely in the space of $\hat{\mathsf{B}}_N$, and the truncation error will not be identically zero. Since the last two terms are not in $\hat{\mathsf{B}}_N$, projecting $R_N(x,t)$ to zero in $\hat{\mathsf{B}}_N$ yields

$$
\frac{da_n(t)}{dt} - \frac{n-1}{2} a_{n-1}(t) + \frac{n+1}{2} a_{n+1}(t) = 0,
$$

with $a_{-(N/2+1)}(t) = a_{N/2+1}(t) = 0$.

In these two examples we illustrated the fact that the Fourier–Galerkin method involves solving the equations in the mode space rather than in physical space. Thus, for each problem we need to derive the equations for the expansion coefficients of the numerical solution. While this was relatively easy for the particular variable coefficient case considered in the previous example, this may be more complicated in general, as seen in the next example.

Example 3.3 Consider the nonlinear problem

$$
\frac{\partial u(x,t)}{\partial t} = u(x,t)\frac{\partial u(x,t)}{\partial x},
$$

with smooth, periodic initial conditions. The solution of such a problem develops discontinuities, which may lead to stability problems; however, the construction of the Fourier–Galerkin method is not affected by this.

As usual, we seek a solution of the form

$$u_N(x,t) = \sum_{|n|\le N/2} a_n(t)e^{inx},$$

and require that the residual

$$R_N(x,t) = \frac{\partial u_N(x,t)}{\partial t} - u_N(x,t)\frac{\partial}{\partial x}u_N(x,t),$$

be orthogonal to $\hat{\mathsf{B}}_N$.

The second term is

$$\begin{aligned} u_N(x,t)\frac{\partial}{\partial x}u_N(x,t) &= \sum_{|l|\le N/2}\sum_{|k|\le N/2} a_l(t)(ik)a_k(t)e^{i(l+k)x} \\ &= \sum_{|k|\le N/2}\sum_{n=-N/2+k}^{N/2+k} (ik)a_{n-k}(t)a_k(t)e^{inx}. \end{aligned}$$

As a result of the nonlinearity, the residual $R_N(x,t) \in \hat{\mathsf{B}}_{2N}$, and not $\hat{\mathsf{B}}_N$. Projecting this onto $\hat{\mathsf{B}}_N$ and setting equal to zero we obtain a set of $(N+1)$ ODEs

$$\frac{da_n(t)}{dt} = \sum_{|k|\le N/2} ika_{n-k}(t)a_k(t), \quad \forall |n| \le \frac{N}{2}.$$

In this example, we obtain the Fourier–Galerkin equation with relative ease due to the fact that the nonlinearity was only quadratic. Whereas quadratic nonlinearity is quite common in the equations of mathematical physics, there are many cases in which the nonlinearity is of a more complicated form, and the derivation of the Fourier–Galerkin equations may become untenable, as in the next example.

Example 3.4 Consider the strongly nonlinear problem

$$\frac{\partial u(x,t)}{\partial t} = e^{u(x,t)}\frac{\partial u(x,t)}{\partial x},$$

where the initial conditions are, as usual, smooth and periodic.

We seek a numerical solution of the form

$$u_N(x,t) = \sum_{|n|\le N/2} a_n(t)e^{inx},$$

and require that the residual

$$R_N(x,t) = \frac{\partial u_N(x,t)}{\partial t} - e^{u_N(x,t)}\frac{\partial}{\partial x}u_N(x,t),$$

is orthogonal to $\hat{\mathsf{B}}_N$. It is exceedingly difficult to obtain the analytic form of the resulting system of ODEs. Hence, it is untenable to formulate the Fourier–Galerkin scheme.

To summarize, the Fourier–Galerkin method is very efficient for linear, constant coefficient problems, but tends to become complicated for variable coefficient and nonlinear problems. The main drawback of the method is the need to derive and solve a different system of governing ODEs for each problem. This derivation may prove very difficult, and even impossible.

3.2 Fourier–collocation methods

We can circumvent the need for evaluating the inner products, which caused us such difficulty in the previous section, by using quadrature formulas to approximate the integrals. This amounts to using the interpolating operator $\mathcal{I}_N$ instead of the orthogonal projection operator $\mathcal{P}_N$, and is called the Fourier–collocation method. This is also known as the pseudospectral method.

When forming the Fourier–Galerkin method we require that the orthogonal projection of the residual onto $\hat{\mathsf{B}}_N$ vanishes. To form the Fourier–collocation method we require, instead, that the residual vanishes identically on some set of gridpoints y_j. We refer to this grid as the *collocation grid*, and note that this grid need not be the same as the *interpolation grid* which we have been using up to now, comprising the points x_j.

In the following, we deal with approximations based on the interpolation grid

$$x_j = \frac{2\pi}{N} j, \quad j \in [0, \ldots, N-1],$$

where N is even. However, the discussion holds true for approximations based on an odd number of points, as well.

We assume that the solution, $u(x,t) \in L^2[0, 2\pi]$, is periodic and consider, once again, the general problem

$$\begin{aligned}
\frac{\partial u(x,t)}{\partial t} &= \mathcal{L}u(x,t), \quad && x \in [0, 2\pi], \quad t \geq 0, \\
u(x,0) &= g(x), && x \in [0, 2\pi], \quad t = 0.
\end{aligned}$$

In the Fourier–collocation method we seek solutions,

$$u_N \in \tilde{\mathsf{B}}_N = \text{span}\{(\cos(nx), 0 \leq n \leq N/2) \cup (\sin(nx), 1 \leq n \leq N/2 - 1)\},$$

of the form

$$u_N(x,t) = \sum_{|n|\le N/2} a_n(t)e^{inx},$$

This trigonometric polynomial can also be expressed

$$u_N(x,t) = \sum_{j=0}^{N-1} u_N(x_j,t)g_j(x),$$

where $g_j(x)$ is the Lagrange interpolation polynomial for an even number of points.

Now the difference between the Fourier–Galerkin and the Fourier–collocation method will appear: we require that the residual

$$R_N(x,t) = \frac{\partial u_N(x,t)}{\partial t} - \mathcal{L}u_N(x,t),$$

vanish at the grid points, y_j, i.e.,

$$R_N(y_j,t) = 0, \quad \forall j \in [0,\ldots,N-1]. \tag{3.2}$$

This yields N equations to determine the N point values, $u_N(x_j,t)$, of the numerical solution. In other words, the pseudospectral approximation u_N satisfies the equation

$$\frac{\partial u_N(x,t)}{\partial t} - \mathcal{I}_N\mathcal{L}u_N(x,t) = 0.$$

Next, we will revisit some of the examples for the Fourier–Galerkin method. For simplicity, the collocation grid will be the same as the interpolation grid, except when stated explicitly.

Example 3.5 Consider first the linear constant coefficient problem

$$\frac{\partial u(x,t)}{\partial t} = c\frac{\partial u(x,t)}{\partial x} + \epsilon\frac{\partial^2 u(x,t)}{\partial x^2},$$

with the assumption that $u(x,t) \in C_p^\infty[0,2\pi]$, and c and ϵ are constants.

We seek solutions of the form

$$u_N(x,t) = \sum_{|n|\le N/2} a_n(t)e^{inx} = \sum_{j=0}^{N-1} u_N(x_j,t)g_j(x), \tag{3.3}$$

such that the residual

$$R_N(x,t) = \frac{\partial u_N(x,t)}{\partial t} - c\frac{\partial}{\partial x}u_N(x,t) - \epsilon\frac{\partial^2}{\partial x^2}u_N(x,t),$$

vanishes at a specified set of grid points, y_j. In this case we choose $y_j = x_j$. This results in N ordinary differential equations for $u_N(x_j, t)$,

$$\begin{aligned}\frac{du_N(x_j,t)}{dt} &= c\mathcal{I}_N\frac{\partial}{\partial x}\mathcal{I}_N u_N(x_j,t) + \epsilon\mathcal{I}_N\frac{\partial^2}{\partial x^2}\mathcal{I}_N u_N(x_j,t)\\ &= \sum_{k=0}^{N-1}\left(c\mathrm{D}^{(1)}_{jk} + \epsilon\mathrm{D}^{(2)}_{jk}\right)u_N(x_k,t),\end{aligned}$$

where $\mathrm{D}^{(1)}$ and $\mathrm{D}^{(2)}$ are the differentiation matrices. Consequently, the scheme consists of solving the ODEs only at the grid points. Note that in the case of variable coefficient $c = c(x)$, the derivation above remains the same, except that c is replaced by $c(x_k)$.

In the collocation method the only use we make of the Fourier approximation is in obtaining the derivatives of the numerical approximation in physical space. As we mentioned in the last chapter, this can be done in two mathematically identical, though computionally different, ways. One way uses the Fourier series and possibly a fast Fourier transform (FFT), while the other employs the direct matrix-vector multiplication.

If we require that the residual vanishes at a set of grid points, y_j, which is different from x_j, we get N equations of the form

$$\frac{du_N(y_j,t)}{dt} = c\sum_{i=0}^{N-1} u_N(x_i,t)\left.\frac{dg_i}{dx}\right|_{y_j} + \epsilon\sum_{i=0}^{N-1} u_N(x_i,t)\left.\frac{d^2g_i}{dx^2}\right|_{y_j},$$

where, as in the previous chapter, g_i are the interpolating functions based on the points x_j.

In the simple linear case above, the formulation of the collocation method is straightforward. However, the Galerkin method was not complicated for this simple linear case either. It is in the realm of nonlinear problems that the advantage of the collocation method becomes apparent.

Example 3.6 Consider the nonlinear problem

$$\frac{\partial u(x,t)}{\partial t} = u(x,t)\frac{\partial u(x,t)}{\partial x},$$

where the initial conditions are given and the solution and all its derivatives are smooth and periodic over the time-interval of interest.

We construct the residual

$$R_N(x,t) = \frac{\partial u_N(x,t)}{\partial t} - u_N(x,t)\frac{\partial u_N(x,t)}{\partial x},$$

where, as before, u_N is a trigonometric polynomial of degree N. The residual is required to vanish at the gridpoints x_j, $j = 0, \ldots, N-1$, leading to

$$\frac{du_N(x_j, t)}{dt} - u_N(x_j, t)\frac{\partial u_N(x, t)}{\partial x}\bigg|_{x=x_j} = 0,$$

i.e.,

$$\frac{du_N(x_j, t)}{dt} - u_N(x_j, t)\sum_{k=0}^{N-1} D_{jk}u_N(x_k, t) = 0.$$

Note that obtaining the equations is equally simple for a nonlinear problem as for a constant coefficient linear problem. This is in marked contrast to the Galerkin case.

Finally, we revisit the problem where the nonlinearity was so strong that formulating the Fourier–Galerkin method was untenable.

Example 3.7 Consider the strongly nonlinear problem

$$\frac{\partial u(x, t)}{\partial t} = e^{u(x,t)}\frac{\partial u(x, t)}{\partial x}.$$

We seek solutions of the form

$$u_N(x, t) = \sum_{|n|\le N/2} a_n(t)e^{inx} = \sum_{j=0}^{N-1} u_N(x_j, t)g_j(x),$$

by requiring that the residual

$$R_N(x, t) = \frac{\partial u_N(x, t)}{\partial t} - e^{u_N(x,t)}\frac{\partial u_N(x, t)}{\partial x},$$

vanishes at the grid points, x_j,

$$R_N(x_j, t) = \frac{du_N(x_j, t)}{dt} - e^{u_N(x_j,t)}\frac{\partial u_N(x, t)}{\partial x}\bigg|_{x=x_j} = 0.$$

Once again, the differentiation can be carried out by Fourier series (using FFT) or by matrix-vector multiplication. The nonlinear term is trivial to evaluate, because this is done in physical space.

Thus, the application of the Fourier–collocation method is easy even for problems where the Fourier–Galerkin method fails. This is due to the fact that we can easily evaluate the nonlinear function, $F(u)$, in terms of the point values of $u(x)$, while it may be very hard, and in some cases impossible, to express the Fourier coefficients of $F(u)$ in terms of the expansion coefficients of $u(x)$. In other words: projection is hard, interpolation is easy.

3.3 Stability of the Fourier–Galerkin method

In Chapter 2, we analyzed the truncation error of the Fourier series. This was done for the continuous approximation and for the discrete approximation, which is mainly relevant to the Galerkin and collocation methods, respectively. In this section and the next, we discuss the stability of Fourier spectral methods. This, too, will be done in two parts: the continuous case, which is the Galerkin method, will be addressed in this section; and the discrete case, which is the collocation method, in the next. This analysis is for the semidiscrete form, where only the spatial components are discretized.

We consider the initial boundary value problem

$$\frac{\partial u}{\partial t} = \mathcal{L}u, \tag{3.4}$$

with proper initial data. The solution $u(x, t)$ is in some Hilbert space with the scalar product $(\cdot, \cdot)$, e.g $L^2[0, 2\pi]$.

A special case of a well posed problem is if $\mathcal{L}$ is semi-bounded in the Hilbert space scalar product, i.e., $\mathcal{L} + \mathcal{L}^* \leq 2\alpha \mathrm{I}$ for some constant α. In this special case, the Fourier–Galerkin method is stable.

First, let's show that Equation (3.4) with a semi-bounded operator $\mathcal{L}$ is well posed. To show this, we estimate the derivative of the norm by considering

$$\begin{aligned} \frac{d}{dt}\|u\|^2 &= \frac{d}{dt}(u, u) = \left(\frac{du}{dt}, u\right) + \left(u, \frac{du}{dt}\right) \\ &= (\mathcal{L}u, u) + (u, \mathcal{L}u) = (u, \mathcal{L}^*u) + (u, \mathcal{L}u) \\ &= (u, (\mathcal{L} + \mathcal{L}^*)u). \end{aligned} \tag{3.5}$$

Since $\mathcal{L} + \mathcal{L}^* \leq 2\alpha \mathrm{I}$, we have $\frac{d}{dt}\|u\|^2 \leq 2\alpha\|u\|^2$ and so $\frac{d}{dt}\|u\| \leq \alpha\|u\|$, which means that the norm is bounded

$$\|u(t)\| \leq e^{\alpha t}\|u(0)\|,$$

and the problem is well posed.

In the following, we consider two specific examples of semi-bounded operators in L^2.

Example 3.8 Consider the operator

$$\mathcal{L} = a(x)\frac{\partial}{\partial x},$$

operating on the Hilbert space $L^2[0, 2\pi]$ where $a(x)$ is a real periodic function with periodic bounded derivative. The adjoint operator is obtained by integration by parts:

$$
\begin{aligned}
(\mathcal{L}u, v)_{L^2[0,2\pi]} &= \int_0^{2\pi} a(x)\frac{\partial u}{\partial x}\overline{v}\,dx \\
&= -\int_0^{2\pi} u\frac{\partial}{\partial x}(a(x)\overline{v})\,dx \\
&= \left(u, \left[-a(x)\frac{\partial}{\partial x} - \frac{da(x)}{dx}\right]v\right)_{L^2[0,2\pi]}.
\end{aligned}
$$

Thus,

$$\mathcal{L}^* = -\frac{\partial}{\partial x}a(x)\mathrm{I} = -a(x)\frac{\partial}{\partial x} - a'(x)\mathrm{I}.$$

This means that

$$\mathcal{L} + \mathcal{L}^* = -a'(x)\mathrm{I},$$

and since the derivative of $a(x)$ is bounded $|a'(x)| \le 2\alpha$, we have

$$\mathcal{L} + \mathcal{L}^* \le 2\alpha\mathrm{I}.$$

Example 3.9 Consider the operator

$$\mathcal{L} = \frac{\partial}{\partial x}b(x)\frac{\partial}{\partial x},$$

where, once again, $\mathcal{L}$ operates on the Hilbert space $L^2[0, 2\pi]$ and $b(x)$ is a real periodic nonnegative function with a periodic bounded derivative.

As above,

$$(u, \mathcal{L}u)_{L^2[0,2\pi]} = \left(u, \frac{\partial}{\partial x}b(x)\frac{\partial}{\partial x}u\right)_{L^2[0,2\pi]} = \left(-b(x)\frac{\partial}{\partial x}u, \frac{\partial}{\partial x}u\right)_{L^2[0,2\pi]}.$$

Thus,

$$(u, (\mathcal{L} + \mathcal{L}^*)u)_{L^2[0,2\pi]} = -2\left(b(x)\frac{\partial}{\partial x}u, \frac{\partial}{\partial x}u\right)_{L^2[0,2\pi]} \le 0.$$

The nice thing about the Fourier–Galerkin method is that it is stable provided only that the operator is semi-bounded.

Theorem 3.10 *Given the problem $\partial u/\partial t = \mathcal{L}u$, where the operator $\mathcal{L}$ is semi-bounded in the usual $L^2[0, 2\pi]$ scalar product, then the Fourier–Galerkin method is stable.*

Proof: First, we show that $\mathcal{P}_N = \mathcal{P}_N^*$. We begin with the simple observation that,

$$(u, \mathcal{P}_N v) = (\mathcal{P}_N u, \mathcal{P}_N v) + ((\mathrm{I} - \mathcal{P}_N)u, \mathcal{P}_N v).$$

The second term is the scalar product of the projection of v on the space $\hat{\mathsf{B}}_N$ with the projection of u on the complement of the space $\hat{\mathsf{B}}_N$. Clearly, this is zero, i.e., $(u, \mathcal{P}_N v) = (\mathcal{P}_N u, \mathcal{P}_N v)$. By the same argument, we find that $(\mathcal{P}_N u, v) = (\mathcal{P}_N u, \mathcal{P}_N v)$. Therefore, $(u, \mathcal{P}_N v) = (\mathcal{P}_N u, v)$, which means that $\mathcal{P}_N = \mathcal{P}_N^*$.

Now, we notice that the Fourier–Galerkin method involves seeking the trigonometric polynomial u_N such that

$$\frac{\partial u_N}{\partial t} = \mathcal{P}_N \mathcal{L} \mathcal{P}_N u_N = \mathcal{L}_N u_N.$$

Notice that

$$\begin{aligned}\mathcal{L}_N + \mathcal{L}_N^* &= \mathcal{P}_N \mathcal{L} \mathcal{P}_N + \mathcal{P}_N \mathcal{L}^* \mathcal{P}_N \\ &= \mathcal{P}_N (\mathcal{L} + \mathcal{L}^*) \mathcal{P}_N \leq 2\alpha \mathcal{P}_N.\end{aligned}$$

Following Equation (3.5) this leads to the stability estimate

$$\|u_N(t)\| \leq e^{\alpha t} \|u_N(0)\|,$$

provided that $\mathcal{L}$ is semi-bounded.

QED

It is important to note that for the collocation method it is *not* true that $\mathcal{I}_N = \mathcal{I}_N^*$ in the usual L^2 scalar product. This occurs because of the aliasing error. Thus, the above proof breaks down for the Fourier–collocation method and different techniques are needed to establish stability.

3.4 Stability of the Fourier–collocation method for hyperbolic problems I

In the Galerkin case, the fact that the operator is semi-bounded was enough to guarantee stability of the numerical method. However, this is not at all the case for the collocation method. While the aliasing error in the collocation method does not cause problems for the accuracy of the method, it significantly affects the stability of the method.

To establish stability of the pseudospectral method, we can proceed along two different avenues, exploiting the properties of the differentiation matrices or the exactness of quadrature rules for trigonometric polynomials.

Whereas for the continuous Galerkin method we use the L^2 continuous inner product norm, for the discrete collocation method we define the discrete inner product

$$(f_N, g_N)_N = \frac{1}{N+1} \sum_{j=0}^{N} f_N(x_j) \bar{g}_N(x_j),$$

and the associated energy norm

$$\| f_N \|_N^2 = (f_N, f_N)_N$$

where $f_N, g_N \in \hat{\mathrm{B}}_N$ and there are an odd number of grid points x_j, $j = 0, \ldots, N$. As a consequence of the exactness of the quadrature rule for trigonometric functions (see Theorem 2.9) we have

$$(f_N, g_N)_N = \frac{1}{2\pi} \int_0^{2\pi} f_N \bar{g}_N \, dx, \quad \| f_N \|_{L^2[0,2\pi]} = \| f_N \|_N.$$

Hence, in $\hat{\mathrm{B}}_N$, the continuous and discrete inner product are the same.

The situation is different when we discuss an even number of grid points. If $f_N, g_N \in \tilde{\mathrm{B}}_N$ and we have an even number of grid points x_j, the discrete inner product

$$(f_N, g_N)_N = \frac{1}{N} \sum_{j=0}^{N-1} f_N(x_j) \bar{g}_N(x_j), \; \| f_N \|_N^2 = (f_N, f_N)_N$$

is not equal to the continuous inner product. However, using the fact that $f_N \in L^2[0, 2\pi]$ it can be shown that there exists a $K > 0$ such that

$$K^{-1} \| f_N \|_{L^2[0,2\pi]}^2 \leq \| f_N \|_N^2 \leq K \| f_N \|_{L^2[0,2\pi]}^2. \tag{3.6}$$

Hence, the continuous and discrete norms are uniformly equivalent.

In the following, we attempt to derive bounds on the energy. We shall use the fact that the differentiation matrices are all symmetric or skew-symmetric. Alternatively, the quadrature rules may, under certain circumstances, allow us to pass from the semidiscrete case with summations to the continuous case with integrals and, thus, simplify the analysis. Unless otherwise stated we focus our attention on the Fourier–collocation methods based on an even number of grid points, and we generally take the collocation grid to be the same as the interpolation grid i.e., $y_j = x_j$. However, most of the results can be generalized.

Let us consider the stability of the pseudospectral Fourier approximation to the periodic hyperbolic problem

$$\frac{\partial u}{\partial t} + a(x) \frac{\partial u}{\partial x} = 0, \tag{3.7}$$
$$u(x, 0) = g(x).$$

We first assume that $0 \leq 1/k \leq |a(x)| \leq k$ for some k; it is critical for stability that $a(x)$ is bounded away from zero. We go on to prove stability for the case in which $a(x)$ is strictly positive. Similar results can be obtained in the same way if $a(x)$ is strictly negative. However, the case in which $a(x)$ passes through zero may not be stable as will be discussed in Section 3.5.

Theorem 3.11 *Stability via method 1: the pseudospectral Fourier approximation to the variable coefficient hyperbolic problem, Equation (3.7), with $0 \leq 1/k \leq |a(x)| \leq k$ is stable:*

$$\sum_{j=0}^{N-1} u_N^2(x_j, t) \leq k^2 \sum_{j=0}^{N-1} u_N^2(x_j, 0).$$

Proof: In the collocation method, the approximant $u_N \in \tilde{\mathrm{B}}_N$ satisfies

$$\left.\frac{\partial u_N}{\partial t}\right|_{x_j} + a(x_j) \left.\frac{\partial u_N}{\partial x}\right|_{x_j} = 0, \tag{3.8}$$

where x_j represents the even grid points and we require that the equation be satisfied on these points also. Since $a(x)$ is uniformly bounded away from zero, $a(x)^{-1}$ exists. Multiplying Equation (3.8) by $a(x_j)^{-1} u_N(x_j)$ and summing over all collocation points we obtain

$$\frac{1}{2} \frac{d}{dt} \sum_{j=0}^{N-1} \frac{1}{a(x_j)} u_N^2(x_j, t) = - \sum_{j=0}^{N-1} u_N(x_j, t) \left.\frac{\partial u_N}{\partial x}\right|_{x_j}.$$

Using the exactness of the quadrature formula, and the fact that u_N is periodic,

$$\frac{1}{2} \frac{d}{dt} \sum_{j=0}^{N-1} \frac{1}{a(x_j)} u_N^2(x_j, t) = - \frac{N}{2\pi} \int_0^{2\pi} u_N(x) \frac{\partial u_N}{\partial x} \, dx = 0.$$

Thus, the summation is not changing over time:

$$\sum_{j=0}^{N-1} \frac{1}{a(x_j)} u_N^2(x_j, t) = \sum_{j=0}^{N-1} \frac{1}{a(x_j)} u_N^2(x_j, 0).$$

Finally, we observe that, since $0 \leq 1/k \leq |a(x)| \leq k$, we have

$$\begin{aligned} \frac{1}{k} \sum_{j=0}^{N-1} u_N^2(x_j, t) \leq \sum_{j=0}^{N-1} \frac{1}{a(x_j)} u_N^2(x_j, t) &= \sum_{j=0}^{N-1} \frac{1}{a(x_j)} u_N^2(x_j, 0) \\ &\leq k \sum_{j=0}^{N-1} u_N^2(x_j, 0), \end{aligned}$$

and so

$$\sum_{j=0}^{N-1} u_N^2(x_j, t) \leq k^2 \sum_{j=0}^{N-1} u_N^2(x_j, 0).$$

From this we conclude that,

$$\|u_N(x, t)\|_N \leq k \|u_N(x, 0)\|_N.$$

The proof for an odd number of points is identical.

QED

Another method of proving stability is based on the matrix representation of the Fourier–collocation method. The Fourier–collocation approximation to the variable coefficient hyperbolic problem can be written as

$$\frac{d\boldsymbol{u}_N(t)}{dt} + \mathrm{AD}\boldsymbol{u}_N(t) = 0, \tag{3.9}$$

where $\boldsymbol{u}_N$ is the vector of unknowns $u_N(x_j, t)$, and A is a diagonal matrix with entries $\mathrm{A}_{jj} = a(x_j)$. The matrix D is our differentiation matrix, Equation (2.19), which is skew-symmetric $\mathrm{D}^T = -\mathrm{D}$. This fact is the crux of the stability proof, and is true regardless of the use of an even or odd number of collocation points.

The solution to Equation (3.9) is given by

$$\boldsymbol{u}(t) = e^{-\mathrm{AD}t}\boldsymbol{u}(0).$$

In this context, stability means that the matrix exponential is bounded independent of the number of points N:

$$\|e^{-\mathrm{AD}t}\| \leq K(t),$$

in the L^2 spectral norm. In other words,

$$e^{-\mathrm{AD}t}e^{-(\mathrm{AD})^T t} \leq K^2(t).$$

Note that if $a(x_j) = a$, a constant, then the matrix $\mathrm{A} = a\mathrm{I}$ and this simplifies to

$$e^{-a\mathrm{D}t}e^{(-a\mathrm{D})^T t} = e^{-a(\mathrm{D}+\mathrm{D}^T)t} = \mathrm{I},$$

since D is skew-symmetric and commutes with itself.

Theorem 3.12 *Stability via method 2: the pseudospectral Fourier approximation to the variable coefficient hyperbolic problem, Equation (3.7) with $0 < 1/k \leq |a(x)| \leq k < \infty$ is stable:*

$$\|e^{-\mathrm{AD}t}\| \leq k.$$

Proof: Once again, we consider the case $a(x) > 0$. We rewrite

$$-\mathrm{AD} = -\mathrm{A}^{\frac{1}{2}}\mathrm{A}^{\frac{1}{2}}\mathrm{D}\mathrm{A}^{\frac{1}{2}}\mathrm{A}^{-\frac{1}{2}}, \quad \text{and so} \quad e^{-\mathrm{AD}t} = \mathrm{A}^{\frac{1}{2}}e^{-\mathrm{A}^{\frac{1}{2}}\mathrm{D}\mathrm{A}^{\frac{1}{2}}t}\mathrm{A}^{-\frac{1}{2}},$$

by using the definition of the matrix exponential $e^A = \sum_{n=0}^{\infty}\frac{1}{n!}A^n$. Since the matrix is skew-symmetric,

$$\left(\mathrm{A}^{\frac{1}{2}}\mathrm{D}\mathrm{A}^{\frac{1}{2}}\right)^T = \mathrm{A}^{\frac{1}{2}}\mathrm{D}^T\mathrm{A}^{\frac{1}{2}} = -\mathrm{A}^{\frac{1}{2}}\mathrm{D}\mathrm{A}^{\frac{1}{2}},$$

we have $\|e^{-A^{\frac{1}{2}}DA^{\frac{1}{2}}t}\| = 1$. Now,

$$\begin{aligned}\|e^{-ADt}\| &\leq \|A^{\frac{1}{2}}\| \|e^{-A^{\frac{1}{2}}DA^{\frac{1}{2}}t}\| \|A^{-\frac{1}{2}}\| \\ &= \|A^{\frac{1}{2}}\| \|A^{-\frac{1}{2}}\| \leq \frac{\sqrt{k}}{\frac{1}{\sqrt{k}}} = k.\end{aligned}$$

We note that the same exact stability proof holds for e^{ADt}, i.e., for the equation $\partial u/\partial t - a(x)\partial u/\partial x = 0$. This establishes the proof for the case $a(x) < 0$ as well. The same stability proof will also hold for e^{-DAt}, which is the solution of $\frac{\partial u}{\partial t} + \frac{\partial a(x)u}{\partial x} = 0$.

QED

3.5 Stability of the Fourier–collocation method for hyperbolic problems II

In the general case where $a(x)$ changes sign, a mild instability can occur. The solution operator, rather than being bounded independently of N, grows linearly with N,

$$\|e^{ADt}\| \sim N.$$

In a well resolved problem, this mild growth is absorbed in the trucation error, and the numerical solution is still accurate. However, if the problem is only marginally resolved, or if there are other perturbations, this linear growth leads to full-blown instability.

Although the Fourier–collocation approximation to Equation (3.7), in the case where $a(x)$ vanishes somewhere in the domain, is not stable in general, there are several known techniques of stabilizing the method.

The most straightforward way to derive a stable pseudospectral Fourier approximation to Equation (3.7) with $|a_x(x)|$ bounded, is to consider the skew-symmetric form of the equation

$$\frac{\partial u}{\partial t} + \frac{1}{2}a(x)\frac{\partial u}{\partial x} + \frac{1}{2}\frac{\partial a(x)u}{\partial x} - \frac{1}{2}a_x(x)u(x,t) = 0. \tag{3.10}$$

The Fourier–collocation approximation to *this* equation is stable.

Theorem 3.13 *The pseudospectral Fourier approximation to the variable coefficient hyperbolic equation, Equation (3.10) is stable:*

$$\|u_N(t)\|_N \leq e^{\alpha t}\|u_N(0)\|_N, \tag{3.11}$$

where

$$\alpha = \frac{1}{2} \max_x |a_x(x)|.$$

Proof: In the Fourier collocation method we seek a polynomial satisfying the equation

$$\left.\frac{\partial u_N}{\partial t}\right|_{x_j} + \frac{1}{2}a(x_j)\left.\frac{\partial u_N}{\partial x}\right|_{x_j} + \frac{1}{2}\left.\frac{\partial \mathcal{J}_N[a(x)u_N]}{\partial x}\right|_{x_j} - \frac{1}{2}a_x(x_j)u_N(x_j) = 0 \tag{3.12}$$

at all grid points, x_j. Multiplying by $u_N(x_j)$ and summing over all the collocation points we obtain

$$\begin{aligned}\frac{1}{2}\frac{d}{dt}\sum_{j=0}^{N-1} u_N^2(x_j) = &-\frac{1}{2}\sum_{j=0}^{N-1} a(x_j)u_N(x_j)\left.\frac{\partial u_N}{\partial x}\right|_{x_j} \\ &-\frac{1}{2}\sum_{j=0}^{N-1} u_N(x_j)\left.\frac{\partial \mathcal{J}_N[a(x)u_N]}{\partial x}\right|_{x_j} + \frac{1}{2}\sum_{j=0}^{N-1} a_x(x_j)u_N^2(x_j).\end{aligned}$$

Observe that in the second term $\mathcal{J}_N \partial \mathcal{J}_N[a(x)u_N(x)]/\partial x \in \hat{\mathsf{B}}_N$, so the quadrature rule in Theorem 2.9 is exact. Thus,

$$\begin{aligned}\frac{1}{2}\sum_{j=0}^{N-1} u_N(x_j)\left.\frac{\partial \mathcal{J}_N[a(x)u_N(x)]}{\partial x}\right|_{x_j} &= \frac{N}{4\pi}\int_0^{2\pi} u_N(x,t)\mathcal{J}_N\frac{\partial \mathcal{J}_N[a(x)u_N(x)]}{\partial x}\,dx \\ &= -\frac{N}{4\pi}\int_0^{2\pi} \mathcal{J}_N[a(x)u_N(x)]\mathcal{J}_N\frac{\partial u_N(x)}{\partial x}\,dx \\ &= -\frac{1}{2}\sum_{j=0}^{N-1} a(x_j)u_N(x_j)\left.\frac{\partial u_N(x)}{\partial x}\right|_{x_j}.\end{aligned}$$

Hence,

$$\begin{aligned}\frac{1}{2}\frac{d}{dt}\|u_N\|_N^2 &= \frac{1}{2}\sum_{j=0}^{N-1} a_x(x_j)u_N^2(x_j,t) \\ &\le \frac{1}{2}\max_x |a_x(x)| \sum_{j=0}^{N-1} u_N^2(x_j,t) \\ &\le \frac{1}{2}\max_x |a_x(x)|\,\|u_N\|_N^2,\end{aligned}$$

leading to the stability estimate of the theorem.

QED

Note that the approximation in Equation (3.12) can be written as

$$\frac{\partial u_N}{\partial t} + \frac{1}{2}\mathcal{J}_N a(x)\frac{\partial u_N}{\partial x} + \frac{1}{2}\frac{\partial}{\partial x}\mathcal{J}_N[a(x)u_N] - \frac{1}{2}\mathcal{J}_N(a_x u_N) = 0.$$

If we write the last term as

$$\mathcal{J}_N(a_x u_N) = \mathcal{J}_N\frac{\partial}{\partial x}(a(x)u_N) - \mathcal{J}_N a(x)\frac{\partial u_N}{\partial x},$$

the approximation becomes

$$\frac{\partial u_N}{\partial t} + \mathcal{J}_N a(x)\frac{\partial u_N}{\partial x} + A_N = 0$$

which is the usual (unstable) Fourier–collocation approximation to Equation (3.7) plus the term

$$A_N = \frac{1}{2}\left(\frac{\partial}{\partial x}\mathcal{J}_N[a(x)u_N] - \mathcal{J}_N\frac{\partial}{\partial x}[a(x)u_N]\right).$$

Thus it is this addition of the aliasing term A_N which stabilizes the method.

Note that, by the aliasing theorem

$$\|A_N\|_{L^2[0,2\pi]} \leq N^{1-2s}\left\|(a(x)u_N)^{(2s)}\right\|_{L^2[0,2\pi]} \leq CN^{1-2s}\left\|u_N^{(2s)}\right\|_{L^2[0,2\pi]}.$$

This suggests a different way of stabilizing the Fourier–collocation method by adding a superviscosity term

$$\frac{\partial u_N}{\partial t} + \mathcal{J}_N a(x)\frac{\partial u_N}{\partial x} = (-1)^{s+1}\frac{\epsilon}{N^{2s-1}}\frac{\partial^{2s} u_N}{\partial x^{2s}}. \tag{3.13}$$

Theorem 3.14 *The superviscosity approximation, Equation (3.13), to Equation (3.7) is stable provided that ϵ is large enough.*

Proof: We rewrite Equation (3.13) as

$$\frac{\partial u_N}{\partial t} + \mathcal{J}_N\left(a(x)\frac{\partial u_N}{\partial x}\right) + A_N = (-1)^{s+1}\frac{\epsilon}{N^{2s-1}}\frac{\partial^{2s} u_N}{\partial x^{2s}} + A_N. \tag{3.14}$$

Multiplying by u_N and summing, we get

$$\begin{aligned}\left(u_N, \frac{\partial u_N}{\partial t}\right)_N &= -\left(u_N, \mathcal{J}_N\left(a(x)\frac{\partial u_N}{\partial x}\right)\right)_N - (u_N, A_N)_N \\ &\quad + \frac{(-1)^{s+1}}{N^{2s-1}}\epsilon\left(u_N, \frac{\partial^{2s} u_N}{\partial x^{2s}}\right)_N + (u_N, A_N)_N,\end{aligned}$$

since the quadrature formula is exact, all the scalar products are, in fact, integrals.

Looking at the first term on the right hand side of the equation, we observe that

$$-\left(u_N, \mathcal{J}_N\left(a(x)\frac{\partial u_N}{\partial x}\right)\right)_N - (u_N, A_N)_N = -\frac{1}{2}(u_N, a_x u_N)_N$$
$$\leq \frac{1}{2} max|a_x|(u_N, u_N)_N.$$

We proceed to show that the next term is negative: first, note that integrating by parts s times yields

$$(u_N, A_N)_N = \frac{1}{2\pi}\int_0^{2\pi} u_N A_N dx = \frac{1}{2\pi}(-1)^s \int_0^{2\pi} u_N^{(s)} A_N^{(-s)} dx \leq c\frac{\left\|u_N^{(s)}\right\|^2}{N^{2s-1}}$$

and

$$\left(u_N, \frac{\partial^{(2s)} u_N}{\partial x^{(2s)}}\right)_N = \frac{1}{2\pi}\int_0^{2\pi} u_N \frac{\partial^{(2s)} u_N}{\partial x^{(2s)}} dx = (-1)^s \left\|u_N^{(s)}\right\|^2.$$

Putting these together, we obtain

$$\left(u_N, \frac{\partial u_N}{\partial t}\right)_N = -\left(u_N, \mathcal{J}_N\left(a(x)\frac{\partial u_N}{\partial x}\right)\right)_N - (u_N, A_N)_N$$
$$+ \frac{(-1)^{s+1}}{N^{2s-1}}\epsilon\left(u_N, \frac{\partial^{2s} u_N}{\partial x^{2s}}\right)_N + (u_N, A_N)_N$$
$$\leq \frac{1}{2} max|a_x|(u_N, u_N)_N + \frac{(-1)^{s+1}}{N^{2s-1}}\epsilon(-1)^s\left\|u_N^{(s)}\right\|^2 + c\frac{\left\|u_N^{(s)}\right\|^2}{N^{2s-1}}$$
$$= \frac{1}{2} max|a_x|(u_N, u_N)_N + (c-\epsilon)\frac{\left\|u_N^{(s)}\right\|^2}{N^{2s-1}}.$$

If $\epsilon > c$ this last term is negative, and thus we conclude that the scheme is stable:

$$\left(u_N, \frac{\partial u_N}{\partial t}\right)_N \leq \frac{1}{2} max|a_x|(u_N, u_N)_N.$$

QED

We note that we are considering the scheme based on an odd number of points. This is only done for simplicity, as it allows us to pass to the integrals without complications. However, the conclusions we reach remain valid for the even case, also.

The advantage of working with the skew-symmetric form is in the simplicity of the stability proof. However, there is a significant practical drawback in that it increases the complexity of the scheme by doubling the number of derivatives.

It would seem that this is also the case in the superviscosity method, but in fact there is a way to use the superviscosity method with no extra cost.

Consider the equation

$$\frac{\partial u_N}{\partial t} = (-1)^{s+1} \frac{\epsilon}{N^{2s-1}} \frac{\partial^{2s}}{\partial x^{2s}} u_N.$$

Using a Fourier–Galerkin method we solve for the coefficients of u_N to get

$$\frac{da_k(t)}{dt} = -\epsilon \frac{k^{2s}}{N^{2s-1}} a_k(t).$$

Thus,

$$a_k(t + \Delta t) = e^{-\epsilon \Delta t N (\frac{k}{N})^{2s}} a_k(t).$$

The coefficients $a_k(t)$ are advanced by the PDE over one time step and then multiplied by an exponential filter. The addition of the superviscosity term is the equivalent of using an exponential filter of the above form. Low pass filters stabilize the Fourier–collocation method at no extra cost.

3.6 Stability for parabolic equations

Let us now consider the question of stability for the strongly parabolic prototype problem

$$\frac{\partial u}{\partial t} = b(x) \frac{\partial^2 u}{\partial x^2}, \tag{3.15}$$
$$u(x, 0) = g(x),$$

where $b(x) > 0$ for wellposedness, and $u(x, t)$ as well as $g(x)$ are assumed to be periodic and smooth.

As before we shall derive two equivalent estimates which address the question of stability for Equation (3.15).

Theorem 3.15 *The Fourier–collocation method approximation to the strongly parabolic problem, Equation (3.15), is stable:*

$$\|u_N(t)\|_N \leq \sqrt{\frac{\max b(x)}{\min b(x)}} \|u_N(0)\|_N.$$

Proof:

Method 1 In matrix form, the Fourier–collocation approximation to the parabolic problem is

$$\frac{d\boldsymbol{u}(t)}{dt} = \mathrm{BD}^{(2)} \boldsymbol{u}(t),$$

where $\boldsymbol{u}$ is the grid vector, whose components are $u_N(x_j, t)$, and x_j represent the grid points. B is the diagonal positive matrix with entries $\mathrm{B}_{jj} = b(x_j)$. Furthermore, $\mathrm{D}^{(2)}$ represents the second derivative differentiation matrix discussed in Section 2.2.4. We need to be careful, however, when defining this matrix because when using an even number of grid points, $\mathrm{D}^{(2)} \neq \mathrm{D} \cdot \mathrm{D}$. The difference between the two formulations stems from the fact that $\mathrm{D}^{(2)}\boldsymbol{u} \in \tilde{\mathrm{B}}_N$ while $(\mathrm{D} \cdot \mathrm{D})\boldsymbol{u} \in \hat{\mathrm{B}}_{N-1}$, i.e., the latter reduces the order of the polynomial. To ensure stability of the Fourier–collocation scheme we choose the latter definition,

$$\mathrm{D}^{(2)} \equiv \mathrm{D} \cdot \mathrm{D}.$$

Note that this problem does not arise when using a method based on an odd number of grid points.

We continue by multiplying with $\boldsymbol{u}^T \mathrm{B}^{-1}$ from the left,

$$\begin{aligned}
\boldsymbol{u}^T \mathrm{B}^{-1} \frac{d}{dt}\boldsymbol{u} &= \boldsymbol{u}^T \mathrm{D}^{(2)} \boldsymbol{u} = \boldsymbol{u}^T \mathrm{D}\mathrm{D}\boldsymbol{u} \\
&= (\mathrm{D}^T \boldsymbol{u})^T (\mathrm{D}\boldsymbol{u}) = -(\mathrm{D}\boldsymbol{u})^T (\mathrm{D}\boldsymbol{u}) \leq 0,
\end{aligned}$$

where we use the fact that D is skew-symmetric. Thus we can say that

$$\frac{d}{dt}\boldsymbol{u}^T \mathrm{B}^{-1} \boldsymbol{u} \leq 0,$$

and so

$$\frac{1}{\max_x b(x)} \|u_N(t)\|_N^2 \leq \boldsymbol{u}^T(t) \mathrm{B}^{-1} \boldsymbol{u}(t) \leq \boldsymbol{u}^T(0) \mathrm{B}^{-1} \boldsymbol{u}(0) \leq \frac{1}{\min_x b(x)} \|u_N(0)\|_N^2.$$

Method 2 The same result may be obtained in scalar form, using the quadrature rules. As usual,

$$u_N(x, t) = \sum_{j=0}^{N-1} u_N(x_j, t) g_j(x),$$

and we require that

$$\left.\frac{\partial u_N(x, t)}{\partial t}\right|_{x_j} = b(x_j) \left.\frac{\partial^2 u_N(x, t)}{\partial x^2}\right|_{x_j}.$$

Multiply by $b(x_j)^{-1} u_N(x_j, t)$ and sum over the collocation points to obtain

$$\frac{1}{2}\frac{d}{dt} \sum_{j=0}^{N-1} \frac{1}{b(x_j)} u_N^2(x_j, t) = \sum_{j=0}^{N-1} u_N(x_j, t) \left.\frac{\partial^2 u_N(x, t)}{\partial x^2}\right|_{x_j}.$$

We realize that the summation on the right hand side is a polynomial of order $2N$, for which the quadrature rule based on an even number of grid points x_j is not exact, and so we cannot immediately pass to the integral. As before, we circumvent this problem by defining the second-order derivative

$$\mathcal{I}_N \frac{d}{dx} \mathcal{I}_N \frac{d}{dx} \mathcal{I}_N.$$

This ensures that

$$\frac{\partial^2 u_N(x,t)}{\partial x^2} = \mathcal{I}_N \frac{\partial}{\partial x} \mathcal{I}_N \frac{\partial}{\partial x} u_N(x,t) \in \hat{\mathrm{B}}_{N-1}.$$

Now the quadrature rule is exact and we have

$$\begin{aligned}
\frac{1}{2}\frac{d}{dt}\sum_{j=0}^{N-1} \frac{1}{b(x_j)} u_N^2(x_j,t) &= \sum_{j=0}^{N-1} u_N(x_j,t)\left(\frac{\partial}{\partial x}\mathcal{I}_N \frac{\partial u_N}{\partial x}\bigg|_{x_j}\right) \\
&= \frac{N}{2\pi}\int_0^{2\pi} u_N(x,t)\mathcal{I}_N\left(\frac{\partial}{\partial x}\mathcal{I}_N\frac{\partial u_N}{\partial x}\right) dx \\
&= -\frac{N}{2\pi}\int_0^{2\pi} \mathcal{I}_N \frac{\partial u_N}{\partial x}\mathcal{I}_N\frac{\partial u_N}{\partial x}\, dx \\
&= -\sum_{j=0}^{N-1}\left(\frac{\partial u_N}{\partial x}\bigg|_{x_j}\right)^2 \le 0.
\end{aligned}$$

The proof continues as in the matrix case. Once again, the case of an odd number of points is simpler since here the quadrature rule is exact for polynomials of order $2N$ and no special definition of the second derivative is necessary.

QED

3.7 Stability for nonlinear equations

In Section 3.5, we showed that the addition of a superviscosity term stabilizes the Fourier–collocation methods for the linear equations. For nonlinear equations, similar techniques can be used to stabilize the methods. Consider the nonlinear hyperbolic system

$$\frac{\partial U}{\partial t} + \frac{\partial f(U)}{\partial x} = 0. \tag{3.16}$$

The Fourier–collocation method can be written as

$$\frac{\partial u_N}{\partial t} + \mathcal{I}_N \frac{\partial \mathcal{I}_N f(u_N)}{\partial x} = 0. \tag{3.17}$$

The method is unstable in general but can be stabilized by either of the following methods:

In the ***Spectral viscosity method*** **(SV)**, Equation (3.17) is modified by

$$\frac{\partial u_N}{\partial t} + \mathcal{I}_N \frac{\partial \mathcal{I}_N f(u_N)}{\partial x} = \epsilon_N (-1)^{s+1} \frac{\partial^s}{\partial x^s}\left[Q_m(x,t) * \frac{\partial^s u_N}{\partial x^s}\right]$$

where the operator Q_m keeps only the high modes of the viscosity term $\frac{\partial^2}{\partial x^2} u_n$:

$$\epsilon_N (-1)^{s+1} \frac{\partial^s}{\partial x^s} \left[Q_m(x,t) * \frac{\partial^s u_N}{\partial x^s} \right] \sim \epsilon \sum_{m<|k|<N} (ik)^{2s} \hat{Q}_k \hat{u}_k e^{ikx}$$

with

$$\begin{aligned} &\epsilon \sim CN^{2s-1}; \\ &m \sim N^{\theta}, \ \theta < \frac{2s-1}{2s}; \\ &1 - \left(\frac{m}{|k|}\right)^{\frac{2s-1}{\theta}} \leq \hat{Q}_k \leq 1. \end{aligned}$$

The most commonly used SV scheme uses $s = 1$.

A better way to stabilize the scheme is the ***Super spectral viscosity*** **(SSV)** method.

$$\frac{\partial u_N}{\partial t} + \mathcal{I}_N \frac{\partial \mathcal{I}_N f(u_N)}{\partial x} = \epsilon_N (-1)^{s+1} \frac{\partial^{2s}}{\partial x^{2s}} u_N. \tag{3.18}$$

Note that for spectral accuracy the order s must be proportional to N, the number of trigonometric polynomials (or grid points) in the approximation. Thus viscosity changes with mesh refinement. As we showed in Section 3.5 the SV and SSV techniques are equivalent to filtering.

The theory developed by Tadmor demonstrates that both the SV and SSV methods converge to the correct entropy solution for stable Fourier approximations to scalar nonlinear hyperbolic equations. It has been proven that even for systems, if the solution converges, it converges to the correct entropy solution (the Lax–Wendroff theorem).

3.8 Further reading

The analysis of stability of the Fourier collocation methods for linear problems was largely completed in the 1970's with contributions by Fornberg (1973), Kreiss and Oliger (1972), Majda et al (1978) and Goodman et al (1979), with the latter highlighting the difficulties posed by variable coefficient problem. A good review of these results are given by Tadmor (1987), including a more thorough discussion of skew-symmetric forms. The exponential convergence for analytic functions is discussed in detail by Tadmor (1986). For the introduction of the vanishing viscosity methods, we refer to the series of papers by Maday and Tadmor (1989) and Tadmor (1989, 1990, 1993), as well as the work by Schochet (1990). The more direct use of filters for stabilization as discussed here was introduced in Gottlieb and Hesthaven (2001).

4
Orthogonal polynomials

At the heart of spectral methods is the fact that any nice enough function $u(x)$ can be expanded in the form

$$u(x) = \sum_{|n|<\infty} \hat{u}_n \phi_n(x),$$

where $\phi_n(x)$ are the global basis functions. Fourier spectral methods, which use the periodic basis $\phi_n(x) = e^{inx}$, perform well and deliver highly accurate results for smooth periodic problems. However, exponential accuracy of the scheme is achieved only when the solution and its derivatives are periodic. Lacking such higher-order periodicity globally impacts the convergence rate of the Fourier series, and reduces the spatial accuracy of the spectral scheme to that of a finite difference scheme. If the function $u(x)$ or any of its derivatives are not periodic, then it makes sense to use a non-periodic basis, such as a polynomial basis. It is convenient to focus on polynomials which are eigensolutions to Sturm–Liouville problems, since this class of polynomials has been extensively studied and has nice convergence properties.

The underlying assumption is that $u(x)$ can be well approximated in the finite dimensional subspace of

$$\mathsf{B}_N = \text{span}\{x^n\}_{n=0}^N,$$

that satisfies the boundary conditions. The emphasis shall be on polynomial approximations of continuous functions, $u(x) \in C^0[a, b]$, where the interval $[a, b]$ could be bounded or unbounded. For simplicity, we will focus on problems defined on a bounded interval, keeping in mind that similar results can be obtained also for problems defined on the semi-infinite interval, $[0, \infty)$, as well as the infinite interval, $(-\infty, \infty)$.

4.1 The general Sturm–Liouville problem

The Sturm–Liouville operator is given by

$$\mathcal{L}\phi(x) = -\frac{d}{dx}\left(p(x)\frac{d\phi(x)}{dx}\right) + q(x)\phi(x) = \lambda w(x)\phi(x), \tag{4.1}$$

subject to the boundary conditions

$$\begin{aligned} \alpha_-\phi(-1) + \beta_-\phi'(-1) = 0, \quad \alpha_-^2 + \beta_-^2 \neq 0, \\ \alpha_+\phi(1) + \beta_+\phi'(1) = 0, \quad \alpha_+^2 + \beta_+^2 \neq 0. \end{aligned} \tag{4.2}$$

We restrict our attention to the interval $[-1, 1]$ for simplicity. In Equation (4.1), we have the real functions $p(x)$, $q(x)$, and $w(x)$. Note that $p(x) \in C^1[-1, 1]$ and is strictly positive in $(-1, 1)$; $q(x) \in C^0[-1, 1]$, is non-negative and bounded; and $w(x) \in C^0[-1, 1]$ is the non-negative continuous weight function.

Assuming that $\alpha_-\beta_- \leq 0$ and $\alpha_+\beta_+ > 0$ one can show that the solutions to the Sturm–Liouville eigenvalue problem are unique sets of eigenfunctions, $\phi_n(x)$, and eigenvalues, λ_n. The eigenfunctions, $\phi_n(x)$, form an $L^2[-1, 1]$-complete basis while the nonnegative and unique eigenvalues form an unbounded sequence,

$$0 \leq \lambda_0 < \lambda_1 < \lambda_2 \ldots$$

Based on this, it is customary to order the eigensolutions in unique pairs (λ_n, ϕ_n). These eigensolutions have the asymptotic behavior

$$\lambda_n \simeq (n\pi)^2 \left(\int_{-1}^{1} \sqrt{\frac{w(x)}{p(x)}}\, dx\right)^{-2} \quad \text{as} \quad n \to \infty,$$

and

$$\phi_n(x) \simeq A_n \sin\left[\sqrt{\lambda_n} \int_1^x \sqrt{\frac{w(x)}{p(x)}}\, dx\right] \quad \text{as} \quad n \to \infty.$$

Furthermore, since $\mathcal{L}$ is self-adjoint, the eigenfunctions of the Sturm–Liouville problem are orthogonal in the $\|\cdot\|_{L^2_w[-1,1]}$-norm,

$$(\phi_n, \phi_m)_{L^2_w[-1,1]} = (\phi_n, \phi_m)_w = \int_{-1}^{1} \phi_n(x)\phi_m(x)w(x)\, dx = \gamma_n\delta_{nm},$$

where $\gamma_n = (\phi_n, \phi_n)_{L^2_w[-1,1]}$. These eigenfunctions are *complete* in L^2_w, so that any function $u(x) \in L^2_w[-1, 1]$ may be represented as

$$u(x) = \sum_{n=0}^{\infty} \hat{u}_n \phi_n(x).$$

A natural approximation to $u(x)$ is then the truncated series

$$\mathcal{P}_N u(x) = \sum_{n=0}^{N} \hat{u}_n \phi_n(x),$$

where the continuous expansion coefficients are obtained from the orthogonality condition

$$\hat{u}_n = \frac{1}{\gamma_n}(u, \phi_n)_{L^2_w[-1,1]}, \quad \gamma_n = \|\phi_n\|^2_{L^2_w[-1,1]}.$$

The Parseval identity

$$\int_{-1}^{1} u^2(x) w(x)\, dx = (u, u)_{L^2_w[-1,1]} = \sum_{n=0}^{\infty} \gamma_n \hat{u}_n^2$$

leads to a truncation error of the form

$$\left\| u(x) - \sum_{n=0}^{N} \hat{u}_n \phi_n(x) \right\|^2_{L^2_w[-1,1]} = \sum_{n=N+1}^{\infty} \gamma_n \hat{u}_n^2.$$

Convergence thus depends solely on the decay of the expansion coefficients, $\hat{u}_n$, similar to the expansions based on trigonometric polynomials.

The decay of the expansion coefficients, $\hat{u}_n$, can be estimated as

$$\begin{aligned}
\hat{u}_n &= \frac{1}{\gamma_n}(u, \phi_n)_{L^2_w[-1,1]} = \frac{1}{\gamma_n}\int_{-1}^{1} u(x)\phi_n(x) w(x)\, dx \\
&= \frac{1}{\gamma_n \lambda_n}\int_{-1}^{1} u(x) \mathcal{L}\phi_n(x)\, dx = \frac{1}{\gamma_n \lambda_n}\int_{-1}^{1} u[-(p\phi_n')' + q\phi_n]\, dx \\
&= \frac{1}{\gamma_n \lambda_n}[-up\phi_n']_{-1}^{1} + \frac{1}{\gamma_n \lambda_n}\int_{-1}^{1} [u' p \phi_n' + qu\phi_n]\, dx \\
&= \frac{1}{\gamma_n \lambda_n}[p(u'\phi_n - u\phi_n')]_{-1}^{1} + \frac{1}{\gamma_n \lambda_n}\int_{-1}^{1} [\mathcal{L}u(x)]\, \phi_n(x)\, dx \\
&= \frac{1}{\gamma_n \lambda_n}[p(u'\phi_n - u\phi_n')]_{-1}^{1} + \frac{1}{\gamma_n \lambda_n}(u_{(1)}, \phi_n)_{L^2_w[-1,1]},
\end{aligned} \tag{4.3}$$

where we have introduced the symbol

$$u_{(m)}(x) = \frac{1}{w(x)}\mathcal{L}u_{(m-1)}(x) = \left(\frac{\mathcal{L}}{w(x)}\right)^m u(x),$$

and $u_{(0)}(x) = u(x)$. This derivation is valid provided $u_{(1)}(x) \in L^2_w[-1, 1]$, i.e., $u^{(2)}(x) \in L^2_w[-1, 1]$ and $w(x)^{-1} \in L^1[-1, 1]$.

In the **singular Sturm–Liouville problem**, $p(x)$ vanishes at the boundary points, i.e., $p(\pm 1) = 0$. Therefore the boundary term in Equation (4.3) vanishes, and repeatedly integrating by parts we obtain

$$|\hat{u}_n| \simeq C\frac{1}{(\lambda_n)^m}\left\|u_{(m)}\right\|_{L^2_w[-1,1]},$$

for $u_{(m)}(x) \in L^2_w[a,b]$, requiring that $u^{(2m)}(x) \in L^2_w[a,b]$. Consequently, if the function $u(x) \in C^\infty[a,b]$ we recover spectral decay of the expansion coefficients, i.e., $|\hat{u}_n|$ decays faster than any algebraic order of λ_n. This result is valid independent of specific boundary conditions on $u(x)$.

This suggests that the eigensolutions of the singular Sturm–Liouville problem are well suited for expanding arbitrary functions defined in the finite interval as the eigenfunctions form a complete, orthogonal basis family, and the expansion is accurate independent of the boundary terms.

It is interesting to note what happens in the **regular Sturm–Liouville problem**. In this case, $p(x)$ and the weight function, $w(x)$, are both strictly positive, and $p(x)$ does not, in general, vanish at both boundaries. Thus the boundary term in Equation (4.3) does not vanish and forms the major contribution to the error. However, for periodic problems this term vanishes by periodicity, thus justifying the use of solutions of the regular Sturm–Liouville problem for periodic problems. The Fourier basis exemplifies this; it is a solution of a regular Sturm–Liouville problem, and is ideal for periodic problems. However, if the problem is not periodic, the decay rate plummets, demonstrating the effects of the first term in Equation (4.3).

4.2 Jacobi polynomials

Spectral methods typically use special cases of Jacobi polynomials, which are the polynomial eigenfunctions of the singular Sturm–Liouville problem.

Theorem 4.1 *The eigensolutions to the singular Sturm–Liouville problem with*

$$p(x) = (1-x)^{\alpha+1}(1+x)^{\beta+1}, \quad w(x) = (1-x)^\alpha(1+x)^\beta, \quad q(x) = cw(x),$$

for $\alpha, \beta > -1$, are polynomials of order n and are known as the Jacobi polynomials $P_n^{(\alpha,\beta)}(x)$. The associated eigenvalues are

$$\lambda_n = n(n+\alpha+\beta+1) - c.$$

A convenient form of the Jacobi polynomial is given by Rodrigues' formula

$$(1-x)^\alpha(1+x)^\beta P_n^{(\alpha,\beta)}(x) = \frac{(-1)^n}{2^n n!}\frac{d^n}{dx^n}[(1-x)^{\alpha+n}(1+x)^{\beta+n}], \qquad (4.4)$$

for integer n. An explicit formula is given by

$$P_n^{(\alpha,\beta)}(x) = \frac{1}{2^n} \sum_{k=0}^{n} \binom{n+\alpha}{k} \binom{n+\beta}{n-k} (x-1)^{n-k}(x+1)^k. \tag{4.5}$$

Furthermore,

$$\frac{d}{dx} P_n^{(\alpha,\beta)}(x) = \frac{1}{2}(n+\alpha+\beta+1)\, P_{n-1}^{(\alpha+1,\beta+1)}(x). \tag{4.6}$$

The Jacobi polynomials are normalized such that

$$P_n^{(\alpha,\beta)}(1) = \binom{n+\alpha}{n} = \frac{\Gamma(n+\alpha+1)}{n!\Gamma(\alpha+1)}. \tag{4.7}$$

An important consequence of the symmetry of the weight function $w(x)$, and the orthogonality of the Jacobi polynomials, is the symmetry relation

$$P_n^{(\alpha,\beta)}(x) = (-1)^n P_n^{(\beta,\alpha)}(-x), \tag{4.8}$$

i.e., the Jacobi polynomials are even and odd depending on the order of the polynomial.

The expansion of functions $u(x) \in L_w^2[-1,1]$, using the Jacobi polynomial, $P_n^{(\alpha,\beta)}(x)$, takes the form

$$u(x) = \sum_{n=0}^{\infty} \hat{u}_n P_n^{(\alpha,\beta)}(x),$$

where the continuous expansion coefficients $\hat{u}_n$ are found through the weighted inner product

$$\begin{aligned} \hat{u}_n &= \frac{1}{\gamma_n} \left(u, P_n^{(\alpha,\beta)}\right)_{L_w^2[-1,1]} \\ &= \frac{1}{\gamma_n} \int_{-1}^{1} u(x) P_n^{(\alpha,\beta)}(x)(1-x)^\alpha (1+x)^\beta \, dx, \end{aligned} \tag{4.9}$$

with the normalizing constant, γ_n, being

$$\begin{aligned} \gamma_n &= \left\| P_n^{(\alpha,\beta)} \right\|_{L_w^2[-1,1]}^2 \\ &= \frac{2^{\alpha+\beta+1}}{(2n+\alpha+\beta+1)n!} \frac{\Gamma(n+\alpha+1)\Gamma(n+\beta+1)}{\Gamma(n+\alpha+\beta+1)}. \end{aligned} \tag{4.10}$$

In practice, we compute the Jacobi polynomials using a variety of recurrence relations. These are now presented in the following theorems:

Theorem 4.2 *All Jacobi polynomials, $P_n^{(\alpha,\beta)}(x)$, satisfy a three-term recurrence relation of the form*

$$x P_n^{(\alpha,\beta)}(x) = a_{n-1,n}^{(\alpha,\beta)} P_{n-1}^{(\alpha,\beta)}(x) + a_{n,n}^{(\alpha,\beta)} P_n^{(\alpha,\beta)}(x) + a_{n+1,n}^{(\alpha,\beta)} P_{n+1}^{(\alpha,\beta)}(x),$$

where $a^{(\alpha,\beta)}$ are given by

$$\begin{aligned} a_{n-1,n}^{(\alpha,\beta)} &= \frac{2(n+\alpha)(n+\beta)}{(2n+\alpha+\beta+1)(2n+\alpha+\beta)}, \\ a_{n,n}^{(\alpha,\beta)} &= -\frac{\alpha^2-\beta^2}{(2n+\alpha+\beta+2)(2n+\alpha+\beta)}, \\ a_{n+1,n}^{(\alpha,\beta)} &= \frac{2(n+1)(n+\alpha+\beta+1)}{(2n+\alpha+\beta+2)(2n+\alpha+\beta+1)}, \end{aligned} \tag{4.11}$$

for $n \geq 0$, where for $n = 0$, $a_{-1,0}^{(\alpha,\beta)} = 0$. To start the recurrence, the first two polynomials are

$$P_0^{\alpha,\beta}(x) = 1, \quad P_1^{\alpha,\beta}(x) = \frac{1}{2}(\alpha+\beta+2)x + \frac{1}{2}(\alpha-\beta).$$

Using this recurrence relation, all Jacobi polynomials can be evaluated at any $x \in [-1, 1]$ and for any order polynomial. Another useful recurrence formula is given in the next theorem.

Theorem 4.3 *All Jacobi polynomials, $P_n^{(\alpha,\beta)}(x)$, satisfy a three-term recurrence relation of the form*

$$P_n^{(\alpha,\beta)}(x) = b_{n-1,n}^{(\alpha,\beta)} \frac{dP_{n-1}^{(\alpha,\beta)}(x)}{dx} + b_{n,n}^{(\alpha,\beta)} \frac{P_n^{(\alpha,\beta)}(x)}{dx} + b_{n+1,n}^{(\alpha,\beta)} \frac{P_{n+1}^{(\alpha,\beta)}(x)}{dx},$$

where

$$\begin{aligned} b_{n-1,n}^{(\alpha,\beta)} &= -\frac{1}{n+\alpha+\beta} a_{n-1,n}^{(\alpha,\beta)}, \\ b_{n,n}^{(\alpha,\beta)} &= -\frac{2}{\alpha+\beta} a_{n,n}^{(\alpha,\beta)}, \\ b_{n+1,n}^{(\alpha,\beta)} &= \frac{1}{n+1} a_{n+1,n}^{(\alpha,\beta)}. \end{aligned} \tag{4.12}$$

Yet another important recurrence property of the Jacobi polynomials is

Theorem 4.4 (Christoffel–Darboux) *For any Jacobi polynomial, $P_n^{(\alpha,\beta)}(x)$, we have*

$$\sum_{n=0}^{N} \frac{1}{\gamma_n} P_n^{(\alpha,\beta)}(x) P_n^{(\alpha,\beta)}(y) = \frac{a_{N+1,N}^{(\alpha,\beta)}}{\gamma_N} \frac{P_{N+1}^{(\alpha,\beta)}(x) P_N^{(\alpha,\beta)}(y) - P_N^{(\alpha,\beta)}(x) P_{N+1}^{(\alpha,\beta)}(y)}{x-y},$$

where

$$\frac{a_{N+1,N}^{(\alpha,\beta)}}{\gamma_N} = \frac{2^{-(\alpha+\beta)}}{2N+\alpha+\beta+2} \frac{\Gamma(N+2)\Gamma(N+\alpha+\beta+2)}{\Gamma(N+\alpha+1)\Gamma(N+\beta+1)},$$

using Equations (4.10) and (4.11).

A consequence of this result is the following theorem.

Theorem 4.5 *All Jacobi polynomials, $P_n^{(\alpha,\beta)}(x)$, satisfy a three-term recurrence relation of the form*

$$(1-x^2)\frac{dP_n^{(\alpha,\beta)}(x)}{dx} = c_{n-1,n}^{(\alpha,\beta)} P_{n-1}^{(\alpha,\beta)}(x) + c_{n,n}^{(\alpha,\beta)} P_n^{(\alpha,\beta)}(x) + c_{n+1,n}^{(\alpha,\beta)} P_{n+1}^{(\alpha,\beta)}(x),$$

where

$$c_{n-1,n}^{(\alpha,\beta)} = \frac{2(n+\alpha)(n+\beta)(n+\alpha+\beta+1)}{(2n+\alpha+\beta)(2n+\alpha+\beta+1)}, \tag{4.13}$$

$$c_{n,n}^{(\alpha,\beta)} = \frac{2n(\alpha-\beta)(n+\alpha+\beta+1)}{(2n+\alpha+\beta)(2n+\alpha+\beta+2)},$$

$$c_{n+1,n}^{(\alpha,\beta)} = -\frac{2n(n+1)(n+\alpha+\beta+1)}{(2n+\alpha+\beta+1)(2n+\alpha+\beta+2)}.$$

Clearly, there is a wide variety of Jacobi polynomials to choose from. Which specific polynomial family, among all the Jacobi polynomials, is best suited for the approximation of functions defined on the finite interval? As can only be expected, the answer to this question depends on what we mean by "best"; in other words, on how the error is measured.

4.2.1 Legendre polynomials

A natural choice of basis functions are those that are orthogonal in $L^2[-1,1]$. In this case, $w(x)=1$ and $\alpha=\beta=0$. These polynomials are known as Legendre polynomials, $P_n(x)$, and are eigensolutions to the Sturm–Liouville problem

$$\frac{d}{dx}(1-x^2)\frac{dP_n(x)}{dx} + n(n+1)P_n(x) = 0,$$

with eigenvalues $\lambda_n = n(n+1)$.

The Rodrigues formula for the Legendre polynomials simplifies to

$$P_n(x) = \frac{(-1)^n}{2^n n!}\frac{d^n}{dx^n}\left[(1-x^2)^n\right]. \tag{4.14}$$

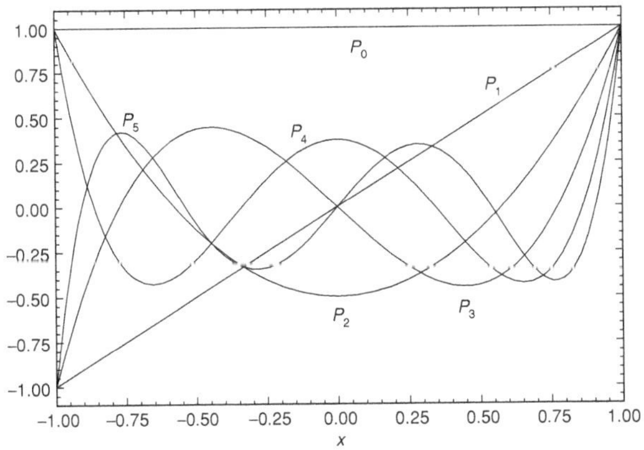

Figure 4.1 Plot of the first 5 Legendre polynomials.

An explicit formula can be recovered from Equation (4.5):

$$P_n(x) = \frac{1}{2^n} \sum_{k=0}^{[n/2]} (-1)^k \binom{n}{k} \binom{2n-2k}{n} x^{n-2k}, \tag{4.15}$$

where $[n/2]$ refers to the integer part of the fraction.

It can be proven that the Legendre polynomials are bounded

$$|P_n(x)| \le 1, \quad |P_n'(x)| \le \frac{1}{2} n(n+1),$$

with boundary values

$$P_n(\pm 1) = (\pm 1)^n, \quad P_n'(\pm 1) = \frac{(\pm 1)^{n+1}}{2} n(n+1). \tag{4.16}$$

The first few Legendre polynomials are

$$P_0(x) = 1, \quad P_1(x) = x, \quad P_2(x) = \frac{1}{2}(3x^2 - 1), \quad P_3(x) = \frac{1}{2}(5x^3 - 3x),$$

and the first 5 Legendre polynomials are shown in Figure 4.1.

The three-term recurrence relation (Theorem 4.2) for the Legendre polynomials,

$$x P_n(x) = \frac{n}{2n+1} P_{n-1}(x) + \frac{n+1}{2n+1} P_{n+1}(x), \tag{4.17}$$

yields a direct way to evaluate legendre polynomials of arbitrary order. Theorem 4.3 yields the recurrence relation

$$P_n(x) = -\frac{1}{2n+1} P_{n-1}'(x) + \frac{1}{2n+1} P_{n+1}'(x). \tag{4.18}$$

Many other properties of the Legendre polynomials are listed in Appendix B.

4.2.2 Chebyshev polynomials

While the Legendre polynomials were the best from the point of view of minimizing the L^2 error, the Chebyshev polynomial family has a simple explicit formula and is straightforward to compute.

The Chebyshev polynomial, $T_n(x)$, is a solution to the singular Sturm–Liouville problem with $p(x) = \sqrt{1-x^2}$, $q(x) = 0$ and the weight function, $w(x) = (\sqrt{1-x^2})^{-1}$, i.e.,

$$\frac{d}{dx}\left(\sqrt{1-x^2}\frac{dT_n(x)}{dx}\right) + \frac{\lambda_n}{\sqrt{1-x^2}}T_n(x) = 0, \tag{4.19}$$

with $\lambda_n = n^2$. The Chebyshev polynomials are related to the Jacobi polynomials with $\alpha = \beta = -\frac{1}{2}$:

$$T_n(x) = \frac{(n!2^n)^2}{(2n)!}P_n^{(-\frac{1}{2},-\frac{1}{2})}(x) = \left[\binom{n-\frac{1}{2}}{n}\right]^{-1} P_n^{(-\frac{1}{2},-\frac{1}{2})}(x).$$

Note that the transformation $x = \cos(\xi)$ in Equation (4.19) leads to the constant coefficient problem

$$\frac{d^2T_n(\xi)}{d\xi^2} + \lambda_n T_n(\xi) = 0$$

so that

$$T_n(\xi) = \cos(n\xi)$$

is the correct solution. Therefore

$$T_n(x) = \cos(n \arccos x)$$

is a convenient alternative way to write the Chebyshev polynomials.

The Rodrigues' formula for Chebyshev polynomials is obtained directly from Equation (4.4) by normalizing appropriately;

$$T_n(x) = \frac{(-1)^n n!2^n}{(2n)!}\sqrt{1-x^2}\frac{d^n}{dx^n}\{(1-x^2)^{n-\frac{1}{2}}\}, \tag{4.20}$$

The definition of the Chebyshev polynomials yields the bounds

$$|T_n(x)| \leq 1, \quad |T_n'(x)| \leq n^2,$$

with the boundary values

$$T_n(\pm 1) = (\pm 1)^n, \quad T_n'(\pm 1) = (\pm 1)^{n+1}n^2, \tag{4.21}$$

due to the normalization employed.

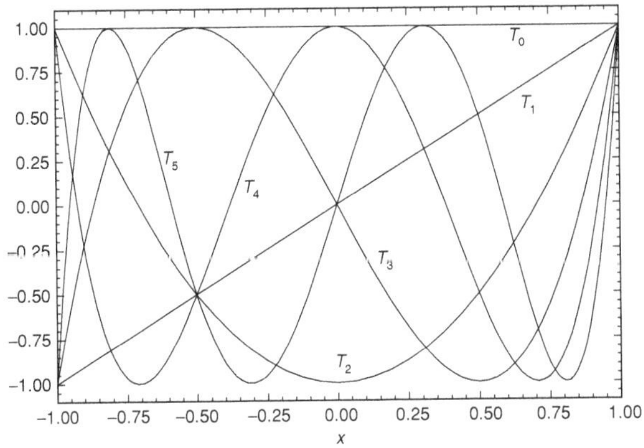

Figure 4.2 Plot of the first 5 Chebyshev polynomials.

For the Chebyshev polynomials, the three-term recurrence relation derived in Theorem 4.2 yields

$$xT_n(x) = \frac{1}{2}T_{n-1}(x) + \frac{1}{2}T_{n+1}(x), \tag{4.22}$$

using the normalized recurrence coefficients in Equation (4.11). In this way, it is simple to compute the first few Chebyshev polynomials

$$T_0(x) = 1,\ \ T_1(x) = x,\ \ T_2(x) = 2x^2 - 1,\ \ T_3(x) = 4x^3 - 3x.$$

The first five Chebyshev polynomials are illustrated in Figure 4.2.

Likewise, from Theorem 4.3 we obtain the recurrence relation

$$T_n(x) = -\frac{1}{2(n-1)}T'_{n-1}(x) + \frac{1}{2(n+1)}T'_{n+1}(x), \tag{4.23}$$

while the recurrence of Theorem 4.5 yields a relation of the form

$$(1-x^2)T'_n(x) = \frac{n}{2}T_{n-1}(x) - \frac{n}{2}T_{n+1}(x). \tag{4.24}$$

In Appendix B, we summarize the properties of Chebyshev polynomials in general, including several results not given here.

A fascinating result is that of all polynomials of degree n, the best approximant (in the maximum norm) to the function x^{n+1} is the nth-order Chebyshev polynomial $T_n(x)$. The Chebyshev polynomial is, in this sense, the optimal choice when the error is measured in the maximum norm.

4.2.3 Ultraspherical polynomials

The Legendre and Chebyshev polynomials are related to a subclass of Jacobi polynomials known as the ultraspherical polynomials. The ultraspherical polynomials are simply Jacobi polynomials with $\alpha = \beta$, and normalized differently:

$$P_n^{(\alpha)}(x) = \frac{\Gamma(\alpha+1)\Gamma(n+2\alpha+1)}{\Gamma(2\alpha+1)\Gamma(n+\alpha+1)} P_n^{(\alpha,\alpha)}(x). \tag{4.25}$$

The relation between Legendre and Chebyshev polynomials and the ultraspherical polynomials is

$$P_n(x) = P_n^{(0)}(x), \quad T_n(x) = n \lim_{\alpha \to -\frac{1}{2}} \Gamma(2\alpha+1) P_n^{(\alpha)}(x). \tag{4.26}$$

Note that $\Gamma(2\alpha+1)$ has a pole for $\alpha = -1/2$, and must be treated carefully. The ultraspherical polynomials, $P_n^{(\alpha)}(x)$, are also known as the Gegenbauer polynomials, $C_n^{(\lambda)}(x)$, of the form

$$C_n^{(\lambda)}(x) = P_n^{(\lambda-\frac{1}{2})}(x).$$

The ultraspherical polynomials, $P_n^{(\alpha)}(x)$, appear as the solution to the Sturm–Liouville problem

$$\frac{d}{dx}(1-x^2)^{\alpha+1}\frac{dP_n^{(\alpha)}(x)}{dx} + n(n+2\alpha+1)(1-x^2)^{\alpha} P_n^{(\alpha)}(x) = 0, \tag{4.27}$$

which is Equation (4.1), with $\lambda_n = n(n+2\alpha+1)$ and $p(x) = (1-x^2)^{\alpha+1}$, $q(x) = 0$ and $w(x) = (1-x^2)^{\alpha}$.

The Rodrigues' formula for the ultraspherical polynomials is obtained from Equation (4.4) and Equation (4.25);

$$(1-x^2)P_n^{(\alpha)}(x) = \frac{\Gamma(\alpha+1)\Gamma(n+2\alpha+1)}{\Gamma(2\alpha+1)\Gamma(n+\alpha+1)} \frac{(-1)^n}{2^n n!} \frac{d^n}{dx^n}\{(1-x^2)^{n+\alpha}\}, \tag{4.28}$$

while an explicit formula, as in Equation (4.5) is given by

$$P_n^{(\alpha)}(x) = \frac{1}{\Gamma\left(\alpha+\frac{1}{2}\right)} \sum_{k=0}^{[n/2]} (-1)^k \frac{\Gamma\left(\alpha+\frac{1}{2}+n-k\right)}{k!(n-2k)!} (2x)^{n-2k} \tag{4.29}$$

$$= \frac{\Gamma(\alpha+1)\Gamma(2n+2\alpha+1)}{2^n n! \Gamma(2\alpha+1)\Gamma(n+\alpha+1)} \left[x^n + \cdots\right],$$

where $[n/2]$ refers to the integer part of the fraction.

The relation between different ultraspherical polynomials is described by

$$\frac{d}{dx} P_n^{(\alpha)}(x) = (2\alpha+1)\, P_{n-1}^{(\alpha+1)}(x), \tag{4.30}$$

while the value of $P_n^{(\alpha)}(x)$ at the boundary is

$$P_n^{(\alpha)}(\pm 1) = (\pm 1)^n \binom{n+2\alpha}{n}, \tag{4.31}$$

from Equation (4.7) and Equation (4.25).

Using Equation (4.28) or Equation (4.29) we recover the first two polynomials $P_0^{(\alpha)}(x) = 1$ and $P_1^{(\alpha)}(x) = (2\alpha+1)x$, while the subsequent polynomials can be computed through the recurrence relation (in Theorem 4.2) of the form

$$x P_n^{(\alpha)}(x) = a_{n-1,n}^{(\alpha)} P_{n-1}^{(\alpha)}(x) + a_{n+1,n}^{(\alpha)} P_{n+1}^{(\alpha)}(x), \tag{4.32}$$

where the recurrence coefficients, obtained by normalizing Equation (4.11) appropriately, become

$$a_{n-1,n}^{(\alpha)} = \frac{n+2\alpha}{2n+2\alpha+1}, \quad a_{n+1,n}^{(\alpha)} = \frac{n+1}{2n+2\alpha+1}. \tag{4.33}$$

The symmetry of the ultraspherical polynomials is emphasized by the relation

$$P_n^{(\alpha)}(x) = (-1)^n P_n^{(\alpha)}(-x). \tag{4.34}$$

The recurrence relation of the form given in Theorem 4.3 becomes

$$P_n^{(\alpha)}(x) = b_{n-1,n}^{(\alpha)} \frac{dP_{n-1}^{(\alpha)}(x)}{dx} + b_{n+1,n}^{(\alpha)} \frac{dP_{n+1}^{(\alpha)}(x)}{dx}, \tag{4.35}$$

where the recurrence coefficients are obtained directly from Equation (4.12) as

$$b_{n-1,n}^{(\alpha)} = -\frac{1}{2n+2\alpha+1}, \quad b_{n+1,n}^{(\alpha)} = \frac{1}{2n+2\alpha+1}. \tag{4.36}$$

Let us finally also give the result of Theorem 4.5 for the ultraspherical polynomials as

$$(1-x^2)\frac{dP_n^{(\alpha)}(x)}{dx} = c_{n-1,n}^{(\alpha)} P_{n-1}^{(\alpha)}(x) + c_{n+1,n}^{(\alpha)} P_{n+1}^{(\alpha)}(x), \tag{4.37}$$

with the coefficients being

$$c_{n-1,n}^{(\alpha)} = \frac{(n+2\alpha+1)(n+2\alpha)}{2n+2\alpha+1}, \quad c_{n+1,n}^{(\alpha)} = -\frac{n(n+1)}{2n+2\alpha+1}, \tag{4.38}$$

from Equation (4.13).

In Chapter 1 we introduced the concept of points-per-wavelength required to accurately represent a wave, e.g., for the Fourier spectral method we found that we needed the minimum value of only two points to represent the wave exactly.

It is illustrative to consider this question also for polynomial expansions as it provides guidelines of how fine a grid, or how many modes, one needs to accurately represent a wave.

Example 4.6 Consider the case in which the plane wave

$$u(x) = e^{i\pi kx},$$

is approximated using a Gegenbauer expansion

$$u_N(x) = \sum_{n=0}^{N} \hat{u}_n C_n^\lambda(x).$$

This approximation is given by

$$u_N(x) = \Gamma(\lambda)\left(\frac{\pi k}{2}\right)^{-\lambda} \sum_{n=0}^{N} i^n (n+\lambda) J_{n+\lambda}(\pi k) C_n^{(\lambda)}(x),$$

where $J_n(k)$ is the Bessel function of the first kind. We have

$$\hat{u}_n = \Gamma(\lambda)\left(\frac{\pi k}{2}\right)^{-\lambda} i^n (n+\lambda) J_{n+\lambda}(\pi k).$$

The decay rate of $\hat{u}_n$ as n gets large depends on the asymptotic behavior of the Bessel function, which is

$$J_{n+\lambda}(\pi k) \simeq \left(\frac{e\pi k}{2(n+\lambda)}\right)^{n+\lambda}.$$

For the Bessel function to decay, we must have $e\pi k/2(n+\lambda) < 1$, which requires $e\pi/2 < (n+\lambda)/k$. For large values of k, we need

$$\frac{n}{k} > \frac{e\pi}{2} \approx 4,$$

so about 4 modes-per-wave, and thus points-per-wavelength, are required to obtain exponential convergence. While this is twice the number required for the Fourier case, it is still dramatically less than needed for the low-order finite difference schemes discussed in Chapter 1.

4.3 Further reading

The subject of orthogonal polynomials as discussed here is classical material. Most results and many generalizations can be found in the texts by Szego (1930) and Funaro (1992).

5
Polynomial expansions

In the last chapter, we showed that smooth functions can be well represented by polynomial expansions. We now turn our attention to obtaining a qualitative understanding of approximating functions and their derivatives using truncated continuous and discrete polynomial expansions. Although such approximations to smooth problems can be formulated using any Jacobi polynomials, the most commonly used polynomials in spectral methods are the Legendre and Chebyshev polynomials, and so we focus on these.

5.1 The continuous expansion

Consider the continuous polynomial expansion of a function

$$u(x) = \sum_{n=0}^{\infty} \hat{u}_n P_n^{(\alpha)}(x), \tag{5.1}$$

with expansion coefficients

$$\hat{u}_n = \frac{1}{\gamma_n}\left(u, P_n^{(\alpha)}\right)_{L^2_w[-1,1]} = \frac{1}{\gamma_n}\int_{-1}^{1} u(x)P_n^{(\alpha)}(x)(1-x^2)^{\alpha}\,dx, \tag{5.2}$$

where $w(x)$ is the weight under which the polynomials $P_n^{(\alpha)}(x)$ are orthogonal, and

$$\gamma_n = \left\| P_n^{(\alpha)} \right\|^2_{L^2_w[-1,1]} = 2^{2\alpha+1}\frac{\Gamma^2(\alpha+1)\Gamma(n+2\alpha+1)}{n!(2n+2\alpha+1)\Gamma^2(2\alpha+1)}. \tag{5.3}$$

The qth derivative of $u(x)$ has a similar expansion

$$\frac{d^q u(x)}{dx^q} = \sum_{n=0}^{\infty} \hat{u}_n^{(q)} P_n^{(\alpha)}(x),$$

and so the question is how the expansion coefficients, $\hat{u}_n^{(q)}$, of the derivative are related to the expansion coefficients, $\hat{u}_n$, of the function. When approximating the solution to partial differential equations, this is a central problem. The following recursion relation gives a solution to this problem:

Theorem 5.1 *Given the expansions*

$$u^{(q)}(x) = \sum_{n=0}^{\infty} \hat{u}_n^{(q)} P_n^{(\alpha)}(x),$$

and

$$u^{(q-1)}(x) = \sum_{n=0}^{\infty} \hat{u}_n^{(q-1)} P_n^{(\alpha,\beta)}(x),$$

then the relation between the coefficients (for $n > 0$) is given by

$$\hat{u}_n^{(q-1)} = b_{n,n-1}^{(\alpha)} \hat{u}_{n-1}^{(q)} + b_{n,n+1}^{(\alpha)} \hat{u}_{n+1}^{(q)},$$

where $b_{n,n-1}^{(\alpha)}$ and $b_{n,n+1}^{(\alpha)}$ are defined in Equation (4.36).

Proof: We establish the result by calculating, up to a constant, the expansion coefficients $\hat{u}_n$ of a function $u(x)$ from the expansion coefficients $\hat{u}_n^{(1)}$ of the first derivative of the function. The recurrence relation Equation (4.35) yields

$$\begin{aligned}
\frac{d}{dx}u(x) &= \sum_{n=0}^{\infty} \hat{u}_n^{(1)} P_n^{(\alpha)}(x) \\
&= \sum_{n=0}^{\infty} \hat{u}_n^{(1)} \left(b_{n-1,n}^{(\alpha)} \frac{dP_{n-1}^{(\alpha)}(x)}{dx} + b_{n+1,n}^{(\alpha)} \frac{P_{n+1}^{(\alpha)}(x)}{dx} \right) \\
&= \sum_{m=-1}^{\infty} b_{m,m+1}^{(\alpha)} \hat{u}_{m+1}^{(1)} \frac{dP_m^{(\alpha)}(x)}{dx} + \sum_{m=1}^{\infty} b_{m,m-1}^{(\alpha)} \hat{u}_{m-1}^{(1)} \frac{dP_m^{(\alpha)}(x)}{dx} \\
&= \sum_{m=1}^{\infty} \left[b_{m,m-1}^{(\alpha)} \hat{u}_{m-1}^{(1)} + b_{m,m+1}^{(\alpha)} \hat{u}_{m+1}^{(1)} \right] \frac{dP_m^{(\alpha)}(x)}{dx} \\
&= \sum_{n=0}^{\infty} \hat{u}_n \frac{dP_n^{(\alpha)}(x)}{dx},
\end{aligned}$$

where we have used the fact that $P_0^{(\alpha)}(x)$ is constant and defined the polynomial $P_{-1}^{(\alpha)}(x)$ to be zero. Note that $\hat{u}_0$ remains undetermined as the operation is essentially an integration, which leaves a constant undetermined. The extension to the general case follows directly.

QED

The theorem tells us that we can easily obtain the expansion coefficients of the function from the expansion coefficients of the derivative. However, we have the opposite problem: given the expansion coefficients of the function, we want the expansion coefficients of the derivative. To obtain them, we invert the tridiagonal integration operator in Theorem 5.1 and obtain the operator (for $\alpha \neq -1/2$):

$$\hat{u}_n^{(q)} = (2n + 2\alpha + 1) \sum_{\substack{p=n+1 \\ n+p \text{ odd}}}^{\infty} \hat{u}_p^{(q-1)}. \tag{5.4}$$

At first glance it seems that the computation of $\hat{u}_n^{(q)}$ from $\hat{u}_n^{(q-1)}$ using Equation (5.4) requires $\mathcal{O}(N^2)$ operation. However, in practice this is done more efficiently. Assume that we have the finite expansion

$$\mathcal{P}_N \frac{d^{(q-1)}u(x)}{dx^{(q-1)}} = \sum_{n=0}^{N} \hat{u}_n^{(q-1)} P_n^{(\alpha)}(x),$$

and we seek an approximation to the derivative

$$\mathcal{P}_N \frac{d^{(q)}u(x)}{dx^{(q)}} = \sum_{n=0}^{N} \hat{u}_n^{(q)} P_n^{(\alpha)}(x).$$

Since the projection $\mathcal{P}_N$ of each expansion is a polynomial of degree N, we can write $\hat{u}_n^{(q)} = 0, \forall n > N$ and $\hat{u}_n^{(q-1)} = 0, \forall n > N$. Equation (5.4) implies that $\hat{u}_N^{(q)} = 0$, as well. Theorem 5.1 then suggests the backward recursion formula for $n \in [N, \ldots, 1]$,

$$\hat{u}_{n-1}^{(q)} = (2n + 2\alpha - 1)\left[\frac{1}{2n + 2\alpha + 3}\hat{u}_{n+1}^{(q)} + \hat{u}_n^{(q-1)}\right]. \tag{5.5}$$

Applying the backward recursion, Equation (5.5), for the computation of $\hat{u}_n^{(q)}$, the computational workload is reduced to $\mathcal{O}(N)$ for any finite expansion.

5.1.1 The continuous legendre expansion

The Legendre expansion of a function $u(x) \in L^2[-1, 1]$, is given by

$$u(x) = \sum_{n=0}^{\infty} \hat{u}_n P_n(x), \tag{5.6}$$

with expansion coefficients

$$\hat{u}_n = \frac{1}{\gamma_n}(u, P_n)_{L^2[-1,1]} = \frac{2n+1}{2}\int_{-1}^{1} u(x) P_n(x)\, dx,$$

since $\gamma_n = 2/(2n + 1)$ using Equation (4.10).

Using Theorem 5.1 and the recurrence in Equation (4.18) we obtain the relation between the expansion coefficients, $\hat{u}_n^{(q-1)}$, and those of its derivative, $\hat{u}_n^{(q)}$:

$$\hat{u}_n^{(q-1)} = \frac{1}{2n-1}\hat{u}_{n-1}^{(q)} - \frac{1}{2n+3}\hat{u}_{n+1}^{(q)}. \tag{5.7}$$

The inversion of this tridiagonal integration operator yields the differentiation operator

$$\hat{u}_n^{(q)} = (2n+1) \sum_{\substack{p=n+1\\ n+p \text{ odd}}}^{\infty} \hat{u}_p^{(q-1)} \quad \forall n. \tag{5.8}$$

As before, if we are dealing with a finite approximation we have $\hat{u}_N^{(q)} = \hat{u}_{N+1}^{(q)} = 0$, so we may compute $\hat{u}_n^{(q)}$ for the truncated expansion through the backward recurrence in Equation (5.5),

$$\hat{u}_{n-1}^{(q)} = \frac{2n-1}{2n+3}\hat{u}_{n+1}^{(q)} + (2n-1)\hat{u}_n^{(q-1)} \quad \forall n \in [N, \ldots, 1]. \tag{5.9}$$

5.1.2 The continuous Chebyshev expansion

The continuous Chebyshev expansion of a function $u(x)$,

$$u(x) = \sum_{n=0}^{\infty} \hat{u}_n T_n(x),$$

with expansion coefficients

$$\hat{u}_n = \frac{1}{\gamma_n}(u, T_n)_{L^2_w[-1,1]} = \frac{2}{c_n \pi}\int_{-1}^{1} u(x)T_n(x)\frac{1}{\sqrt{1-x^2}}\, dx,$$

where $w(x)$ is the appropriate weight function,

$$\gamma_n = \int_{-1}^{1} T_n(x)T_n(x)\frac{1}{\sqrt{1-x^2}}\, dx = \frac{\pi}{2}c_n,$$

and

$$c_n = \begin{cases} 2 & n = 0 \\ 1 & n > 0. \end{cases} \tag{5.10}$$

This additional constant is a consequence of the normalization we have chosen, i.e. $T_n(\pm 1) = (\pm 1)^n$.

The connection between the expansion coefficients of $\hat{u}_n^{(q-1)}$ and those of its derivative $\hat{u}_n^{(q)}$ is (for $n > 0$),

$$\hat{u}_n^{(q-1)} = \frac{c_{n-1}}{2n}\hat{u}_{n-1}^{(q)} - \frac{1}{2n}\hat{u}_{n+1}^{(q)}, \tag{5.11}$$

where c_{n-1} enters due to the normalization. Inverting this tridiagonal integration operator yields the differentiation operator

$$\hat{u}_n^{(q)} = \frac{2}{c_n} \sum_{\substack{p=n+1 \\ n+p \text{ odd}}}^{\infty} p\hat{u}_p^{(q-1)} \quad \forall n. \tag{5.12}$$

In the case where $\hat{u}_{N+1}^{(q)} = \hat{u}_N^{(q)} = 0$, an $\mathcal{O}(N)$ method to compute the expansion coefficients for the approximate derivative is given by the backward recurrence in Equation (5.11)

$$c_{n-1}\hat{u}_{n-1}^{(q)} = \hat{u}_{n+1}^{(q)} + 2n\hat{u}_n^{(q-1)} \quad \forall n \in [N, \dots, 1]. \tag{5.13}$$

5.2 Gauss quadrature for ultraspherical polynomials

In the previous section we discussed how to obtain the expansion coefficients of the derivative from the expansion coefficients of the function itself. However, just obtaining the expansion coefficients may be difficult, as this requires an evaluation of an integral. To obviate this problem we turn to quadrature formulas to approximate these integrals, just as we did for the Fourier expansions. In this case, classical Gauss quadratures enable the practical use of polynomial methods for the approximation of general functions.

Gauss–Lobatto quadrature In the Gauss–Lobatto quadrature for the ultraspherical polynomial, $P_N^{(\alpha)}(x)$, we approximate the integral

$$\int_{-1}^{1} p(x)w(x)\,dx = \sum_{j=0}^{N} p(x_j)w_j.$$

The nodes are $-1 = x_0, x_1, \dots, x_{N-1}, x_N = 1$ where the internal points $x_1, \dots, x_{N-1}$ are the roots of the polynomial

$$\frac{d}{dx} P_N^{(\alpha)}(x). \tag{5.14}$$

The weights w_j are given by

$$w_j = \begin{cases} (\alpha+1)\Pi_{N,j}^{(\alpha)} & j = 0, N, \\ \Pi_{N,j}^{(\alpha)} & j \in [1, \dots, N-1], \end{cases} \tag{5.15}$$

where

$$\Pi_{N,j}^{(\alpha)} = 2^{2\alpha+1} \frac{\Gamma^2(\alpha+1)\Gamma(N+2\alpha+1)}{NN!(N+2\alpha+1)\Gamma^2(2\alpha+1)} \left[P_N^{(\alpha)}(x_j)\right]^{-2}.$$

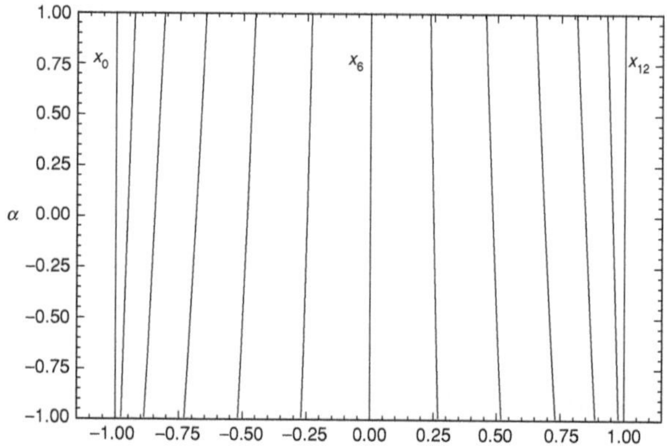

Figure 5.1 The ultraspherical Gauss–Lobatto collocation points, x_i, for $N = 12$ for the polynomial, $P_{12}^{(\alpha)}(x)$, as a function of α.

The Gauss–Lobatto formula is exact for all polynomials $p(x) \in \mathsf{B}_{2N-1}$.

Note that in general, the quadrature nodes are not given explicitly but must be computed by finding the zeroes of the polynomial in Equation (5.14), as discussed in Chapter 11. At this stage it is worth noting that the nodes cluster close to the boundaries with the amount of clustering decreasing as α increases, as illustrated in Figure 5.1.

Gauss–Radau quadrature In the Gauss–Radau formula, only one of the endpoints of the interval $[-1, 1]$ is included as a node in the summation. If we choose the endpoint -1, the remaining quadrature points are the roots of the polynomial

$$q(y) = P_{N+1}^{(\alpha)}(y) + a_N P_N^{(\alpha)}(y), \tag{5.16}$$

with a_N chosen such that $q(-1) = 0$, and the weights are

$$v_j = \begin{cases} (\alpha+1)\Pi_{N,0}^{(\alpha)} & j = 0, \\ \Pi_{N,j}^{(\alpha)} & j \in [1, \dots, N], \end{cases} \tag{5.17}$$

where

$$\Pi_{N,j}^{(\alpha)} = 2^{2\alpha} \frac{\Gamma^2(\alpha+1)\Gamma(N+2\alpha+1)}{N!(N+\alpha+1)(N+2\alpha+1)\Gamma^2(2\alpha+1)} (1-y_j)\left[P_N^{(\alpha)}(y_j)\right]^{-2}.$$

The Gauss–Radau quadrature

$$\int_{-1}^{1} p(y)w(y)\,dy = \sum_{j=0}^{N} p(y_j)v_j,$$

is exact for all polynomials $p(y) \in \mathsf{B}_{2N}$.

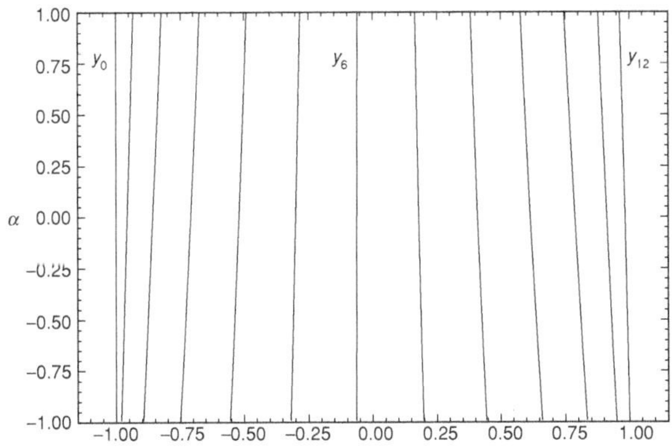

Figure 5.2 The Gauss–Radau collocation points, y_i, for $N = 12$, for the ultraspherical, $P_{12}^{(\alpha)}(x)$, as a function of α with the node $y_0 = -1$ being fixed.

As before, the quadrature nodes y_j must be obtained numerically. In Figure 5.2 we plot the position of the ultraspherical Gauss–Radau quadrature nodes for $N = 12$ and $\alpha \in (-1, 1)$. As in the case of the Gauss–Lobatto quadrature, the nodes cluster close to the boundaries, with the amount of clustering decreasing as α increases and only the left boundary is included in the nodal set.

The formulation of a Gauss–Radau quadrature that includes the right end point follows directly from the above by reflecting the weights v_j, as well as the quadrature nodes y_j, around the center of the interval.

Gauss quadrature The quadrature points for the Gauss formula are all in the interior of the domain $[-1, 1]$. The quadrature points z_j are the roots of the polynomial,

$$q(z) = P_{N+1}^{(\alpha)}(z), \tag{5.18}$$

while the weights, u_j, are

$$u_j = 2^{2\alpha+1} \frac{\Gamma^2(\alpha+1)\Gamma(2\alpha+N+2)}{\Gamma(N+2)\Gamma^2(2\alpha+1)\left(1-z_j^2\right)} \left[\frac{d}{dx} P_{N+1}^{(\alpha)}(z_j)\right]^{-2}, \tag{5.19}$$

for all $j \in [0, \ldots, N]$.

The Gauss formula

$$\int_{-1}^{1} p(z)w(z)\,dz = \sum_{j=0}^{N} p(z_j)u_j,$$

is exact for all $p(x) \in \mathrm{B}_{2N+1}$. This is the classical Gauss quadrature, which provides the rule of exact integration of a polynomial of maximum order.

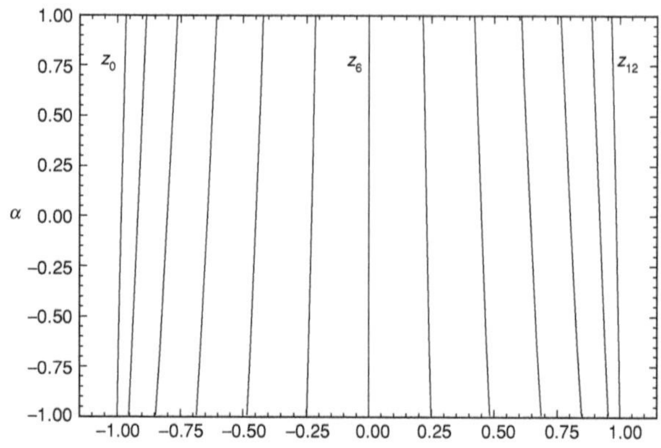

Figure 5.3 The Gauss collocation points, z_i, for $N = 12$, for the ultraspherical polynomial, $P_{12}^{(\alpha)}(x)$, as a function of α.

The quadrature points z_j are generally not given in analytic form and in Figure 5.3 we plot, for illustration, the position of the nodes in the case of the ultraspherical polynomials for $N = 12$ and $\alpha \in (-1, 1)$. We emphasize, as is also evident from Figure 5.3, that the nodal set associated with the Gauss quadrature does not include any of the endpoints of the interval.

5.2.1 Quadrature for Legendre polynomials

The quadrature formulas for the integration of polynomials specified at the Legendre quadrature points can be obtained directly from the formulas derived above by setting $\alpha = 0$. Due to the extensive use of the Legendre polynomials we summarize the expressions below.

Legendre Gauss–Lobatto quadrature The Legendre Gauss–Lobatto quadrature points, x_j, are the roots of the polynomial

$$q(x) = (1 - x^2)\frac{d}{dx} P_N(x). \tag{5.20}$$

The weights, w_j are

$$w_j = \frac{2}{N(N+1)}[P_N(x_j)]^{-2}. \tag{5.21}$$

Legendre Gauss–Radau quadrature The Legendre Gauss–Radau quadrature points, y_j, are the roots of the polynomial

$$q(y) = P_{N+1}(y) + P_N(y), \tag{5.22}$$

assuming that $y = -1$ is included in the set of quadrature points. The weights, v_j, are given as

$$v_j = \frac{1}{(N+1)^2} \frac{1 - y_j}{[P_N(y_j)]^2}. \tag{5.23}$$

Legendre Gauss quadrature The Legendre Gauss quadrature points z_j are the $N+1$ roots of the polynomial

$$q(z) = P_{N+1}(z). \tag{5.24}$$

The weights u_j are obtained directly from Equation (5.19) for $\alpha = 0$,

$$u_j = \frac{2}{[1 - (z_j)^2][P'_{N+1}(z_j)]^2}. \tag{5.25}$$

5.2.2 Quadrature for Chebyshev polynomials

The Chebyshev quadrature is distinguished from the previous cases by its explicit and simple expressions for the quadrature points as well as the corresponding weights. This is a compelling motivation for the use of these polynomials. Note that these expressions are not recovered directly from the general results above when $\alpha = -1/2$, but must be normalized differently, cf. Equation (4.26).

Chebyshev Gauss–Lobatto quadrature The Chebyshev Gauss–Lobatto quadrature points and weights are given explicitly by

$$x_j = -\cos\left(\frac{\pi}{N} j\right), \quad j \in [0, \ldots, N], \tag{5.26}$$

and

$$w_j = \frac{\pi}{c_j N}, \quad c_j = \begin{cases} 2 & j = 0, N, \\ 1 & j \in [1, \ldots, N-1]. \end{cases} \tag{5.27}$$

Chebyshev Gauss–Radau quadrature The Chebyshev Gauss–Radau quadrature points, y_j, have the explicit expression

$$y_j = -\cos\left(\frac{2\pi}{2N+1} j\right), \quad j \in [0, \ldots, N]. \tag{5.28}$$

Assuming that the left endpoint is included in the set of quadrature points, the weights, v_j, are given as

$$v_j = \frac{\pi}{c_j} \frac{1}{2N+1}, \tag{5.29}$$

where c_n takes the usual meaning, Equation (5.10).

Chebyshev Gauss quadrature The Chebyshev Gauss quadrature points and weights are given by:

$$z_j = -\cos\left(\frac{(2j+1)\pi}{2N+2}\right), \quad j \in [0, \ldots, N], \tag{5.30}$$

and

$$u_j = \frac{\pi}{N+1} \quad \forall j \in [0, \ldots, N]. \tag{5.31}$$

Note that this is the only quadrature for which the weights are all the same. In Chapter 11 we will discuss how to best compute the nodes and weights for the quadrature based on the ultraspherical polynomials.

5.3 Discrete inner products and norms

The identification of the quadrature formulas motivates the introduction of discrete versions of the inner product and the corresponding L^2_w-norm. We recall that in the continuous case, the inner product and norm take the form

$$(f, g)_{L^2_w[-1,1]} = \int_{-1}^{1} f(x)g(x)w(x)\,dx, \quad \|f\|^2_{L^2_w[-1,1]} = (f, f)_{L^2_w[-1,1]},$$

for $f, g \in L^2_w[-1, 1]$. Using the quadrature formulas it seems natural to define the corresponding discrete inner product

$$[f, g]_w = \sum_{j=0}^{N} f(x_j)g(x_j)w_j, \quad \|f\|^2_{N,w} = [f, f]_w,$$

where $x_j = (x_j, y_j, z_j)$ can be any of the Gauss quadrature points with the corresponding weights, $w_j = (w_j, v_j, u_j)$. If the product $f(x)g(x) \in \mathsf{B}_{2N}$, then the Gauss–Radau and Gauss quadratures are exact and the discrete inner product based on these quadrature rules is identical to the continuous inner product. This, however, is not true for the Gauss–Lobatto quadrature, since it is only exact for functions in B_{2N-1}.

Recall that the norm, $\tilde{\gamma}_n$, of $P_n^{(\alpha)}(x)$ using Gauss or Gauss–Radau quadrature, is

$$\tilde{\gamma}_n = \left(P_n^{(\alpha)}, P_n^{(\alpha)}\right)_{L^2_w[-1,1]} = 2^{2\alpha+1}\frac{\Gamma^2(\alpha+1)\Gamma(n+2\alpha+1)}{n!(2n+2\alpha+1)\Gamma^2(2\alpha+1)}, \tag{5.32}$$

using Equation (5.2). However, for the Gauss–Lobatto quadrature, the quadrature is inexact for $n = N$. The following result allows us to evaluate the inner product.

Lemma 5.2 *The ultraspherical polynomials, $P_N^{(\alpha)}(x)$, satisfy*

$$\frac{d}{dx} P_{N-1}^{(\alpha)}(x_j) = -N P_N^{(\alpha)}(x_j) \quad \forall j \in [1, \ldots, N-1],$$

where x_j represent the interior Gauss–Lobatto quadrature points.

Using this result and the Gauss–Lobatto quadrature, we obtain

$$\tilde{\gamma}_N = \left\| P_N^{(\alpha)} \right\|_{N,w}^2 = 2^{2\alpha+1} \frac{\Gamma^2(\alpha+1)\Gamma(N+2\alpha+1)}{N N! \Gamma^2(2\alpha+1)}. \tag{5.33}$$

While the discrete Gauss–Lobatto norm is slightly different from the continuous norm, it follows immediately from the above that they are equivalent for any polynomial, $u_N \in \mathsf{P}_N$, since

$$\|u_N\|_{L^2_w[-1,1]} \leq \|u_N\|_{N,\omega} \leq \sqrt{2 + \frac{2\alpha+1}{N}} \|u_N\|_{L^2_w[-1,1]},$$

as $\tilde{\gamma}_N > \gamma_N$ for all values of N and $\alpha > -1$.

Legendre polynomials For the Legendre polynomials, $P_n(x)$, the discrete norm $\tilde{\gamma}_n$ is

$$\tilde{\gamma}_n = \begin{cases} \frac{2}{2n+1} & n < N, \\ \frac{2}{2N+1} & n = N \text{ for Gauss and Gauss–Radau quadrature}, \\ \frac{2}{N} & n = N \text{ for Gauss–Lobatto quadrature}. \end{cases} \tag{5.34}$$

Chebyshev polynomials The results for the Chebyshev polynomials, $T_n(x)$, can likewise be summarized as

$$\tilde{\gamma}_n = \begin{cases} \frac{\pi}{2} c_n & n < N, \\ \frac{\pi}{2} & n = N \text{ for Gauss and Gauss–Radau quadrature}, \\ \pi & n = N \text{ for Gauss–Lobatto quadrature}, \end{cases} \tag{5.35}$$

where c_n is as defined in Equation (5.10).

5.4 The discrete expansion

The quadrature rules provide the tools needed for approximating the expansion coefficients. Recall the definition of the truncated continuous expansion

$$\mathcal{P}_N u(x) = \sum_{n=0}^{N} \hat{u}_n P_n^{(\alpha)}(x), \quad \hat{u}_n = \frac{1}{\gamma_n} \int_{-1}^{1} u(x) P_n^{(\alpha)}(x)(1-x^2)^{\alpha}\, dx,$$

where the normalizing factor, γ_n, is given in Equation (5.3).

Using the Gauss–Lobatto quadrature it is natural to define the discrete approximation

$$\mathcal{I}_N u(x) = \sum_{n=0}^{N} \tilde{u}_n P_n^{(\alpha)}(x), \quad \tilde{u}_n = \frac{1}{\tilde{\gamma}_n} \sum_{j=0}^{N} u(x_j) P_n^{(\alpha)}(x_j) w_j, \tag{5.36}$$

where the nodes, weights, and normalization factors are given in Equations (5.14)–(5.15) and (5.32)–(5.33).

Likewise, the Gauss quadrature gives us the discrete approximation

$$\mathcal{I}_N u(z) = \sum_{n=0}^{N} \tilde{u}_n P_n^{(\alpha)}(z), \quad \tilde{u}_n = \frac{1}{\tilde{\gamma}_n} \sum_{j=0}^{N} u(z_j) P_n^{(\alpha)}(z_j) u_j, \tag{5.37}$$

with the nodes, weights and normalization factors given in Equations (5.18), (5.19), and (5.32).

The approximation of the discrete expansion coefficients using the Gauss quadratures has a striking consequence.

Theorem 5.3 *Let $u(x)$ be a function defined on all points in the interval $[-1, 1]$, and let the discrete expansion coefficients $\tilde{u}_n$, be defined in Equation (5.37). Then $\mathcal{I}_N u$ interpolates u at all the Gauss quadrature points z_j:*

$$\mathcal{I}_N u(z_j) = u(z_j).$$

Proof: If we put the discrete expansion coefficients into the polynomial approximation we recover

$$\begin{aligned}
\mathcal{I}_N u(z) &= \sum_{n=0}^{N} \tilde{u}_n P_n^{(\alpha)}(z) \\
&= \sum_{n=0}^{N} \left(\frac{1}{\tilde{\gamma}_n} \sum_{j=0}^{N} u(z_j) P_n^{(\alpha)}(z_j) u_j \right) P_n^{(\alpha)}(z) \\
&= \sum_{j=0}^{N} u(z_j) \left(u_j \sum_{n=0}^{N} \frac{1}{\tilde{\gamma}_n} P_n^{(\alpha)}(z) P_n^{(\alpha)}(z_j) \right) \\
&= \sum_{j=0}^{N} u(z_j) l_j(z),
\end{aligned}$$

where

$$l_j(z) = u_j \sum_{n=0}^{N} \frac{1}{\tilde{\gamma}_n} P_n^{(\alpha)}(z) P_n^{(\alpha)}(z_j). \tag{5.38}$$

Here u_j are the Gauss quadrature weights (5.19). We note that $l_j(z) \in \mathrm{B}_N$ as expected. We proceed by showing that $l_j(z)$ is the Lagrange interpolation polynomial based on the Gauss quadrature nodes, i.e., $l_j(z_i) = \delta_{ij}$.

Recalling that $\tilde{\gamma}_n = \gamma_n$ in the Gauss quadrature, we apply the Christoffel–Darboux formula in Theorem 4.4 to obtain

$$l_j(z) = u_j 2^{-(2\alpha+1)} \frac{\Gamma(N+2)\Gamma^2(2\alpha+1)}{\Gamma(N+2\alpha+1)\Gamma^2(\alpha+1)} \frac{P_{N+1}^{(\alpha)}(z) P_N^{(\alpha)}(z_j)}{z - z_j},$$

since $P_{N+1}^{(\alpha)}(z_j) = 0$ defines the Gauss points. Clearly $l_j(z_i) = 0$ for $i \neq j$. Using l'Hôpital rule we obtain for $i = j$ that

$$l_j(z_j) = u_j 2^{-(2\alpha+1)} \frac{\Gamma(N+2)\Gamma^2(2\alpha+1)}{\Gamma(N+2\alpha+1)\Gamma^2(\alpha+1)} P_N^{(\alpha)}(z_j) \frac{d}{dx} P_{N+1}^{(\alpha)}(z_j).$$

Introducing the formula for the Gauss weights, Equation (5.19), yields

$$l_j(z_j) = (N+2\alpha+1) P_N^{(\alpha)}(z_j) \left[\left(1 - z_j^2\right) \frac{d}{dx} P_{N+1}^{(\alpha)}(z_j) \right]^{-1} = 1,$$

where the last reduction follows by combining Equations (4.32) and (4.37) and using the definition of the Gauss quadrature points.

QED

Similar results hold for the Gauss–Radau and Gauss–Lobatto grids. Hence, the discrete approximations based on the Gauss–Radau and Gauss–Lobatto quadratures interpolate the undelying function at their respective points. These discrete approximations are thus known as interpolation methods.

The identification of the discrete approximation with the interpolation polynomials suggests a mathematically equivalent but computationally different way of representing the discrete expansion. We can either calculate the discrete expansion coefficients from their formulas (Equations (5.36)–(5.37)), or we can make use of the Lagrange interpolation polynomial $l_j(x)$, directly.

To facilitate the use of Lagrange interpolation, we list the Lagrange polynomials based on the Gauss–Lobatto, Gauss–Radau, and Gauss quadrature points.

Gauss–Lobatto points The Lagrange interpolation polynomial, $l_j(x)$, based on the Gauss–Lobatto points x_j is

$$l_j(x) = \begin{cases} (\alpha+1)\Pi_{N,j}^{(\alpha)}(x) & j = 0, N, \\ \Pi_{N,j}^{(\alpha)}(x) & \text{otherwise}, \end{cases} \tag{5.39}$$

where

$$\Pi_{N,j}^{(\alpha)}(x) = -\frac{1}{N(N+2\alpha+1)} \frac{(1-x^2)\left(P_N^{(\alpha)}\right)'(x)}{(x-x_j) P_N^{(\alpha)}(x_j)}.$$

Gauss–Radau points The Lagrange interpolation polynomial, $l_j(y)$, based on the Gauss–Radau points, y_j, is

$$l_j(y) = \begin{cases} (\alpha+1)\Pi^{(\alpha)}_{N,j}(y) & j = 0, N, \\ \Pi^{(\alpha)}_{N,j}(y) & \text{otherwise}, \end{cases} \tag{5.40}$$

where

$$\Pi^{(\alpha)}_{N,j}(y) = \frac{1}{2(N+\alpha+1)(N+2\alpha+1)} \frac{(1-y_j)}{P^{(\alpha)}_N(y_j)} \times \frac{(N+1)P^{(\alpha)}_{N+1}(y) + (N+2\alpha+1)P^{(\alpha)}_N}{y-y_j}.$$

Gauss points The Lagrange interpolation polynomial, $l_j(z)$, based on the Gauss quadrature points, z_j, is

$$l_j(z) = \frac{P^{(\alpha)}_{N+1}(z)}{(z-z_j)\left(P^{(\alpha)}_{N+1}\right)'(z)}. \tag{5.41}$$

Let us now return to the issue of how to obtain an approximation to the derivative of a function once the discrete approximation, expressed by using the discrete expansion coefficients or the Lagrange polynomials, is known.

First we focus on the approximation of the derivative resulting from the computation of the discrete expansion coefficients by quadrature. If we consider the Gauss–Lobatto approximation

$$\mathcal{I}_N u(x) = \sum_{n=0}^{N} \tilde{u}_n P^{(\alpha)}_n(x), \quad \tilde{u}_n = \frac{1}{\tilde{\gamma}_n} \sum_{j=0}^{N} u(x_j) P^{(\alpha)}_n(x_j) w_j,$$

we obtain an approximation to the derivative of $u(x)$

$$\mathcal{I}_N \frac{d}{dx} u(x) = \sum_{n=0}^{N} \tilde{u}^{(1)}_n P^{(\alpha)}_n(x),$$

where the new expansion coefficients, $\tilde{u}^{(1)}_n$, are obtained by using the backward recursion

$$\tilde{u}^{(1)}_n = (2n+2\alpha+1)\left[\frac{1}{2n+2\alpha+5}\tilde{u}^{(1)}_{n+2} + \tilde{u}_{n+1}\right],$$

from Equation (5.5) and initialized by $\tilde{u}^{(1)}_{N+1} = \tilde{u}^{(1)}_N = 0$ as $\mathcal{I}_N u(x) \in \mathrm{B}_N$. Higher derivatives are computed by applying the recurrence relation repeatedly. For the Chebyshev case, use Equation (5.13).

Of course, the use of the Gauss–Lobatto approximation was just an example; the same procedure can be followed using any of the Gauss quadrature points.

Next, we turn to the mathematically equivalent formulation which utilizes the Lagrange interpolating polynomial

$$\mathcal{I}_N u(x) = \sum_{j=0}^{N} u(x_j) l_j(x).$$

The derivative of $u(x)$ at any point is approximated simply by differentiating the global polynomial, $l_j(x)$. In particular, if one evaluates it at the quadrature points one obtains

$$\mathcal{I}_N \frac{d}{dx} \mathcal{I}_N u(x_i) = \sum_{j=0}^{N} u(x_j) \left. \frac{dl_j(x)}{dx} \right|_{x_i} = \sum_{j=0}^{N} \mathrm{D}_{ij} u(x_j).$$

Where D_{ij} are the (ij)-th elements of the $(N+1) \times (N+1)$ differentiation matrix, D.

The differentiation matrix, D, can be written explicitly.

Gauss–Lobatto quadrature points

$$\mathrm{D}_{ij} = \begin{cases} \frac{\alpha - N(N+2\alpha+1)}{2(\alpha+2)} & i = j = 0, \\ \frac{\alpha x_j}{1-(x_i)^2} & i = j \in [1, \dots, N-1], \\ \frac{\alpha+1}{x_i - x_j} \frac{P_N^{(\alpha)}(x_i)}{P_N^{(\alpha)}(x_j)} & i \neq j,\ j = 0, N, \\ \frac{1}{x_i - x_j} \frac{P_N^{(\alpha)}(x_i)}{P_N^{(\alpha)}(x_j)} & i \neq j,\ j \in [1, \dots, N-1], \\ -\mathrm{D}_{00} & i = j = N. \end{cases} \tag{5.42}$$

Gauss quadrature points

$$\mathrm{D}_{ij} = \begin{cases} \frac{(\alpha+1)z_i}{1-(z_i)^2} & i = j, \\ \frac{\left(P_{N+1}^{(\alpha)}\right)'(z_i)}{(z_i - z_j)\left(P_{N+1}^{(\alpha)}\right)'(z_j)} & i \neq j. \end{cases} \tag{5.43}$$

For an approximation based on the use of the Gauss–Radau quadrature points one can obtain an equivalent expression by using the associated Lagrange interpolation polynomial.

Some of these differentiation matrices D have interesting properties which shall be useful. In particular, the differentiation matrix, D, derived from the Lagrange interpolation polynomial based on any of the Gauss quadrature points, is nilpotent. This property is hardly a surprise, i.e., by differentiating a polynomial the order of the polynomial is reduced by one order and after $N+1$ differentiations the polynomial vanishes identically. The reason this works is that since the interpolating polynomial of degree N is exact for polynomials of degree N, the differentiation operator D is also exact. D has $(N+1)$ zero eigenvalues and only 1 eigenvector, i.e. it has a size $(N+1)$ Jordan block.

Another interesting property is limited to the differentiation matrix, D, based on the Gauss or the Gauss–Lobatto quadrature points. In these cases, the differentiation matrix D is centro-antisymmetric,

$$\mathrm{D}_{ij} = -\mathrm{D}_{N-i,N-j}.$$

This property follows from the expressions for the entries of D in Equation (5.42) and Equation (5.43), the even-odd symmetry of the ultraspherical polynomials, Equation (4.34), and, as a reflection of this, the symmetry of the quadrature points around $x = 0$. As we shall see in Chapter 11, this subtle symmetry enables a factorization of the differentiation matrices that ultimately allows for the computation of a derivative at a reduced cost. It is worth emphasizing that the differentiation matrix based on Gauss–Radau quadrature points does not possess the centro-antisymmetric property due to the lack of symmetry in the grid points.

The computation of higher derivatives follows the approach for the computation of the first derivative. One may compute entries of the qth-order differentiation matrix, $\mathrm{D}^{(q)}$ by evaluating the qth derivative of the Lagrange interpolation polynomial at the quadrature points. Alternatively, one obtains the qth-order differentiation matrix by simply multiplying the first-order differentiation matrices, i.e.,

$$\mathrm{D}^{(q)} = (\mathrm{D})^q,$$

where $q \leq N$. Although this latter approach certainly is appealing in terms of simplicity, we shall see later that the first method is preferable, in terms of accuracy.

5.4.1 The discrete Legendre expansion

The Legendre case (obtained by setting $\alpha = 0$ or $\lambda = 1/2$) is one of the most frequently used; thus, we devote this section to summarizing the differentiation formulas for Legendre.

Legendre Gauss–Lobatto In this case we consider

$$\mathcal{I}_N u(x) = \sum_{n=0}^{N} \tilde{u}_n P_n(x), \quad \tilde{u}_n = \frac{1}{\tilde{\gamma}_n} \sum_{j=0}^{N} u(x_j) P_n(x_j) w_j, \qquad (5.44)$$

where $\tilde{\gamma}_n$ is given in Equation (5.34) and the quadrature points, x_j, and the weights, w_j, are given as the solution to Equation (5.20) and in Equation (5.21), respectively. Computation of the expansion coefficients for the derivatives is done using the backward recurrence relation given in Equation (5.9).

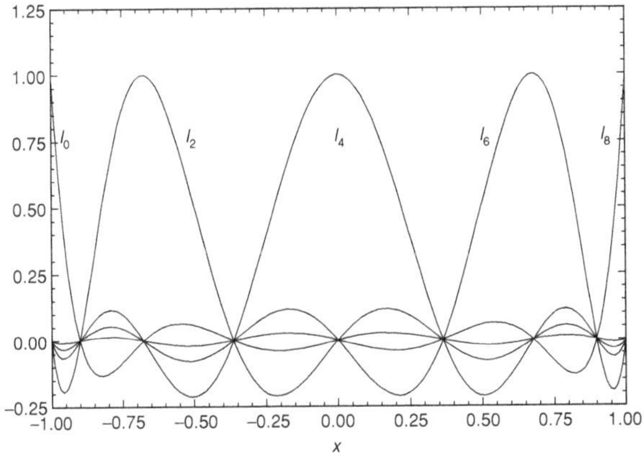

Figure 5.4 The interpolating Lagrange polynomial, $l_j(x)$, based on the Legendre Gauss–Lobatto quadrature points with $N = 8$.

Since $\mathcal{I}_N u(x)$ is the interpolant of $u(x)$ at the Legendre Gauss–Lobatto quadrature points, we may express the approximation as

$$\mathcal{I}_N u(x) = \sum_{j=0}^{N} u(x_j) l_j(x), \tag{5.45}$$

where the Lagrange interpolation polynomial, obtained directly from Equation (5.39) with $\alpha = 0$, takes the form

$$l_j(x) = \frac{-1}{N(N+1)} \frac{(1-x^2)P_N'(x)}{(x-x_j)P_N(x_j)}. \tag{5.46}$$

Examples of the Lagrange polynomials based on the Legendre Gauss–Lobatto points are shown in Figure 5.4.

The differentiation matrix is given by

$$\mathrm{D}_{ij} = \begin{cases} -\frac{N(N+1)}{4} & i = j = 0, \\ 0 & i = j \in [1, \ldots, N-1], \\ \frac{P_N(x_i)}{P_N(x_j)} \frac{1}{x_i - x_j} & i \neq j, \\ \frac{N(N+1)}{4} & i = j = N. \end{cases} \tag{5.47}$$

Legendre Gauss The discrete Legendre expansion based on the Legendre Gauss approximation to the continuous expansion coefficients is

$$\mathcal{I}_N u(z) = \sum_{n=0}^{N} \tilde{u}_n P_n(z), \quad \tilde{u}_n = \frac{1}{\tilde{\gamma}_n} \sum_{j=0}^{N} u(z_j) P_n(z_j) u_j. \tag{5.48}$$

The normalization constant, $\tilde{\gamma}_n$, is given in Equation (5.34), the quadrature points, z_j, are found as the roots of the polynomial, Equation (5.24), and the weights, u_j, are given in Equation (5.25).

We express the interpolation as

$$\mathcal{I}_N u(z) = \sum_{j=0}^{N} u(z_j) l_j(z), \tag{5.49}$$

with

$$l_j(z) = \frac{P_{N+1}(z)}{(z - z_j) P'_{N+1}(z_j)}, \tag{5.50}$$

Leading to the differentiation matrix

$$\mathrm{D}_{ij} = \begin{cases} \frac{z_i}{1-z_i^2} & i = j, \\ \frac{P'_{N+1}(z_i)}{(z_i - z_j) P'_{N+1}(z_j)} & i \neq j. \end{cases} \tag{5.51}$$

5.4.2 The discrete Chebyshev expansion

Methods based on Chebyshev polynomials continue to play a key role in the context of spectral methods. Their widespread use can be traced back to a number of reasons. Not only are the polynomials given in a simple form but all the Gauss quadrature nodes and the associated weights are also given in closed form. Furthermore, the close relationship between Chebyshev expansions and Fourier series allows for the fast evaluation of derivatives.

Chebyshev Gauss–Lobatto The discrete expansion is

$$\mathcal{I}_N u(x) = \sum_{n=0}^{N} \tilde{u}_n T_n(x), \quad \tilde{u}_n = \frac{2}{N \bar{c}_n} \sum_{j=0}^{N} \frac{1}{\bar{c}_j} u(x_j) T_n(x_j), \tag{5.52}$$

where

$$\bar{c}_n = \begin{cases} 2 & n = 0, N, \\ 1 & n \in [1, \dots, N-1]. \end{cases}$$

The Chebyshev Gauss–Lobatto quadrature points are

$$x_j = -\cos\left(\frac{\pi}{N} j\right),$$

which allows us to express the computation of the interpolating polynomial at the quadrature points as

$$\mathcal{I}_N u(x_j) = \sum_{n=0}^{N} \tilde{u}_n \cos\left(\frac{\pi}{N} nj\right), \quad \tilde{u}_n = \frac{2}{N \bar{c}_n} \sum_{j=0}^{N} \frac{1}{\bar{c}_j} u(x_j) \cos\left(\frac{\pi}{N} nj\right),$$

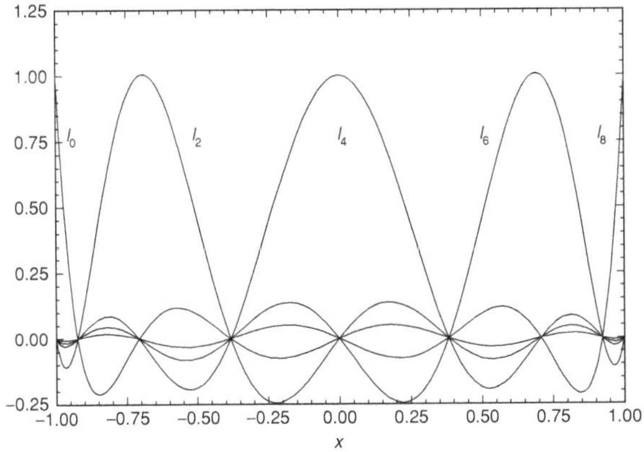

Figure 5.5 The Lagrange interpolation polynomial, $l_j(x)$, based on the Chebyshev Gauss–Lobatto quadrature points with $N = 8$.

which is the discrete cosine expansion. As we will see in Chapter 11, fast Fourier transform techniques can be used to evaluate $\tilde{u}_n$.

Alternatively, we can use the Lagrange polynomial formulation

$$\mathcal{I}_N u(x) = \sum_{j=0}^{N} u(x_j) l_j(x), \tag{5.53}$$

where the Lagrange interpolation polynomial is

$$l_j(x) = \frac{(-1)^{N+j+1}(1-x^2)T_N'(x)}{\bar{c}_j N^2 (x - x_j)}. \tag{5.54}$$

Leading to the differentiation matrix

$$\mathrm{D}_{ij} = \begin{cases} -\frac{2N^2+1}{6} & i = j = 0, \\ \frac{\bar{c}_i}{\bar{c}_j}\frac{(-1)^{i+j}}{x_i - x_j} & i \neq j, \\ -\frac{x_i}{2(1-x_i^2)} & i = j \in [1, \ldots, N-1], \\ \frac{2N^2+1}{6} & i = j = N. \end{cases} \tag{5.55}$$

In Figure 5.5 we show examples of the Lagrange polynomials, $l_j(x)$, based on the Chebyshev Gauss–Lobatto nodes. Comparing these with the polynomials based on the Legendre Gauss–Lobatto nodes in Figure 5.4 we observe only small differences; this is a natural consequence of the grid points being qualitatively the same.

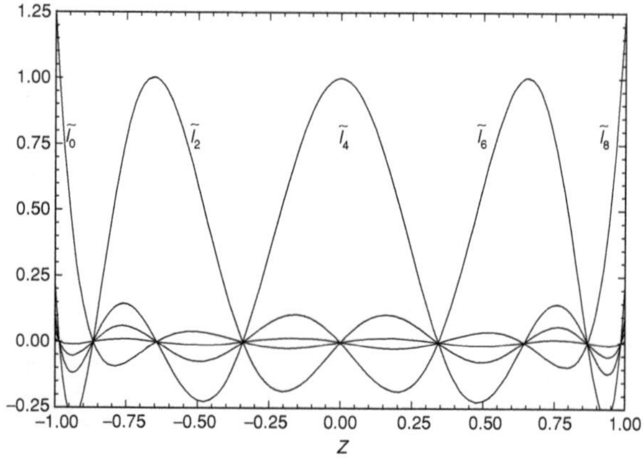

Figure 5.6 The Lagrange interpolation polynomial, $l_j(z)$, based on the Chebyshev Gauss quadrature points with $N = 8$.

Chebyshev Gauss In the Chebyshev Gauss method,

$$\mathcal{I}_N u(z) = \sum_{n=0}^{N} \tilde{u}_n T_n(z), \quad \tilde{u}_n = \frac{2}{c_n(N+1)} \sum_{j=0}^{N} u(z_j) T_n(z_j), \tag{5.56}$$

where c_n is defined in Equation (5.10). Since the Chebyshev Gauss quadrature points are

$$z_j = -\cos\left(\frac{(2j+1)\pi}{2N+2}\right),$$

the Chebyshev Gauss discrete expansion coefficients may be obtained using a modified Fast Fourier Transform.

Alternatively, using the Lagrange interpolation polynomial approach,

$$\mathcal{I}_N u(z) = \sum_{j=0}^{N} u(z_j) l_j(z), \tag{5.57}$$

with

$$l_j(z) = \frac{T_{N+1}(z)}{(z - z_j) T'_{N+1}(z_j)}, \tag{5.58}$$

leads to the differentiation matrix

$$\mathrm{D}_{ij} = \begin{cases} \frac{z_i}{2(1-z_i^2)} & i = j, \\ \frac{T'_{N+1}(z_i)}{(z_i - z_j) T'_{N+1}(z_j)} & i \neq j. \end{cases} \tag{5.59}$$

For the purpose of illustration, in Figure 5.6 we plot the Lagrange polynomials based on the Gauss quadrature points. We note, in particular, the different behavior at the boundaries of the domain as compared to polynomials based on the Gauss–Lobatto points and shown in Figure 5.5. Once again, we defer the discussion of the computational aspects of computing the nodes and weights to Chapter 11.

5.4.3 On Lagrange interpolation, electrostatics, and the Lebesgue constant

Up to this point, we have considered the approximation of functions and their derivatives using both discrete expansions and Lagrange interpolation polynomials. We observed that for certain choices of grid points and quadrature rules, these representations are identical. In this section, we will consider approximations

$$\mathcal{I}_N u(x) = \sum_{j=0}^{N} u(x_j) l_j(x),$$

where $l_j(x)$ are the Lagrange interpolation polynomials based on the grid points x_j,

$$l_j(x) = \frac{q_N(x)}{(x - x_j) q_N'(x_j)}, \quad q_N(x) = \prod_{j=0}^{N} (x - x_j). \tag{5.60}$$

Following the approach in previous sections, we obtain the entries of the differentiation matrix

$$\mathrm{D}_{ij} = \frac{1}{q_N'(x_j)} \begin{cases} q_N'(x_i)(x_i - x_j)^{-1} & i \neq j, \\ \frac{1}{2} q_N''(x_i) & i = j. \end{cases}$$

Now, we must decide how to specify the grid points x_j. Although we are free to chose any set of grid points, the following examples illustrate that we must choose x_j very carefully.

Example 5.4 We approximate the analytic function,

$$u(x) = \frac{1}{1 + 16x^2}, \quad x \in [-1, 1],$$

by $\mathcal{I}_N u(x)$ using Lagrange interpolants as above. We compare the interpolation using the equidistant points

$$x_j = \frac{2}{N} j - 1, \quad j \in [0, \ldots, N],$$

with that based on the Chebyshev Gauss–Lobatto quadrature points

$$x_j = -\cos\left(\frac{\pi}{N}j\right), \quad j \in [0, \ldots, N].$$

In Figure 5.7 we observe the dramatic differences between the resulting interpolations. We note that while the one based on the Chebyshev Gauss–Lobatto grid points seems to converge as expected, the interpolation polynomial based on the equidistant grid is divergent as N increases. Clearly, the choice of the grid points matters when considering the quality of the global interpolation.

This wildly oscillatory and divergent behavior close to the limits of the domain is known as the Runge phenomenon.

In the previous example we could easily see which interpolation was good and which was not, but in general the quality of the interpolation needs to be quantified. A useful measure of the quality of the interpolation is introduced in the following theorem:

Theorem 5.5 *Assume that $u(x) \in C^0[-1, 1]$ with $\mathcal{I}_N u(x)$ being the corresponding Nth-order polynomial interpolation based on the grid points, x_j. Then*

$$\|u - \mathcal{I}_N u\|_\infty \leq [1 + \Lambda_N] \|u - p^*\|_\infty,$$

where p^* is the best approximating Nth-order polynomial, and

$$\Lambda_N = \max_{x\in[-1,1]} \lambda_N(x), \quad \lambda_N(x) = \sum_{j=0}^{N} |l_j(x)|,$$

are the Lebesgue constant and the Lebesgue function, respectively.

Proof: As $u(x) \in C^0[-1, 1]$, the best approximating polynomial, p^*, exists and we immediately have

$$\|u - \mathcal{I}_N u\|_\infty \leq \|u - p^*\|_\infty + \|p^* - \mathcal{I}_N u\|_\infty.$$

However, since both p^* and $\mathcal{I}_N u$ are Nth-order polynomials,

$$\|p^* - \mathcal{I}_N u\|_\infty \leq \Lambda_N \|u - p^*\|_\infty,$$

where

$$\Lambda_N = \max_{x\in[-1,1]} \lambda_N(x), \quad \lambda_N^{(x)} = \sum_{j=0}^{N} |l_j(x)|.$$

QED

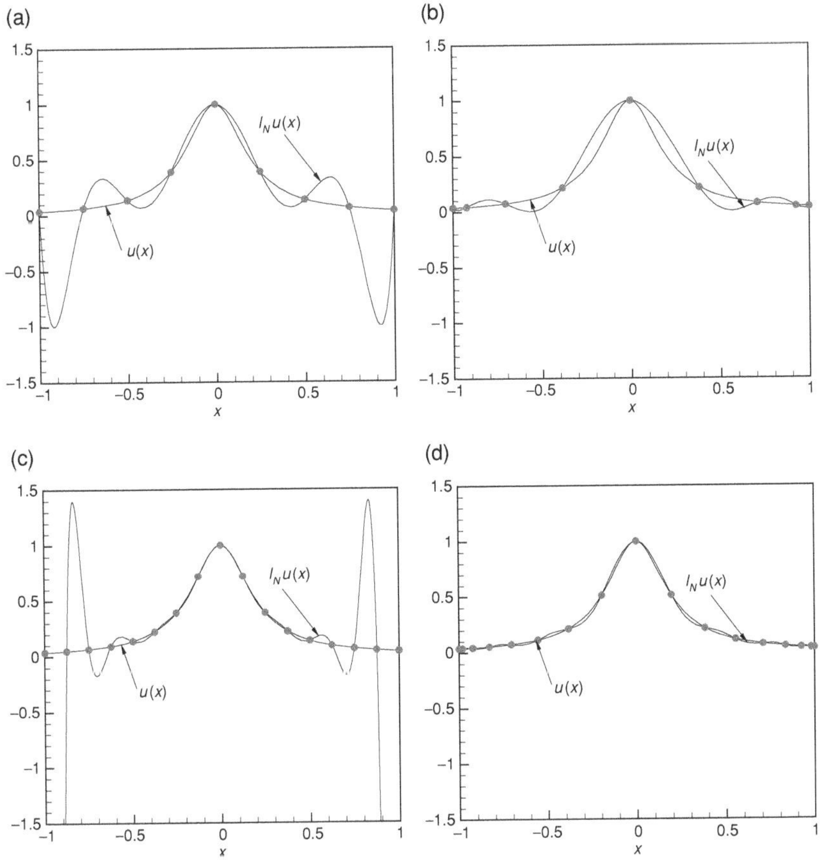

Figure 5.7 (a) Interpolation of $u(x)$ using $N = 8$ equidistant grid points. (b) Interpolation of $u(x)$ using $N = 8$ Chebyshev Gauss–Lobatto distributed grid points. (c) Interpolation of $u(x)$ using $N = 16$ equidistant grid points. (d) Interpolation of $u(x)$ using $N = 16$ Chebyshev Gauss–Lobatto distributed grid points.

The Lebesgue function and the Lebesgue constant both depend only on the choice of x_j. Theorem 5.5 tells us that a good set of grid points is one which minimizes the Lebesgue constant.

On a more practical matter, the Lebesgue function gives us insight into how computational issues such as rounding errors can impact the accuracy of the interpolation. As an example, assume that $u_\varepsilon(x)$ represents a perturbed version of $u(x)$

$$\|u - u_\varepsilon\|_\infty \leq \varepsilon.$$

The difference between the two polynomial representations is then

$$\|\mathcal{I}_N u - \mathcal{I}_N u_\varepsilon\|_\infty \leq \varepsilon \Lambda_N.$$

Clearly, if the Lebesgue constant Λ_N is large enough the interpolation is illposed and the impact of the rounding is very severe.

It is thus worthwhile to look at the behavior of the Lebesgue function and the value of Λ_N for various choices of grid points. Indeed, one could hope to identify families of grid points, x_j, for which Λ_N remains a constant. A seminal result in approximation theory, however, rules out the existence of such a set of grid points.

Theorem 5.6 *For any set of $(N+1)$ distinct grid points, $x_j \in [-1, 1]$, and all values of N, the Lebesgue constant is bounded as*

$$\Lambda_N \geq \frac{2}{\pi} \log(N+1) + A,$$

where

$$A = \frac{2}{\pi}\left(\gamma + \log \frac{4}{\pi}\right),$$

in the limit of large N. Here $\gamma = 0.577221566$ is Euler's constant.

In other words, the Lebesgue constant grows at least logarithmically with N. This has the unfortunate consequence that for any given set of grid points there exist continuous functions for which the polynomial representations will exhibit non-uniform convergence. On the other hand, one can also show that for any given continuous function one can always construct a set of grid points that will result in a uniformly convergent polynomial representation.

Thus, we can not in general seek one set of grid points, x_j, that will exhibit optimal behavior for all possible interpolation problems. However, the behavior of the Lebesgue constant can serve as a guideline to understand whether certain families of grid points are likely to result in well behaved interpolation polynomials.

Let's consider how the Lebesgue constant varies for different choices of x_j.

Theorem 5.7 *Assume that the interpolation is based on the equidistributed set of grid points*

$$x_j = -1 + \frac{2j}{N}, \quad j \in [0, \ldots, N].$$

Then the corresponding Lebesgue constant, Λ_N^{eq} is bounded for $N \geq 1$,

$$\frac{2^{N-2}}{N^2} \leq \Lambda_N^{\text{eq}} \leq \frac{2^{N+3}}{N}.$$

Asymptotically,

$$\Lambda_N^{\mathrm{eq}} \simeq \frac{2^{N+1}}{eN(\log N + \gamma)}.$$

This is clearly far from optimal, and for large values of N one can not expect anything meaningful from the interpolation based on the equidistant set of grid points, i.e., for $N \geq 65$, $\Lambda_N^{\mathrm{eq}} \sim 10^{16}$ and the interpolation is very illposed.

How can we identify which grid points lead to well behaved interpolations? To understand this, consider the Cauchy remainder for the interpolation,

$$u(z) - \mathcal{I}_N u(z) = R_N(z) = \frac{u^{(N+1)}(\xi)}{(N+1)!} q_N(z),$$

where ξ refers to some position in $[-1, 1]$, $q_N(z)$ is as defined in Equation (5.60) and z is the complex extension of x. Note that the grid points, x_j, remain real. Considering $q_N(z)$ it follows directly that

$$\log |q_N(z)| = \sum_{j=0}^{N} \log |z - x_j| = -(N+1)\phi_N(z), \tag{5.61}$$

where $\phi_N(z)$ can be interpreted as the electrostatic energy associated with $N+1$ unit mass, unit charge particles interacting according to a logarithmic potential. In the limit of $N \to \infty$ it is natural to model this as

$$\phi_\infty(z) = \lim_{N\to\infty} \frac{1}{N+1} \sum_{j=0}^{N} \log |z - x_j| = \int_{-1}^{1} \rho(x) \log |z - x| \, dx,$$

where $\rho(x)$ represents a normalized charge density, reflecting an asymptotic measure of the grid point distribution. This implies that

$$\lim_{N\to\infty} |q(z)|^{1/N} = e^{-\phi_\infty(z)},$$

for large values of N. Understanding the second part of the remainder, associated with the particular function being interpolated, is a bit more difficult. Using complex analysis allows one to derive that

$$\lim_{N\to\infty} \left| \frac{u^{(N+1)}(\xi)}{(N+1)!} \right|^{1/N} = e^{\phi_\infty(z_0)},$$

where z_0 represents the radius of the largest circle within which $u(z)$ is analytic. Hence, we recover that

$$\lim_{N\to\infty} |R(z)|^{1/N} = e^{\phi_\infty(z_0) - \phi_\infty(z)}.$$

Clearly, if $\phi_\infty(z_0) - \phi_\infty(z)$ is less than zero throughout the interval $[-1, 1]$, we can expect exponential convergence in N. On the other hand, if $\phi_\infty(z_0) - \phi_\infty(z)$ exceeds zero anywhere in the unit interval we expect exponential divergence of the remainder and thus of the error.

Let us first return to the equidistant charge distribution, in which case $\rho(x) = \frac{1}{2}$ and the electrostatic energy becomes

$$\phi_\infty(z) = 1 + \frac{1}{2}\mathrm{Re}\left[|z-1|\log|z-1| - |z+1|\log|z+1|\right].$$

We first of all note that while $\phi_\infty(0) = 1$, we have $\phi_\infty(\pm 1) = 1 - \log 2$, i.e., we should expect to see the most severe problems of divergence closer to the limits of the domain, as observed in Figure 5.7. In fact, if we consider the $u(x)$ in Example 5.4, it has a pole at $z_0 = \pm i/4$ and

$$\phi_\infty(\pm i/4) = 1 - \frac{1}{2}\left[\log\frac{17}{16} + \tfrac{1}{2}\arctan 4\right] \simeq 0.63823327\ldots$$

Thus, the remainder, and hence the polynomial representation of $u(x)$ will diverge in regions close to $[-1, 1]$ as $\phi_\infty(\pm i/4) - \phi_\infty(z)$ will exceed zero. Finding the exact point of divergence yields $x \simeq \pm 0.794226$. A close inspection of the results in Figure 5.7 confirms this.

An equidistant distribution of grid points is evidently not a good choice for high-order polynomial interpolation of analytic functions defined on the interval. On the other hand, there is evidence in the above that the main problems are close to the limits of the domain and that clustering the grid points relieves these problems as illustrated in Example 5.4.

To study this further let us begin by realizing that the continuous charge distribution leading to the Chebyshev Gauss–Lobatto nodes used in Example 5.4 takes the form

$$\rho(x) = \frac{1}{\pi\sqrt{1-x^2}}, \tag{5.62}$$

since we have

$$j = N\int_{-1}^{x_j} \frac{1}{\pi\sqrt{1-x^2}}dx \;\Rightarrow\; x_j = -\cos\left(\frac{\pi}{N}j\right),$$

where $j \in [0, \ldots, N]$ is the charge number. With this, we have the corresponding electrostatic energy of the form

$$\phi_\infty(z) = -\log\frac{|z - \sqrt{z^2-1}|}{2}.$$

Inspection reveals that $\phi_\infty(z)$ are level curves associated with an ellipsoid with foci at ± 1. Furthermore, as z becomes purely real, this level curve collapses to a ridge spanning the domain of $[-1, 1]$ along which $\phi_\infty(z)$ takes on its maximum value. Hence, there are no restrictions on the position of the poles of the function, $u(x)$, being approximated, as we can guarantee that $\phi_\infty(z_0) - \phi_\infty(x)$ is negative for any value of z_0. This collapse of the level curve of $\phi_\infty(z)$ to a perfectly flat ridge clearly represents the optimal choice when one considers the electrostatic analysis.

Having identified grid point distributions which lead to well behaved Lagrange interpolations, let us now return to the evaluation of these interpolations in terms of the Lebesgue constant.

For the symmetric Chebyshev Gauss and Chebyshev Gauss–Lobatto, both having the same asymptotic distribution given in Equation (5.62), we have the Lebesgue constant, Λ_N^{CG}, of the former as

$$\Lambda_N^{\mathrm{CG}} \leq \frac{2}{\pi}\log(N+1) + A + \frac{2}{\pi}\log 2.$$

The constant A is given in Theorem 5.6. The Lebesgue constant, Λ_N^{CGL} of the latter set of grid points is bounded

$$\Lambda_N^{\mathrm{CGL}} \leq \Lambda_{N-1}^{\mathrm{CG}},$$

i.e., the Gauss–Lobatto points are, when measured by the growth of the Lebesgue constant, superior to the Gauss points and very close to the theoretical optimum given in Theorem 5.6.

The particular characteristic of the Chebyshev distributed grid points that gives the well behaved Lagrange polynomials is the quadratic clustering of the grid points close to the ends of the domain. This quality is, however, shared among the zeros of all the ultraspherical polynomials as they all have a minimum grid size of the kind

$$\Delta x_{\min} = 1 - cN^2,$$

where the constant c depends on the particular polynomial. This difference, however, vanishes as N approaches infinity as the grid distributions of the zeros of the ultraspherical polynomials all share the limiting continuous charge distribution, Equation (5.62).

With this result, it becomes clear that when choosing grid points well suited for high-order polynomial interpolation, we need only impose structural conditions on the position of the grid points, e.g., close to the boundaries of the domain the grid points must cluster quadratically. This means that in terms of interpolation, the exact position of the grid points is immaterial, e.g., Legendre

Gauss–Lobatto points are as good as the Chebyshev Gauss–Lobatto points as the basis of the Lagrange polynomials. This is also reflected in the associated Lebesgue constant of the form

$$\Lambda_N^{\text{LGL}} \leq \frac{2}{\pi}\log(N+1) + 0.685\ldots$$

for the Legendre Gauss–Lobatto grid. Having realized, however, that it is the quadratic behavior of the grid points close to the end of the interval that is the key to high-order convergence, there is nothing that prohibits us from seeking grid point distributions with this particular quality. Indeed, the closest to optimal grid point distribution as measured through the Lebesgue constant, and for which a simple formula is known, is defined as

$$x_j^{\text{ECG}} = -\frac{\cos\left(\frac{2j+1}{2N+2}\pi\right)}{\cos\left(\frac{\pi}{2N+2}\right)},$$

known as the extended Chebyshev Gauss grid points. These are not zeros of any ultraspherical polynomial, yet they have a Lebesgue constant, Λ_N^{ECG}, bounded as

$$\Lambda_N^{\text{ECG}} = \frac{2}{\pi}\log(N+1) + A + \frac{2}{\pi}\left(\log 2 - \frac{2}{3}\right),$$

which is very close to the optimal set of grid points and for all practical purposes can serve as that.

The above discussion of the Lebesgue constant and the electrostatic approach revolves, to a large extent, around the behavior of the interpolation as it depends on the choice of the grid points in the limit where N is very large. It is, however, illustrative and useful to realize that there is a close connection between the zeros of the ultraspherical polynomials and the solution to a slightly modified version of the finite dimensional electrostatic problem, Equation (5.61).

Let us define the electrostatic energy, $E(x_0, \ldots, x_N)$, as

$$E(x_0, \ldots, x_N) = -\frac{1}{2}\sum_{i=0}^{N}\sum_{\substack{j=0 \\ j\neq i}}^{N} \log|x_i - x_j|,$$

for the $N+1$ unit mass, unit charge particles interacting according to a logarithmic potential and consider the problem as an N-body problem for which we seek the steady state, minimum energy solution if it exists. For this definition, however, the dynamics of the problem are such that all charges would move to infinity as that would be the minimum energy solution. Let us therefore consider the slightly modified problem

$$E(p, x_0, \ldots, x_N) = -\sum_{i=0}^{N}\left[p\log\left(1-x_i^2\right)+\frac{1}{2}\sum_{\substack{j=0\\ j\neq i}}^{N}\log|x_i-x_j|\right]. \quad (5.63)$$

This corresponds to forcing the $N+1$ charges with an exterior field corresponding to two charges, positioned at ± 1, of strength $p > 0$. If we now assume that all charges initially are positioned in the interior of $[-1, 1]$, they are confined there and nontrivial steady-state minimum energy solutions can be sought.

Considering the gradient of E, we find that a condition for minimum energy is

$$\frac{\partial E}{\partial x_i} = \frac{1}{2}\sum_{\substack{j=0\\ j\neq i}}^{N}\frac{1}{x_i-x_j} - \frac{2x_i p}{1-x_i^2} = 0.$$

Using $q_N(x)$ as defined in Equation (5.60) we recover

$$\frac{1}{2}\frac{q_N''(x_i)}{q_N'(x_i)} - \frac{2x_i p}{1-x_i^2} = 0,$$

or equivalently

$$\left(1-x_i^2\right)q_N''(x_i) - 4px_i q_N'(x_i) = 0.$$

Since this is a polynomial of order $N+1$ with $N+1$ point constraints where it vanishes, it must, due to the definition of $q_N(x)$, be proportional to $q_N(x)$ it self. By matching coefficients we recover

$$(1-x^2)q_N''(x) - 4xpq_N'(x) + (N+1)(N+4p)q_N(x) = 0. \quad (5.64)$$

The polynomial solution, $q_N \in \mathrm{B}_{N+1}$, of Equation (5.64) is the optimal solution to the electrostatic problem, Equation (5.63), as its $N+1$ roots. If, however, we take $\alpha = 2p-1$ and multiply Equation (5.63) by $(1-x^2)^\alpha$ we recover

$$\frac{d}{dx}(1-x^2)^{\alpha+1}\frac{dq_N}{dx} + (N+1)(N+2\alpha+2)(1-x^2)^\alpha q_N = 0,$$

which we may recognize as the Sturm–Liouville problem defining the general ultraspherical polynomial, $P_{N+1}^{(\alpha)}(x)$, i.e., $q_N(x) = P_{N+1}^{(\alpha)}(x)$.

Hence, a further manifestation of the close relation between grid points, well suited for interpolation, and the solution to problems of electrostatics is realized by observing that the minimum energy steady state charge distribution solution to the N-body problem stated in Equation (5.63) is exactly the Gauss quadrature points of the ultraspherical polynomial, $P_N^{(2p-1)}(x)$. Using the simple relation

$$2\frac{dP_N^{(\alpha)}}{dx} = (N+1+2\alpha)P_{N-1}^{(\alpha+1)}(x),$$

we see that the interior part of the Gauss–Lobatto points can be found by taking $\alpha = 2(p-1)$, e.g., the Chebyshev Gauss–Lobatto grid appears as the steady state solution for $p = 3/4$.

It should be noted that by allowing an asymmetric exterior field in Equation (5.63) one can recover the Gauss quadrature nodes for all the Jacobi polynomials, $P_N^{(\alpha,\beta)}(x)$, in Equation (5.63).

5.5 Further reading

The discrete expansions and differential operators are discussed in several references, e.g., Gottlieb et al (1983), Canuto et al (1986, 2006), and Funaro (1992), albeit in slightly different forms. The connection between interpolation and electrostatics is pointed out in Szego (1930) and further elaborated in Hesthaven (1998). A good discussion of the Runge phenomenon and its explanation through potentials can be found in the texts by Fornberg (1996), Boyd (2000), and Trefethen (2000). In the latter there is also a brief discussion of results related to the Lebesgue constants.

6
Polynomial approximation theory for smooth functions

Since the orthogonal polynomials are solutions of a singular Sturm–Liouville problem, we expect the polynomial expansion of smooth functions to converge at a rate depending only on the smoothness of the function being approximated. In particular, when approximating a C^∞ function by such a polynomial expansion up to order N, we expect the convergence rate to be spectral, i.e., faster than any algebraic order of N. In this chapter we will review the major approximation results for functions of finite regularity. We restrict our attention to those results which are most useful in the context of polynomial spectral methods: namely, those involving ultraspherical polynomials, and in particular, Chebyshev and Legendre polynomials.

6.1 The continuous expansion

Our aim is to estimate the distance between $u(x)$ and its spectral expansion

$$\mathcal{P}_N u(x) = \sum_{n=0}^{N} \hat{u}_n P_n^{(\alpha)}(x), \quad \hat{u}_n = \frac{1}{\gamma_n}\left(u, P_n^{(\alpha)}\right)_w,$$

as measured in the weighted Sobolev norm $\|\cdot\|_{H_w^m[-1,1]}$, where $w(x)$ is the weight under which the family of polynomials $P_n^{(\alpha)}(x)$ is orthogonal.

The following theorem provides the basic approximation results for ultraspherical polynomial expansions.

Theorem 6.1 *For any $u(x) \in H_w^p[-1,1]$, $p \geq 0$, there exists a constant C, independent of N, such that*

$$\|u - \mathcal{P}_N u\|_{L_w^2[-1,1]} \leq C N^{-p} \|u\|_{H_w^p[-1,1]}.$$

Proof: Parseval's identity implies

$$\|u - \mathcal{P}_N u\|^2_{L^2_w[-1,1]} = \sum_{n=N+1}^{\infty} \gamma_n |\hat{u}_n|^2,$$

where the expansion coefficients, $\hat{u}_n$, are

$$\hat{u}_n = \frac{1}{\gamma_n} \int_{-1}^{1} u(x) P_n^{(\alpha)}(x)(1-x^2)^\alpha \, dx,$$

and $P_n^{(\alpha)}(x)$ satisfies the Sturm–Liouville equation

$$[\mathcal{Q} + \lambda_n] \, P_n^{(\alpha)} = 0,$$

where

$$\mathcal{Q}u = (1-x^2)\frac{d^2u}{dx^2} - 2x(1+\alpha)\frac{du}{dx}.$$

Here $\lambda_n = n(n + 2\alpha + 1)$ is the nth eigenvalue.

Repeated integration by parts of $\hat{u}_n$ yields

$$\hat{u}_n = \frac{(-1)^m}{\gamma_n \lambda_n^m} \int_{-1}^{1} [\mathcal{Q}^m u(x)] P_n^{(\alpha)}(x) w(x) \, dx,$$

and so the Cauchy–Schwartz inequality implies

$$|\hat{u}_n|^2 \leq C \frac{1}{\lambda_n^{2m}} \|\mathcal{Q}^m u\|^2_{L^2_w[-1,1]}.$$

Since $|x| \leq 1$, we have

$$\|\mathcal{Q}u\|_{L^2_w[-1,1]} \leq C \|u\|_{H^2_w[-1,1]}.$$

By induction,

$$\|\mathcal{Q}^m u\|_{L^2_w[-1,1]} \leq C \|u\|_{H^{2m}_w[-1,1]}.$$

Combining the above results,

$$\|u - \mathcal{P}_N u\|^2_{L^2_w[-1,1]} \leq C \|u\|^2_{H^{2m}_w[-1,1]} \sum_{n=N+1}^{\infty} \gamma_n \lambda_n^{-2m} \leq C N^{-4m} \|u\|^2_{H^{2m}_w[-1,1]}.$$

Taking $p = 2m$ establishes the theorem.

QED

This theorem demonstrates that the error, as measured in the weighted L^2 norm, decays spectrally. In the next theorem we show that this is also the case when the error is measured in the corresponding weighted Sobolev norm.

Theorem 6.2 *Let $u(x) \in H_w^p[-1,1]$; there exists a constant C, independent of N, such that*

$$\left\| \mathcal{P}_N \frac{du}{dx} - \frac{d}{dx}\mathcal{P}_N u \right\|_{H_w^q[-1,1]} \le C N^{2q-p+3/2} \|u\|_{H_w^p[-1,1]},$$

where $1 \le q \le p$.

Proof: Without loss of generality, we consider the case where N is even. Let us first of all recall that

$$u(x) = \sum_{n=0}^{\infty} \hat{u}_n P_n^{(\alpha)}(x), \quad \text{and} \quad \mathcal{P}_N u(x) = \sum_{n=0}^{N} \hat{u}_n P_n^{(\alpha)}(x),$$

and so,

$$\frac{d}{dx}\mathcal{P}_N u(x) = \sum_{n=0}^{N} \hat{u}_n \left(P_n^{(\alpha)}(x)\right)'.$$

On the other hand, since

$$\frac{d}{dx} u(x) = \sum_{n=0}^{\infty} \hat{u}_n \left(P_n^{(\alpha)}(x)\right)',$$

we have that

$$\mathcal{P}_N \frac{d}{dx} u(x) = \sum_{n=0}^{N+1} \hat{u}_n \left(P_n^{(\alpha)}(x)\right)'.$$

Comparing these two expressions, we see that

$$\mathcal{P}_N \frac{du}{dx} - \frac{d}{dx}\mathcal{P}_N u = \hat{u}_{N+1} P_{N+1}^{(\alpha)}.$$

Theorem 6.1 tells us that

$$|\hat{u}_{N+1}|^2 \le \|u - \mathcal{P}_N u\|^2_{L_w^2[-1,1]} \le C N^{-p} \|u\|^2_{H_w^p[-1,1]}.$$

Furthermore, using orthogonality of $P_k^{(\alpha)}(x)$ and the fact that that γ_k is bounded in k provided $|\alpha| \le 1/2$, we get

$$\begin{aligned}
\left\|\left(P_N^{(\alpha)}\right)'\right\|^2_{L_w^2[-1,1]} &= \left\| \sum_{\substack{k=0 \\ k+N \text{ odd}}}^{N-1} (2k+2\alpha+1) P_k^{(\alpha)} \right\|^2_{L_w^2[-1,1]} \\
&\le \sum_{\substack{k=0 \\ k+N \text{ odd}}}^{N-1} (2k+2\alpha+1)^2 \gamma_k \le C N^3.
\end{aligned}$$

Combining these estimates yields

$$\left\| \mathcal{P}_N \frac{du}{dx} - \frac{d}{dx}\mathcal{P}_N u \right\|_{L^2_w[-1,1]} \leq C N^{\frac{3}{2}-p} \|u\|_{H^p_w[-1,1]}.$$

The Sturm–Liouville equations and the Poincaré inequality lead to

$$\left| \frac{d^m}{dx^m} P_n^{(\alpha)}(x) \right| \leq C N^{2m} \left| P_n^{(\alpha)}(x) \right|,$$

so that any further differentiation causes a loss of N^2 in the error bound. Thus, the error bound in the norm dependent on the qth derivative becomes

$$\left\| \mathcal{P}_N \frac{du}{dx} - \frac{d}{dx}\mathcal{P}_N u \right\|_{H^q_w[-1,1]} \leq C N^{2q-p+\frac{3}{2}} \|u\|_{H^p_w[-1,1]}.$$

QED

Consequently, we obtain the following generalization:

Theorem 6.3 *For any $u(x) \in H^p_w[-1,1]$ there exists a constant C, independent of N, such that*

$$\|u - \mathcal{P}_N u\|_{H^q_w[-1,1]} \leq C N^{\sigma(q,p)} \|u\|_{H^p_w[-1,1]},$$

where

$$\sigma(q,p) = \begin{cases} \frac{3}{2}q - p & 0 \leq q \leq 1, \\ 2q - p - \frac{1}{2} & q \geq 1, \end{cases}$$

and $0 \leq q \leq p$.

These results establish the spectral accuracy of the polynomial expansion. Recall that when approximating the derivative by the Fourier method we lose a factor of N in the error estimate. However, as we see in Theorem 6.3, when approximating the derivative by the polynomial method we lose a factor of N^2 in the error estimate. This fact limits the number of converging derivatives for a given polynomial approximation. The following example illustrates the loss of convergence.

Example 6.4 Consider the function

$$u(x) = \sum_{k=1}^{\infty} \frac{P_{k+1}(x) - P_{k-1}(x)}{(2k+1)k^{\epsilon}},$$

where $P_k(x)$ is the kth degree Legendre polynomial, and $\epsilon \ll 1$.

Using the recurrence for the Legendre polynomials, Equation (4.18),

$$\frac{P'_{k+1}(x) - P'_{k-1}(x)}{2k+1} = P_k(x),$$

we obtain

$$\frac{du}{dx} = \sum_{k=1}^{\infty} \frac{P_k(x)}{k^\epsilon}.$$

We note that this series converges in L_2; in fact, since $\int P_k^2(x)dx = \frac{2}{2k+1}$, we have

$$\left\| \frac{du}{dx} \right\|^2 = \sum_{k=1}^{\infty} \int \frac{P_k^2(x)}{k^{2\epsilon}} dx = \sum_{k=1}^{\infty} \frac{1}{k^{2\epsilon}} \frac{2}{2k+1},$$

which converges for $\epsilon > 0$.

Approximating $u(x)$ by the truncated Legendre expansion

$$\mathcal{P}_N u = \sum_{n=0}^{N} \hat{u}_n P_n(x),$$

we observe that

$$\mathcal{P}_N u = \sum_{k=1}^{N} \frac{P_{k+1}(x) - P_{k-1}(x)}{(2k+1)k^\epsilon} - \frac{P_{N+1}(x)}{(2N+1)N^\epsilon},$$

so that

$$\frac{d}{dx}\mathcal{P}_N u = \sum_{k=1}^{N} \frac{P_k(x)}{k^\epsilon} - \frac{P'_{N+1}(x)}{(2N+1)N^\epsilon},$$

while

$$\mathcal{P}_N \frac{du}{dx} = \sum_{k=1}^{N} \frac{P_k(x)}{k^\epsilon}.$$

Thus, the error in approximating the derivative:

$$\frac{d}{dx}\mathcal{P}_N u - \mathcal{P}_N \frac{du}{dx} = -\frac{1}{(2N+1)N^\epsilon} \frac{dP_{N+1}}{dx}.$$

The norm of this error is

$$\begin{aligned} \left\| \frac{d}{dx}\mathcal{P}_N u - \mathcal{P}_N \frac{du}{dx} \right\|^2 &= \frac{1}{(2N+1)^2 N^{2\epsilon}} \int (P'_{N+1})^2 dx \\ &\leq \frac{N^3}{(2N+1)^2 N^{2\epsilon}} \approx N^{1-2\epsilon}, \end{aligned}$$

which diverges as N increases. The divergence is caused by the inability of the derivative of the truncated approximation to approximate the derivative of u, as predicted in Theorem 6.2.

6.2 The discrete expansion

Similar approximation results are obtained for the discrete expansion,

$$\mathcal{I}_N u(x) = \sum_{n=0}^{N} \tilde{u}_n P_n^{(\alpha)}(x) = \sum_{j=0}^{N} u(x_j) l_j(x),$$

though the proofs are more technically complex than for the continuous expansion. The main reason for this added complexity is the aliasing error introduced by using quadrature to approximate the integrals. Under the assumption of sufficient smoothness, e.g., $u(x) \in H_w^1[-1,1]$, the aliasing error is

$$\tilde{u}_n = \hat{u}_n + \frac{1}{\tilde{\gamma}_n} \sum_{k>N}^{\infty} \left[P_n^{(\alpha)}, P_k^{(\alpha)}\right]_w \hat{u}_k,$$

where $[\cdot,\cdot]_w$ is the discrete inner product introduced in Section 5.3. From the orthogonality of the polynomial basis, it follows that

$$\|u - \mathcal{I}_N u\|_{L_w^2[-1,1]}^2 = \|u - \mathcal{P}_N u\|_{L_w^2[-1,1]}^2 + \|\mathcal{A}_N u\|_{L_w^2[-1,1]}^2,$$

where the aliasing error is

$$\mathcal{A}_N u(x) = \sum_{n=0}^{N} \frac{1}{\tilde{\gamma}_n} \left(\sum_{k>N}^{\infty} \left[P_n^{(\alpha)}, P_k^{(\alpha)}\right]_w \hat{u}_k \right) P_n^{(\alpha)}(x).$$

Interchanging the two summations we obtain the simple expression

$$\mathcal{A}_N u(x) = \sum_{k>N}^{\infty} \left(\mathcal{I}_N P_k^{(\alpha)}\right) \hat{u}_k.$$

Thus, the aliasing error can be seen as the error introduced by using the interpolation of the basis, $\mathcal{I}_N P_k^{(\alpha)}$, rather than the basis itself to represent the higher modes. The aliasing error stems from the fact that we cannot distinguish between lower and higher modes on a finite grid.

The discrete analog to Theorem 6.1 is:

Theorem 6.5 *For $u \in H_w^p[-1,1]$ where $p > 1/2 \max(1, 1+\alpha)$, there exists a constant, C, which depends on α and p but not on N, such that*

$$\|u - \mathcal{I}_N u\|_{L_w^2[-1,1]} \leq C N^{-p} \|u\|_{H_w^p[-1,1]},$$

where $\mathcal{I}_N u$ is constructed using ultraspherical polynomials, $P_n^{\alpha}(x)$, with $|\alpha| \leq 1$. This holds for Gauss and Gauss–Lobatto based interpolations.

This result confirms that for well resolved smooth functions the qualitative behavior of the continuous and the discrete expansion is similar for all practical purposes.

To be more specific and obtain results in higher norms we leave the general ultraspherical expansion and consider discrete expansions based on Legendre and Chebyshev polynomials and the associated Gauss-type quadrature points.

Results for the discrete Legendre expansion Consider first the discrete Legendre expansion

$$\mathcal{I}_N u(x) = \sum_{n=0}^{N} \tilde{u}_n P_n(x) = \sum_{j=0}^{N} u(x_j) l_j(x).$$

The most general result is given by the following theorem.

Theorem 6.6 *For any $u(x) \in H^p[-1,1]$ with $p > 1/2$ and $0 \le q \le p$, there exists a positive constant, C, independent of N, such that*

$$\|u - \mathcal{I}_N u\|_{H^q[-1,1]} \le C N^{2q-p+1/2} \|u\|_{H^p[-1,1]}.$$

Results for the discrete Chebyshev expansion The behavior of the discrete Chebyshev expansion

$$\mathcal{I}_N u(x) = \sum_{n=0}^{N} \tilde{u}_n T_n(x) = \sum_{j=0}^{N} u(x_j) l_j(x),$$

can be made clear by using the close connection between the Chebyshev expansion and the Fourier cosine series. If we introduce the transformation, $x = \cos(\theta)$, we can use the orthogonality of the exponential function to get

$$\tilde{p}_n = \frac{1}{\tilde{\gamma}_n} [T_k(x), T_n(x)]_N = \begin{cases} 1 & k = 2Np \pm n \quad p = 0, \pm 1, \pm 2, \ldots, \\ 0 & \text{otherwise}, \end{cases}$$

so that the aliasing error can be expressed simply as

$$\mathcal{A}_N u = \sum_{k>N}^{\infty} (\mathcal{I}_N T_k) \hat{u}_k = \sum_{\substack{p=-\infty \\ p \neq 0}}^{p=\infty} (\hat{u}_{2Np+n} + \hat{u}_{2Np-n}) T_n(x).$$

The following theorem shows us that we gain a factor of $\sqrt{N}$ by using the Chebyshev method, compared to the Legendre method.

Theorem 6.7 *For any $u(x) \in H^p_w[-1,1]$ with $p > 1/2$ and $0 \le q \le p$, there exists a positive constant, C, independent of N, such that*

$$\|u - \mathcal{I}_N u\|_{H^q_w[-1,1]} \le C N^{2q-p} \|u\|_{H^p_w[-1,1]}.$$

When we look at the L^∞-error for the discrete Chebyshev expansion, we note that we lose a factor of $\sqrt{N}$.

Theorem 6.8 *For any $u(x) \in H_w^p[-1, 1]$ with $p > 1/2$, there exists a positive constant, C, independent of N, such that*

$$\|u - \mathcal{I}_N u\|_{L^\infty[-1,1]} \leq C N^{1/2-p} \|u\|_{H_w^p[-1,1]}.$$

6.3 Further reading

The polynomial approximation theory was developed mainly by Canuto and Quarteroni (1981) and Bernardi and Maday (1989, 1992, 1999) with a thorough review and many results given in the overview paper by Bernardi and Maday (1999) and the texts by Canuto et al (1986, 2006). Some new results for the Jacobi case can be found in the text by Funaro (1992).

7

Polynomial spectral methods

In this chapter, we turn our attention to the formulation of polynomial spectral methods for solving partial differential equations, using the Galerkin, tau, and collocation approaches. The presence of non-periodic boundary conditions, which provide the motivation for using polynomial methods, further distinguishes between the methods. Once again we shall restrict ourselves to problems involving smooth solutions and turn to the special issues related to nonsmooth problems in Chapter 9.

Consider the problem

$$
\begin{aligned}
\frac{\partial u(x,t)}{\partial t} &= \mathcal{L}u(x,t), && x \in [a,b], \quad t \geq 0, \\
\mathcal{B}_L u &= 0, && t \geq 0 \\
\mathcal{B}_R u &= 0, && t \geq 0 \\
u(x,0) &= f(x), && x \in [a,b], \quad t = 0.
\end{aligned}
\tag{7.1}
$$

where $\mathcal{B}_{L,R}$ are the boundary operators at $x = a$ and $x = b$, respectively.

For simplicity we restrict much of the discussion to methods based on Legendre or Chebyshev expansions on the finite interval $[-1, 1]$. However, all results extend in a straightforward manner to schemes based on ultraspherical polynomials.

7.1 Galerkin methods

In the Galerkin method we seek solutions, $u_N(x,t) \in \mathrm{B}_N$, of the form

$$
u_N(x,t) = \sum_{n=0}^{N} a_n(t)\phi_n(x),
$$

where $\phi_n(x)$ is a polynomial taken from the space

$$\mathsf{B}_N = \text{span}\left\{\phi_n(x) \in \text{span}\{x^k\}_{k=0}^n \,|\mathcal{B}_L\phi_n = 0, \quad \mathcal{B}_R\phi_n = 0\,\right\}_{n=0}^N.$$

The $N+1$ equations for the unknown expansion coefficients, $a_n(t)$, are obtained from Equation (7.1) by requiring the residual

$$R_N(x,t) = \frac{\partial u_N}{\partial t} - \mathcal{L}u_N(x,t),$$

to be orthogonal to B_N, under some weight function $w(x)$. Using the test-functions, $\phi_k(x)$, this yields

$$\sum_{n=0}^N \mathrm{M}_{kn}\frac{da_n}{dt} = \sum_{n=0}^N \mathrm{S}_{kn}a_n(t) \quad \forall k \in [0,\ldots,N],$$

where

$$\mathrm{M}_{kn} = \frac{1}{\gamma_k}\int_{-1}^1 \phi_k(x)\phi_n(x)w(x)\,dx \quad \forall k,n \in [0,\ldots,N],$$

are entries of the mass matrix M, and

$$\mathrm{S}_{kn} = \frac{1}{\gamma_k}\int_{-1}^1 \phi_k(x)\mathcal{L}\phi_n(x)w(x)\,dx \quad \forall k,n \in [0,\ldots,N]$$

are the entries of the stiffness matrix S. We note that the mass matrix is symmetric and positive definite. The symmetry is obvious, while its positive definiteness can be seen by considering a general non-zero N-vector, $\boldsymbol{u}$, consisting of entries u_j, and using the definition of M to obtain

$$\begin{aligned}
\boldsymbol{u}^T\mathrm{M}\boldsymbol{u} &= \sum_{i,j=0}^N u_iu_j\mathrm{M}_{ij} \\
&= \sum_{i,j=0}^N u_iu_j\int_{-1}^1 \phi_i(x)\phi_j(x)w(x)\,dx \\
&= \int_{-1}^1\left(\sum_{i=0}^N u_i\phi_i(x)\right)^2 w(x)\,dx > 0,
\end{aligned}$$

since $\|\cdot\|_{L^2_w[-1,1]}$ is a norm.

The basis, $\phi_n(x)$, is usually constructed from a linear combination of $P_n^{(\alpha)}(x)$ to ensure that the boundary conditions are satisfied for all $\phi_n(x)$. For example, if the boundary condition is $u(1,t)=0$, we can choose the basis $\{P_n^{(\alpha)}(x) - P_n^{(\alpha)}(1)\}$ which satisfies the boundary conditions. Since we are using the polynomials $P_n^{(\alpha)}(x)$ as a basis it is natural to choose the weight function, $w(x)$, such that $(P_n^{(\alpha)}, P_k^{(\alpha)})_w = \gamma_n\delta_{nk}$.

We illustrate this process with some examples.

Example 7.1 Consider the linear hyperbolic problem

$$\frac{\partial u}{\partial t} = \frac{\partial u}{\partial x},$$

with the boundary condition

$$u(1,t) = 0,$$

and the initial condition $u(x,0) = f(x)$.

In the Chebyshev Galerkin method we define

$$\phi_n(x) = T_n(x) - 1,$$

so that the boundary condition is satisfied by each one of the polynomials. We seek solutions, $u_N(x,t) \in \mathrm{B}_N$, of the form

$$u_N(x,t) = \sum_{n=1}^{N} a_n(t)\phi_n(x) = \sum_{n=1}^{N} a_n(t)(T_n(x) - 1).$$

Note that the sum is from $n = 1$ rather than $n = 0$ since $\phi_0(x) = 0$. We now require that the residual

$$R_N(x,t) = \frac{\partial u_N}{\partial t} - \frac{\partial u_N}{\partial x},$$

is orthogonal to $\phi_n(x)$ in $L^2_w[-1,1]$:

$$\frac{2}{\pi}\int_{-1}^{1} R_N(x,t)\phi_k(x)\frac{1}{\sqrt{1-x^2}}\,dx = 0 \quad \forall k \subset [1,\ldots,N].$$

We have chosen $w(x)$ as the weight for which $T_n(x)$ is orthogonal in order to simplify the scheme. This procedure yields the Chebyshev Galerkin scheme

$$\sum_{n=1}^{N} \mathrm{M}_{kn}\frac{da_n}{dt} = \sum_{n=1}^{N} \mathrm{S}_{kn}a_n(t) \quad \forall k \in [1,\ldots,N],$$

where the mass matrix has the entries

$$\mathrm{M}_{kn} = \frac{2}{\pi}\int_{-1}^{1}(T_k(x)-1)(T_n(x)-1)\frac{1}{\sqrt{1-x^2}}\,dx = 2 + \delta_{kn}.$$

Computing the entries of the stiffness matrix requires a little more work; the entries are given by

$$\mathrm{S}_{kn} = \frac{2}{\pi}\int_{-1}^{1}(T_k(x)-1)\frac{dT_n(x)}{dx}\frac{1}{\sqrt{1-x^2}}\,dx \quad \forall k,n \in [1,\ldots,N].$$

From Equation (4.23),

$$\frac{dT_n(x)}{dx} = 2n \sum_{\substack{p=0 \\ p+n \text{ odd}}}^{n-1} \frac{T_p(x)}{c_p},$$

where $c_0 = 2$ and $c_p = 1$ otherwise. Introducing this into the expression for the stiffness matrix yields

$$\begin{aligned} \mathrm{S}_{kn} &= \frac{2}{\pi} \int_{-1}^{1} (T_k(x) - 1)\, 2n \sum_{\substack{p=0 \\ p+n \text{ odd}}}^{n-1} \frac{T_p(x)}{c_p} \frac{1}{\sqrt{1-x^2}}\, dx \\ &= 2n \sum_{\substack{p=0 \\ p+n \text{ odd}}}^{n-1} (\delta_{kp} - \delta_{0p}). \end{aligned}$$

This yields N ordinary differential equations (ODEs) for the N unknowns, $\boldsymbol{a} = (a_1, \ldots, a_N)^T$, of the form

$$\frac{d\boldsymbol{a}}{dt} = \mathrm{M}^{-1}\mathrm{S}\boldsymbol{a}(t),$$

with the initial conditions

$$a_n(0) = \frac{2}{\pi} \int_{-1}^{1} f(x)\,(T_n(x) - 1) \frac{1}{\sqrt{1-x^2}}\, dx \quad \forall n \in [1, \ldots, N].$$

This example illustrates that the formulation of the Chebyshev Galerkin method is a bit cumbersome and typically involves the inversion of the mass matrix. Fortunately, this mass matrix is invertible, since it is symmetric positive-definite.

Example 7.2 Consider the linear parabolic problem

$$\frac{\partial u}{\partial t} = \frac{\partial^2 u}{\partial x^2},$$

with boundary conditions

$$u(-1, t) = u(1, t) = 0,$$

and initial condition $u(x, 0) = f(x)$.

In the Legendre Galerkin methods we may choose the basis functions in several different ways. A convenient choice could be

$$\phi_n(x) = P_{n+1}(x) - P_{n-1}(x), \quad n \geq 1.$$

An alternative option is

$$\phi_n(x) = \begin{cases} P_n(x) - P_0(x) & n \text{ even}, \\ P_n(x) - P_1(x) & n \text{ odd}. \end{cases}$$

Both these choices have the desired property, $\phi_n(\pm 1) = 0$. Choosing the former, we seek solutions, $u_N(x,t) \in \mathrm{B}_N$, of the form

$$u_N(x,t) = \sum_{n=1}^{N-1} a_n(t)\phi_n(x) = \sum_{n=1}^{N-1} a_n(t)(P_{n+1}(x) - P_{n-1}(x)),$$

and require the residual

$$R_N(x,t) = \frac{\partial u_N}{\partial t} - \frac{\partial^2 u_N}{\partial x^2}$$

to be L^2-orthogonal to $\phi_k(x)$

$$\frac{2k+1}{2}\int_{-1}^{1} R_N(x,t)\phi_k(x)\,dx = 0 \quad \forall k \in [1,\ldots,N-1].$$

This yields the Legendre Galerkin scheme

$$\sum_{n=1}^{N-1} \mathrm{M}_{kn}\frac{da_n}{dt} = \sum_{n=1}^{N-1} \mathrm{S}_{kn}a_n(t) \quad \forall k \in [1,\ldots,N-1],$$

with the mass matrix given by

$$\begin{aligned}\mathrm{M}_{kn} &= \frac{2k+1}{2}\int_{-1}^{1}\phi_k(x)\phi_n(x)\,dx\\ &= \frac{2(2k+1)^2}{(2k-1)(2k+3)}\delta_{kn} - \frac{2k+1}{2k+3}\delta_{k,n+2} - \frac{2k+1}{2k-1}\delta_{k,n-2}.\end{aligned}$$

Note that the mass matrix is tridiagonal, i.e., the computation of M^{-1} is straightforward.

The stiffness matrix is given by

$$\mathrm{S}_{kn} = \frac{2k+1}{2}\int_{-1}^{1}\phi_k(x)\frac{d^2\phi_n(x)}{dx^2}\,dx \quad \forall k,n \in [1,\ldots,N-1].$$

These entries can either be derived by using the properties of the Legendre polynomials or by using the Gauss quadrature.

This procedure yields an ODE system of $(N-1)$ equations with $N-1$ unknowns, $\boldsymbol{a} = (a_1,\ldots,a_{N-1})^T$, of the form

$$\frac{d\boldsymbol{a}}{dt} = \mathrm{M}^{-1}\mathrm{S}\boldsymbol{a}(t),$$

with the initial conditions

$$a_n(0) = \frac{2n+1}{2}\int_{-1}^{1} f(x)(P_{n+1}(x) - P_{n-1}(x))\,dx \quad \forall n \in [1,\ldots,N-1].$$

As we will see in the following example, this process becomes much more complicated in the nonlinear case.

Example 7.3 Consider Burgers' equation

$$\frac{\partial u}{\partial t} + u\frac{\partial u}{\partial x} = \nu\frac{\partial^2 u}{\partial x^2},$$

with the homogeneous boundary conditions

$$u(-1,t) = u(1,t) = 0,$$

and the initial condition $u(x,0) = f(x)$.

To use a Chebyshev Galerkin method we choose the basis

$$\phi_n(x) = T_{n+1}(x) - T_{n-1}(x),$$

(so that $\phi_n(\pm 1) = 0$), and seek solutions, $u_N(x,t) \in \mathrm{B}_N$, of the form

$$u_N(x,t) = \sum_{n=1}^{N-1} a_n(t)\phi_n(x) = \sum_{n=1}^{N-1} a_n(t)(T_{n+1}(x) - T_{n-1}(x)).$$

As usual, we require the residual,

$$R_N(x,t) = \frac{\partial u_N}{\partial t} + u_N\frac{\partial u_N}{\partial x} - \nu\frac{\partial^2 u_N}{\partial x^2}$$

to be orthogonal to $\phi_k(x)$:

$$\frac{2}{\pi}\int_{-1}^{1} R_N(x,t)\phi_k(x)\frac{1}{\sqrt{1-x^2}}\,dx = 0 \quad \forall k \in [1,\ldots,N-1],$$

yielding the Chebyshev Galerkin scheme

$$\sum_{n=1}^{N-1} \mathrm{M}_{kn}\frac{da_n}{dt} = \sum_{n=1}^{N-1} \mathrm{S}_{kn}(\boldsymbol{a}(t))a_n(t) \quad \forall k \in [1,\ldots,N-1],$$

where $\boldsymbol{a}(t)$ is the vector of elements $a_n(t)$. The tridiagonal mass matrix is given by

$$\mathrm{M}_{kn} = \frac{2}{\pi}\int_{-1}^{1} \phi_k(x)\phi_n(x)\frac{1}{\sqrt{1-x^2}}\,dx = 2\delta_{kn} - \delta_{k,n+2} - \delta_{k,n-2}.$$

The nonlinearity complicates the expression of the elements of the stiffness matrix, since the entries of S now consist of contributions from the nonlinear hyperbolic part, S^h, and the parabolic part, S^p:

$$\begin{aligned}
\mathrm{S}_{kn}(\boldsymbol{a}(t)) &= -\mathrm{S}^h_{kn}(\boldsymbol{a}(t)) + \nu\mathrm{S}^p_{kn} \\
&= -\frac{2}{\pi}\int_{-1}^{1} \phi_k(x)\phi_n(x)\sum_{l=1}^{N-1} a_l(t)\frac{d\phi_l(x)}{dx}\frac{1}{\sqrt{1-x^2}}\,dx \\
&\quad + \nu\frac{2}{\pi}\int_{-1}^{1} \phi_k(x)\frac{d^2\phi_n(x)}{dx^2}\frac{1}{\sqrt{1-x^2}}\,dx.
\end{aligned}$$

To simplify the nonlinear hyperbolic part, we may use the definition of $\phi_n(x)$ and the identity

$$2T_n(x)T_l(x) = T_{|n+l|}(x) + T_{|n-l|}(x).$$

However, the computation of the integrals involved in the stiffness matrix is still expensive.

Once the components of the stiffness matrix are calculated, we have $(N-1)$ ODEs for the $(N-1)$ unknowns, $\boldsymbol{a} = (a_1, \ldots, a_{N-1})^T$,

$$\frac{d\boldsymbol{a}}{dt} = \mathrm{M}^{-1}(-\mathrm{S}^h(\boldsymbol{a}) + \nu \mathrm{S}^p)\boldsymbol{a}(t),$$

with initial conditions

$$a_n(0) = \frac{2}{\pi}\int_{-1}^{1} f(x)\,(T_{n+1}(x) - T_{n-1}(x))\,\frac{1}{\sqrt{1-x^2}}\,dx \quad \forall n \in [1, \ldots, N-1].$$

The polynomial Galerkin methods are of considerable complexity, analytically as well as computationally. The requirement that each polynomial from the basis chosen must satisfy the boundary conditions individually often causes the Galerkin formulation to become complicated, particularly when the boundary conditions are time-dependent. Furthermore, although the mass matrix depends only on the basis being used, the computation of the entries of the stiffness matrix has to be completed for each individual problem and is by no means simple even for constant coefficient linear problems. The presence of nonlinear terms further complicates the computation of the stiffness matrix, rendering the Galerkin method extremely complex. A simplification of the Galerkin procedure is known as the tau method, and is presented in the next section.

7.2 Tau methods

In the tau method, we still seek solutions $u_N(x,t) \in \mathsf{B}_N$. However, we do not project the residual onto the space B_N but rather onto the polynomial space

$$\mathsf{P}_{N-k} = \mathrm{span}\{x^n\}_{n=0}^{N-k},$$

where k is the number of boundary conditions. The approximant is a polynomial of degree N but k degrees of freedom are used to enforce the boundary conditions. The following examples illustrate this procedure.

Example 7.4 Consider the linear hyperbolic problem

$$\frac{\partial u}{\partial t} = \frac{\partial u}{\partial x},$$

with the boundary condition

$$u(1,t) = h(t),$$

and the initial condition, $u(x,0) = f(x)$.

In the Legendre tau method we seek solutions, $u_N(x,t) \in \mathrm{B}_N$, of the form

$$u_N(x,t) = \sum_{n=0}^{N} a_n(t) P_n(x),$$

with the additional constraint that

$$\sum_{n=0}^{N} a_n(t) P_n(1) = \sum_{n=0}^{N} a_n(t) = h(t),$$

to ensure that the solution obeys the boundary condition. The first N equations are found by requiring that the residual

$$R_N(x,t) = \frac{\partial u_N}{\partial t} - \frac{\partial u_N}{\partial x},$$

is L^2-orthogonal to P_{N-1}

$$\frac{2k+1}{2} \int_{-1}^{1} R_N(x,t) P_k(x)\, dx = 0 \quad \forall k \in [0,\dots,N-1].$$

This yields

$$\frac{da_n}{dt} = (2n+1) \sum_{\substack{p=n+1 \\ p+n \text{ odd}}}^{N} a_p(t) \quad \forall n \in [0,\dots,N-1].$$

We see that for the linear constant coefficient problem, the derivatives of the first N coefficients are exactly the coefficients of the first derivative. Recall the identity, Equation (5.8),

$$a_n^{(1)}(t) = (2n+1) \sum_{\substack{p=n+1 \\ p+n \text{ odd}}}^{N} a_p(t).$$

The last coefficient $a_N(t)$ can be obtained explicitly from the other coefficients

$$a_N(t) = -\sum_{n=0}^{N-1} a_n(t) - h(t).$$

This reflects the ease by which time dependent boundary conditions are introduced into the tau method.

The initial conditions yield

$$a_n(0) = \frac{2n+1}{2} \int_{-1}^{1} f(x) P_n(x)\, dx \quad \forall n \in [0,\dots,N-1],$$

with the final coefficient

$$a_N(0) = -\sum_{n=0}^{N-1} a_n(0) - h(0).$$

Since we project the residual onto P_{N-1} rather than B_N the mass matrix remains diagonal due to orthogonality, and the stiffness matrix is obtained directly from the properties of the polynomial basis, typically resulting in schemes much simpler than the Galerkin approximation.

Example 7.5 Once again, consider the linear parabolic problem

$$\frac{\partial u}{\partial t} = \frac{\partial^2 u}{\partial x^2},$$

with boundary conditions

$$u(-1,t) = \alpha_1 u(-1,t) - \beta_1 \frac{\partial u(-1,t)}{\partial x} = g(t)$$

$$u(1,t) = \alpha_2 u(1,t) + \beta_2 \frac{\partial u(1,t)}{\partial x} = h(t),$$

and initial condition $u(x,0) = f(x)$.

In the Chebyshev tau method we seek solutions $u_N(x,t) \in \mathrm{B}_N$, of the form

$$u_N(x,t) = \sum_{n=0}^{N} a_n(t) T_n(x),$$

with the additional constraints from the boundary conditions

$$\sum_{n=0}^{N} a_n(t) \left(\alpha_1 T_n(-1) - \beta_1 T_n'(-1)\right) = g(t),$$

$$\sum_{n=0}^{N} a_n(t) \left(\alpha_2 T_n(1) + \beta_2 T_n'(1)\right) = h(t).$$

Using the boundary values of $T_n(x)$ and its derivative

$$T_n(\pm 1) = (\pm 1)^n, \quad T_n'(\pm 1) = (\pm 1)^{n+1} n^2,$$

the constraints become

$$\sum_{n=0}^{N} a_n(t) (\alpha_1 (-1)^n - \beta_1 (-1)^{n+1} n^2) = g(t), \tag{7.2}$$

$$\sum_{n=0}^{N} a_n(t) (\alpha_2 + \beta_2 n^2) = h(t),$$

to ensure that the solution obeys the boundary conditions.

We now require that the residual

$$R_N(x,t) = \frac{\partial u_N}{\partial t} - \frac{\partial^2 u_N}{\partial x^2},$$

is orthogonal to P_{N-2} in $L^2_w[-1,1]$:

$$\frac{2}{\pi c_k}\int_{-1}^{1} R_N(x,t)T_k(x)\frac{1}{\sqrt{1-x^2}}\,dx = 0 \quad \forall k \in [0,\ldots,N-2].$$

Using the identity

$$a_n^{(2)}(t) = \frac{1}{c_n}\sum_{\substack{p=n+2\\ p+n \text{ even}}}^{N} p(p^2-n^2)a_p(t),$$

we obtain the first $(N-1)$ ODEs for the coefficients

$$\frac{da_n}{dt} = \frac{1}{c_n}\sum_{\substack{p=n+2\\ p+n \text{ even}}}^{N} p(p^2-n^2)a_p(t) \quad \forall n \in [0,\ldots,N-2].$$

The system is closed by obtaining $a_{N-1}(t)$ and $a_N(t)$ from the 2-by-2 linear system in Equation (7.2).

The initial conditions for this system of ODEs are

$$a_n(0) = \frac{2}{c_n\pi}\int_{-1}^{1} f(x)T_n(x)\frac{1}{\sqrt{1-x^2}}\,dx \quad \forall n \in [0,\ldots,N-2].$$

Even though we considered very general time-dependent boundary conditions, the resulting Chebyshev tau method remains simple because the effects of the boundary conditions are separated from the approximation of the operator $\mathcal{L}$.

While the emphasis in this text is on schemes for the solution of time-dependent problems we would like to take a small detour to illustrate the efficacy of tau methods for the solution of elliptic problems.

Example 7.6 Consider the elliptic problem

$$\frac{\partial^2 u}{\partial x^2} = f(x),$$

subject to the general boundary conditions

$$\alpha_1 u(-1) + \beta_1\frac{\partial u(-1)}{\partial x} = c_-,$$

$$\alpha_2 u(1) + \beta_2\frac{\partial u(1)}{\partial x} = c_+.$$

We now seek a solution, $u_N(x) \in \mathrm{B}_N$, of the form

$$u_N(x,t) = \sum_{n=0}^{N} a_n(t)T_n(x),$$

with the additional constraints from the boundary conditions

$$\sum_{n=0}^{N} a_n(\alpha_1(-1)^n + \beta_1(-1)^{n+1}n^2) = c_-, \tag{7.3}$$

$$\sum_{n=0}^{N} a_n(\alpha_2 + \beta_2 n^2) = c_+.$$

The residual is

$$R_N(x) = \frac{\partial^2 u_N}{\partial x^2} - f_{N-2}(x),$$

with $f_{N-2}(x) \in \mathrm{P}_{N-2}$ given by

$$f_{N-2}(x) = \sum_{n=0}^{N-2} \hat{f}_n T_n(x),$$

where

$$\hat{f}_n = \frac{2}{c_n \pi}\int_{-1}^{1} f(x)T_n(x)\frac{1}{\sqrt{1-x^2}}\,dx.$$

Requiring the residual to be L^2_w-orthogonal to P_{N-2} yields the first $(N-1)$ equations

$$a_n^{(2)} = \frac{1}{c_n}\sum_{\substack{p=n+2\\ p+n \text{ even}}}^{N} p(p^2-n^2)a_p = \hat{f}_n \quad \forall n \in [0,\ldots,N-2].$$

The system is closed by considering the two boundary conditions in Equation (7.3). Apart from the two equations appearing from the boundary conditions, the matrix is fairly sparse and upper triangular.

The tau formulation of spectral methods may well be the most efficient way to solve linear elliptic problems or eigenvalue problems, as the resulting matrices typically are sparse and very efficient preconditioners can be developed.

Let us finally consider the formulation of a tau method for the solution of Burgers' equation.

Example 7.7 Consider Burgers' equation

$$\frac{\partial u}{\partial t} + u\frac{\partial u}{\partial x} = \nu\frac{\partial^2 u}{\partial x^2},$$

with boundary conditions

$$u(-1,t) = u(1,t) = 0,$$

and initial conditions $u(x,0) = f(x)$.

To solve Burgers' equation using a Chebyshev tau method we seek solutions $u_N(x,t) \in \mathrm{B}_N$ of the form

$$u_N(x,t) = \sum_{n=0}^{N} a_n(t) T_n(x),$$

with the constraints

$$\sum_{n=0}^{N} a_n(t)(-1)^n = \sum_{n=0}^{N} a_n(t) = 0.$$

We require the residual

$$R_N(x,t) = \frac{\partial u_N}{\partial t} + u_N \frac{\partial u_N}{\partial x} - \nu \frac{\partial^2 u_N}{\partial x^2},$$

to be L^2_w-orthogonal to P_{N-2}:

$$\frac{2}{c_k \pi} \int_{-1}^{1} R_N(x,t) T_k(x) \frac{1}{\sqrt{1-x^2}}\, dx = 0 \quad \forall k \in [0, \ldots, N-2],$$

leading to the system of ODEs

$$\frac{da_k}{dt} + \frac{2}{c_k \pi} \int_{-1}^{1} u_N \frac{\partial u_N}{\partial x} T_k(x) \frac{1}{\sqrt{1-x^2}}\, dx = \nu a_k^{(2)}(t) \quad \forall k \in [0, \ldots, N-2].$$

Recall that

$$a_k^{(2)}(t) = \frac{1}{c_k} \sum_{\substack{p=n+2 \\ p+n \text{ even}}}^{N} p(p^2 - k^2) a_p(t).$$

The calculation of the nonlinear term requires a little more work. Introducing the expansions for $u_N(x,t)$ and its derivative we obtain

$$\begin{aligned} &\frac{2}{c_k \pi} \int_{-1}^{1} u_N \frac{\partial u_N}{\partial x} T_k(x) \frac{1}{\sqrt{1-x^2}}\, dx \\ &\quad = \frac{2}{c_k \pi} \int_{-1}^{1} \sum_{n,l=0}^{N} a_n(t) a_l^{(1)}(t) T_n(x) T_l(x) T_k(x) \frac{1}{\sqrt{1-x^2}}\, dx. \end{aligned}$$

The identity for Chebyshev polynomials

$$T_n(x) T_l(x) = \frac{1}{2}\left(T_{n+l}(x) + T_{|n-l|}(x)\right),$$

yields

$$
\begin{aligned}
&\frac{2}{c_k\pi}\int_{-1}^{1} u_N\frac{\partial u_N}{\partial x}T_k(x)\frac{1}{\sqrt{1-x^2}}\,dx\\
&\quad=\frac{1}{2}\frac{2}{c_k\pi}\int_{-1}^{1}\sum_{n,l=0}^{N}a_n(t)a_l^{(1)}(t)\left(T_{n+l}(x)+T_{|n-l|}(x)\right)T_k(x)\frac{1}{\sqrt{1-x^2}}\,dx\\
&\quad=\frac{1}{2}\left(\sum_{\substack{n,l=0\\ n+l=k}}^{N}a_n(t)a_l^{(1)}(t)+\sum_{\substack{k,l=0\\ |n-l|=k}}^{N}a_n(t)a_l^{(1)}(t)\right),
\end{aligned}
$$

where

$$
a_l^{(1)}(t)=\frac{2}{c_l}\sum_{\substack{p=l+1\\ p+l \text{ odd}}}^{N} pa_p(t).
$$

This establishes the equations for the first $N-1$ expansion coefficients, $a_k(t)$. The remaining two are found by enforcing the boundary conditions

$$
\sum_{n=0}^{N}a_n(t)(-1)^n=\sum_{n=0}^{N}a_n(t)=0.
$$

As usual, the initial conditions are

$$
a_n(0)=\frac{2}{c_n\pi}\int_{-1}^{1}f(x)T_n(x)\frac{1}{\sqrt{1-x^2}}\,dx \quad \forall n\in[0,\ldots,N-2].
$$

Although the tau method avoids some of the problems of the Galerkin methods and allows for the development of fairly simple and efficient methods for solving the partial differential equation it is primarily used for the solution of linear elliptic equations. The main reason is that the tau method still requires one to derive equations for the expansion coefficients for each individual problem. For variable coefficient or nonlinear problems this process may be very complicated and in many cases impossible. However, for linear constant coefficient or special cases of variable coefficient nonlinear problems the resulting tau method provides an efficient approach.

7.3 Collocation methods

The projection operators involved in the Galerkin and tau methods lead to complications in deriving the system of ODEs. In both these methods, we require that the projection of the residual onto some polynomial space be zero. In contrast, the collocation method obtains the system of ODEs by requiring that the residual vanish at a certain set of grid points, typically some set of Gauss-type points. As a result, while the Galerkin and tau methods may become untenable for strongly nonlinear problems, the collocation method deals with linear, variable coefficient, and nonlinear problems with equal ease.

Since we are considering initial boundary value problems, the appropriate choice of grid is the set of Gauss–Lobatto quadrature points

$$x_j = \left\{x \,\middle|\, (1-x^2)\left(P_N^{(\alpha)}\right)'(x) = 0\right\} \quad \forall j \in [0, \ldots, N],$$

as this includes the boundaries.

Once again, we begin by seeking an approximation in the space of Nth-order polynomials which satisfy the boundary conditions, $u_N(x,t) \in \mathsf{B}_N$, such that the residual

$$R_N(x,t) = \frac{\partial u_N}{\partial t} - \mathcal{L}u_N(x,t),$$

vanishes at the interior gridpoints,

$$\mathcal{I}_N R_N(x_j, t) = 0 \quad \forall j \in [1, \ldots, N-1],$$

leading to the $N-1$ equations

$$\left.\frac{\partial u_N}{\partial t}\right|_{x_j} = (\mathcal{L}u_N)\,|_{x_j} \quad \forall j \in [1, \ldots, N-1].$$

The last two equations are prescribed by the boundary conditions. The following examples illustrate the simplicity of the collocation method.

Example 7.8 We solve the hyperbolic problem

$$\frac{\partial u}{\partial t} = \frac{\partial f(u)}{\partial x},$$

with boundary condition

$$u(1,t) = h(t),$$

and initial condition $u(x,0) = g(x)$. Assume that $f'(u) > 0$ at the boundaries, so that one boundary condition is sufficient.

In the Chebyshev collocation method, we seek a solution

$$u_N(x,t) = \sum_{n=0}^{N} a_n(t) T_n(x) = \sum_{j=0}^{N} u_N(x_j, t) l_j(x).$$

We need also to define

$$f_N\left(u_N(x,t)\right) = \sum_{n=0}^{N} \tilde{f}_n(t) T_n(x) = \sum_{j=0}^{N} f(u_N(x_j, t)) l_j(x),$$

where the x_j are the Gauss–Lobatto quadrature points

$$x_j = -\cos\left(\frac{\pi}{N} j\right) \quad \forall j \in [0, \ldots, N].$$

$$l_j(x) = \frac{(-1)^{j+1+N}(1-x^2)T'_N(x)}{\overline{c}_j N^2 (x - x_j)}.$$

The coefficients are

$$a_n(t) = \frac{2}{\overline{c}_n N} \sum_{j=0}^{N} \frac{1}{\overline{c}_j} u_N(x_j, t) T_n(x_j),$$

and

$$\tilde{f}_n(t) = \frac{2}{\overline{c}_n N} \sum_{j=0}^{N} \frac{1}{\overline{c}_j} f\left(u_N(x_j, t)\right) T_n(x_j),$$

where $\overline{c}_0 = \overline{c}_N = 2$ and $\overline{c}_n = 1$ otherwise. And l_j are the interpolating Lagrange polynomial

The solution is found by requiring the residual

$$R_N(x,t) = \frac{\partial u_N}{\partial t} - \frac{\partial f_N}{\partial x},$$

to vanish at the collocation points (which in this case are also chosen to be the Gauss–Lobatto points):

$$\mathcal{I}_N R_N(x_j, t) = \left.\frac{\partial u_N}{\partial t}\right|_{x_j} - \mathcal{I}_N \left.\frac{\partial f_N}{\partial x}\right|_{x_j} = 0 \quad \forall j \in [0, \ldots, N-1],$$

yielding N equations

$$\left.\frac{\partial u_N}{\partial t}\right|_{x_j} = \mathcal{I}_N \left.\frac{\partial f_N}{\partial x}\right|_{x_j} \qquad \forall j \in [0, \ldots, N-1].$$

The main computational work is in the derivative evaluation. This can be accomplished in two different ways. One can use

$$\mathcal{I}_N \left.\frac{\partial f_N}{\partial x}\right|_{x_j} = \sum_{j=0}^{N} \tilde{f}_n^{(1)}(t) T_n(x_j),$$

where the discrete expansion coefficients are found through the backward recursion relation

$$c_{n-1} \tilde{f}_{n-1}^{(1)}(t) = \tilde{f}_{n+1}^{(1)}(t) + 2n \tilde{f}_{n-1}(t).$$

Alternatively, we may compute the derivative through the introduction of the differentiation matrix, D,

$$\mathcal{I}_N \frac{\partial f_N}{\partial x}\bigg|_{x_j} = \sum_{i=0}^{N} \mathrm{D}_{ji} f_N(x_i, t).$$

The system is closed through the boundary condition

$$u_N(x_N, t) = h(t),$$

with the initial condition

$$u_N(x_j, 0) = (\mathcal{I}_N g)(x_j) = g(x_j).$$

Note the ease with which the collocation method handles a general nonlinear hyperbolic equation. Now let us consider an example of a polynomial collocation method for the solution of a partial differential equation.

Example 7.9 Consider the linear parabolic problem

$$\frac{\partial u}{\partial t} = \frac{\partial^2 u}{\partial x^2},$$

with boundary conditions

$$u(-1, t) = \alpha_1 u(-1, t) - \beta_1 \frac{\partial u(-1, t)}{\partial x} = g(t),$$
$$u(1, t) = \alpha_2 u(1, t) + \beta_2 \frac{\partial u(1, t)}{\partial x} = h(t),$$

and initial condition $u(x, 0) = f(x)$.

In the Legendre Gauss–Lobatto collocation method we seek solutions, $u_N(x, t) \in \mathrm{B}_N$, of the form

$$u_N(x, t) = \sum_{j=0}^{N} u_N(x_j, t) l_j(x),$$

where $l_j(x)$ is the interpolating Lagrange polynomial based on the Legendre Gauss–Lobatto quadrature points

$$x_j = \left\{x \middle| (1 - x^2) P_N'(x) = 0\right\} \quad \forall j \in [0, \dots, N],$$

and we require the residual

$$R_N(x, t) = \frac{\partial u_N}{\partial t} - \frac{\partial^2 u_N}{\partial x^2},$$

to vanish at the interior points, yielding the $(N - 1)$ equations

$$\frac{du_N(x_j, t)}{dt} = \sum_{i=0}^{N} \mathrm{D}^{(2)}_{ji} u_N(x_i, t) \quad \forall j \in [1, \dots, N - 1].$$

Here $\mathrm{D}^{(2)}$ is the second-order differentiation matrix based on the Legendre Gauss–Lobatto quadrature points.

We close the system by using the boundary conditions,

$$\alpha_1 u_N(x_0,t) - \beta_1 \sum_{j=0}^{N} \mathrm{D}_{0j} u_N(x_j,t) = g(t),$$

$$\alpha_2 u_N(x_N,t) + \beta_2 \sum_{j=0}^{N} \mathrm{D}_{Nj} u_N(x_j,t) = h(t).$$

where D is the first-order differentiation matrix. This yields a 2-by-2 system for the computation of the boundary values

$$(\alpha_1 - \beta_1 \mathrm{D}_{00}) u_N(x_0,t) - \beta_1 \mathrm{D}_{0N} u_N(x_N,t) = g(t) + \beta_1 \sum_{j=1}^{N-1} \mathrm{D}_{0j} u_N(x_j,t),$$

$$\beta_2 \mathrm{D}_{N0} u_N(x_0,t) + (\alpha_2 + \beta_2 \mathrm{D}_{NN}) u_N(x_N,t) = h(t) - \beta_2 \sum_{j=1}^{N-1} \mathrm{D}_{Nj} u_N(x_j,t).$$

In this way the boundary conditions are enforced exactly.

7.4 Penalty method boundary conditions

Up to now, we have been imposing the boundary conditions exactly. However, it is sufficient to impose the boundary conditions only up to the order of the scheme. This procedure is illustrated in the following example.

Example 7.10 Consider the constant coefficient wave equation

$$u_t + a u_x = 0 \quad a > 0$$

with boundary condition $u(-1,t) = h(t)$ and the initial condition $u(x,0) = g(x)$.

First, we define the polynomial $Q^-(x)$ which vanishes on the grid points except the boundary point $x = -1$:

$$Q^-(x) = \frac{(1-x)P_N'(x)}{2P_N'(-1)} = \begin{cases} 1 & x = -1 \\ 0 & x = x_j \neq -1 \end{cases}$$

where x_j refers to the Legendre Gauss–Lobatto points.

In the Legendre collocation method with *weakly imposed* boundary conditions, we seek polynomial solutions of degree N, $u_N(x,t)$, which satisfy

$$\frac{\partial u_N}{\partial t} + a\frac{\partial u_N}{\partial x} = -\tau a Q^-(x)[u_N(-1,t) - h(t)],$$

where τ is a parameter which we adjust for stability. This means that the equation is satisfied exactly at all the grid points except the boundary point, while at the boundary point we have

$$\frac{\partial u_N(-1,t)}{\partial t} + a\frac{\partial u_N}{\partial x}(-1,t) = -\tau a[u_N(-1,t) - h(t)\,].$$

The scheme is consistent since when the exact solution is used, the right-hand-side is zero. We'll show shortly that this method is asymptotically stable, and this fact coupled with stability implies convergence.

8

Stability of polynomial spectral methods

In this chapter, we shall study the stability of Galerkin and collocation polynomial methods for a variety of linear problems. We then discuss the stability of the penalty method approach, and briefly review the stability theory for polynomial spectral methods for nonlinear equations.

8.1 The Galerkin approach

In the Galerkin case, both Fourier and polynomial methods are stable for the equation

$$u_t = \mathcal{L}u,$$

as long as the operator $\mathcal{L}$ is semi-bounded in the norm under which the underlying approximating basis is orthogonal.

Theorem 8.1 *If $\mathcal{L}$ satisfies*

$$\mathcal{L} + \mathcal{L}^* \leq 2\gamma \mathrm{I},$$

then the Galerkin method is stable.

Proof: When the Galerkin method is applied to

$$\frac{\partial u}{\partial t} = \mathcal{L}u,$$

we seek a solution $u_N(x, t) \in \mathrm{B}_N$ such that the residual is orthogonal to the space B_N under the relevant non-negative weight function $w(x)$. Thus, since $u_N \in \mathrm{B}_N$ we have

$$\left(\frac{\partial u_N}{\partial t} - \mathcal{L}u_N, u_N\right)_w = 0,$$

which means

$$\left(\frac{\partial u_N}{\partial t}, u_N\right)_w = (\mathcal{L}u_N, u_N)_w.$$

The left hand side becomes

$$\left(\frac{\partial u_N}{\partial t}, u_N\right)_w = \frac{1}{2}\frac{d}{dt}(u_N, u_N)_w,$$

while the right hand side is

$$(\mathcal{L}u_N, u_N)_w = \frac{1}{2}((\mathcal{L} + \mathcal{L}^*)u_N, u_N)_w.$$

Since $\mathcal{L}$ is semi-bounded, we have:

$$(u_N, u_N)_w \leq e^{\gamma t}(u_N(0), u_N(0))_w.$$

QED

Let's consider a few examples of bounded operators. We begin with the parabolic operator.

Example 8.2 The parabolic operator

$$\mathcal{L}u = u_{xx},$$

with boundary conditions $u(\pm 1, t) = 0$, is semi-bounded under the weight $w(x) = (1 - x^2)^\alpha$, for values of $-1/2 \leq \alpha \leq 1$. In this case, $\mathcal{L} = \mathcal{L}^*$, and we will see that $\mathcal{L} + \mathcal{L}^* \leq 0$.

First, we look at values of $-1/2 \leq \alpha \leq 0$. This case is significant since the two most common spectral methods are Legendre for which $\alpha = 0$, and Chebyshev, for which $\alpha = -1/2$. We rewrite the inner-product

$$(\mathcal{L}u_N, u_N)_w = \int_{-1}^{1} w u u_{xx} dx = -\int_{-1}^{1} (wu)_x u_x dx,$$

by integration by parts and the fact that the boundary values are zero. Now we let $v = wu$, and we have

$$u_x = \frac{(wu)_x}{w} - \frac{w_x u}{w} = \frac{v_x}{w} + v\left(\frac{1}{w}\right)'.$$

Plugging this in and integrating by parts again, we have

$$\begin{aligned}(\mathcal{L}u_N, u_N)_w &= -\int_{-1}^{1} \frac{v_x^2}{w} dx - \int_{-1}^{1} v v_x \left(\frac{1}{w}\right)' dx \\ &= -\int_{-1}^{1} \frac{v_x^2}{w} dx + \int_{-1}^{1} \frac{1}{2} v^2 \left(\frac{1}{w}\right)'' dx.\end{aligned}$$

The first integrand is non-negative, while the second can be shown to be non-positive by observing that

$$\left(\frac{1}{w}\right)'' = \alpha(1-x^2)^{-\alpha-2}(x^2(4\alpha+2)+2) \le 0.$$

Thus, we have $(\mathcal{L}u_N, u_N)_w \le 0$.

Next, we look at values of $0 \le \alpha \le 1$, for which the inner product

$$\begin{aligned}(\mathcal{L}u_N, u_N)_w &= \int_{-1}^{1} wuu_{xx}dx = -\int_{-1}^{1}(wu)_x u_x dx \\ &= -\int_{-1}^{1} w_x u u_x dx - \int_{-1}^{1} wu_x^2 dx \\ &= \frac{1}{2}\int_{-1}^{1} w_{xx}u^2 dx - \int_{-1}^{1} wu_x^2 dx.\end{aligned}$$

The second integrand is strictly positive. We can show that the first is non-positive by observing that

$$w_{xx} = \alpha(1-x^2)^{\alpha-2}(x^2(4\alpha-2)-2) \le 0,$$

and so, since $w(x) \ge 0$, $u^2 \ge 0$, $u_x^2 \ge 0$, and $w_{xx} \le 0$, we have

$$(Lu_N, u_N)_w \le 0.$$

Next, we consider the hyperbolic operator.

Example 8.3 Consider the Legendre Galerkin approximation to the linear hyperbolic problem

$$\frac{\partial u}{\partial t} = \frac{\partial u}{\partial x},$$

with the boundary condition

$$u(1,t) = 0,$$

and the initial condition $u(x,0) = g(x)$.

The operator $\mathcal{L}u = u_x$ is semi-bounded for the Legendre weight $w(x) = 1$:

$$\left(u_N, \frac{\partial u_N}{\partial t}\right)_w = \left(u_N, \frac{\partial u_N}{\partial x}\right)_w$$

the right hand side is, as usual

$$\left(u_N, \frac{\partial u_N}{\partial t}\right)_w = \frac{1}{2}\frac{d}{dt}(u_N, u_N)_w,$$

whereas the left hand side

$$\left(u_N, \frac{\partial u_N}{\partial x}\right)_w = \frac{1}{2}\int_{-1}^{1} \frac{\partial u_N^2}{\partial x} dx$$
$$= -\frac{1}{2} u_N^2(-1) \leq 0.$$

Putting this together, we have

$$\frac{d}{dt}(u_N, u_N)_w \leq 0.$$

The next example shows that this is true for all Jacobi polynomials with $\alpha \geq 0$ and $\beta \leq 0$:

Example 8.4 Consider the operator $\mathcal{L}u = u_x$ and the weight $w(x) = (1+x)^\alpha(1-x)^\beta$. As above, we have

$$\frac{d}{dt}(u_N, u_N)_w = \int_{-1}^{1} w(x)\frac{\partial u_N^2}{\partial x} dx = -\int_{-1}^{1} w'(x) u_N^2 dx.$$

Now we observe that, for $\alpha > 0$ and $\beta < 0$,

$$w'(x) = \alpha(1+x)^{\alpha-1}(1-x)^\beta - \beta(1+x)^\alpha(1-x)^{\beta-1}$$
$$= (1+x)^{\alpha-1}(1-x)^{\beta-1}\left(\alpha(1-x) - \beta(1+x)\right) \geq 0,$$

so that we have

$$\frac{d}{dt}(u_N, u_N)_w \leq 0.$$

This example establishes that the linear hyperbolic equation is wellposed in the Jacobi norm provided $\alpha \geq 0$ and $\beta \leq 0$. The following example demonstrates that this equation is not wellposed in the Chebyshev norm. This implies that we will not be able to prove stability in this norm.

Example 8.5 Consider the linear hyperbolic equation $u_t = u_x$, on $x \epsilon[-1, 1]$ with initial condition

$$u(x, 0) = \begin{cases} 1 - \frac{|x|}{\epsilon} & \text{if } |x| < \epsilon, \\ 0 & \text{if } |x| \geq \epsilon, \end{cases}$$

which has the solution at time $t = 1$

$$u(x, 1) = \begin{cases} 1 - \frac{x+1}{\epsilon} & \text{if } |x+1| < \epsilon, \\ 0 & \text{if } |x+1| \geq \epsilon. \end{cases}$$

The norms of the solution as $\epsilon \to 0^+$ at times $t = 0$ and $t = 1$,

$$\|u(x, 0)\|_w^2 \sim \epsilon \; \|u(x, 1)\|_w^2 \sim \frac{2}{3}\sqrt{2\epsilon},$$

give us insight into the solution's growth:

$$\|e^L\|_w \geq \frac{\|u(x,1)\|_w}{\|u(x,0)\|_w} \sim \left(\frac{8}{9}\right)^{\frac{1}{4}} \epsilon^{-\frac{1}{4}}.$$

In fact, as $\epsilon \to 0^+$, we have $\|e^{Lt}\| \to \infty$ for $0 \leq t \leq 1+\epsilon$, and $\|e^{Lt}\| = 0$ for $t \geq 1+\epsilon$ (i.e., after the initial profile exits the domain), so that this equation is not wellposed in the Chebyshev norm.

Clearly, if the equation itself is not wellposed in this norm, the Galerkin method will certainly not be stable in this norm. However, if we consider the norm based on the weight $w(x) = (1+x)/\sqrt{1-x^2}$, we see that the linear hyperbolic equation is wellposed in *this* norm.

Example 8.6 The scalar linear hyperbolic equation is wellposed under the norm $(\cdot,\cdot)_w$ where $w(x) = (1+x)/\sqrt{1-x^2}$. To see this, note that the weight

$$w(x) = \frac{1+x}{\sqrt{1-x^2}} = (1+x)^{\frac{1}{2}}(1-x)^{-\frac{1}{2}}$$

and recall that we showed in Example 8.4 that in this case $w'(x) \geq 0$. Consequently,

$$\begin{aligned}
\frac{d}{dt}\int_{-1}^{1} wu^2 dx &= 2\int_{-1}^{1} wuu_t dx \\
&= 2\int_{-1}^{1} wuu_x dx = \int_{-1}^{1} w(u^2)_x dx \\
&= -\int_{-1}^{1} w'u^2 dx \leq 0.
\end{aligned}$$

In *this* norm, the Galerkin method for the linear hyperbolic equation is stable for a wide variety of Jacobi polynomials with weight $w(x) = (1+x)^\alpha(1-x)^\beta$.

Example 8.7 The Jacobi polynomial Galerkin method for the scalar hyperbolic equation

$$\frac{\partial u}{\partial t} = \frac{\partial u}{\partial x}$$

with boundary condition

$$u(1,t) = 0$$

is stable under the norm $(\cdot,\cdot)_w$ where $w(x) = (1+x)^\alpha(1-x)^\beta$ for all values of $\alpha \geq -1$ and $\beta \leq 0$.

The approximation, using the polynomial basis $\{P_n(x) - P_n(1)\}$, where $P_n = P_n^{(\alpha,\beta)}$ are the Jacobi polynomials, can be written

$$u_N = \sum_{n=0}^{N} a_n(t)(P_n(x) - P_n(1)),$$

where u_N satisfies

$$\frac{\partial u_N}{\partial t} - \frac{\partial u_N}{\partial x} = \tau(t) R_N(x). \tag{8.1}$$

If we multiply this by $(1+x)w(x)u_N$ and integrate over all x, the left hand side becomes

$$\begin{aligned}
LHS &= \int_{-1}^{1} (1+x)w(x)u_N \left(\frac{\partial u_N}{\partial t} - \frac{\partial u_N}{\partial x} \right) dx \\
&= \int_{-1}^{1} (1+x)w(x)u_N \frac{\partial u_N}{\partial t} dx - \int_{-1}^{1} (1+x)w(x)u_N \frac{\partial u_N}{\partial x} dx \\
&= \frac{1}{2}\frac{d}{dt} \int_{-1}^{1} (1+x)w(x)(u_N)^2 dx - \int_{-1}^{1} (1+x)w(x)u_N \frac{\partial u_N}{\partial x} dx \\
&= \frac{1}{2} \left(\frac{d}{dt} \|u_N\|^2_{\tilde{w}(x)} + \int_{-1}^{1} \tilde{w}'(x)(u_N)^2 \, dx \right)
\end{aligned}$$

where $\tilde{w}(x) = (1+x)^{\alpha+1}(1-x)^{\beta}$. Note that we showed before that $\tilde{w}'(x) \geq 0$ for $\alpha + 1 \geq 0$ and $\beta \leq 0$.

Now, let's look at the right hand side,

$$\begin{aligned}
RHS &= \tau(t) \int_{-1}^{1} (1+x)w(x)u_N R_N dx \\
&= \tau(t) \int_{-1}^{1} w(x)u_N R_N dx + \tau(t) \int_{-1}^{1} x w(x) u_N R_N dx.
\end{aligned}$$

Clearly, the first integral is zero because R_N is orthogonal to the space of u_N under $w(x)$. The second integral is

$$\int_{-1}^{1} x w(x) u_N R_N dx = \int_{-1}^{1} x w(x) \sum_{n=0}^{N} a_n(t)(P_n(x) - P_n(1)) R_N dx.$$

Since each Jacobi polynomial (with parameters α and β) satisfies the recursion in Chapter 4, Theorem 4.2,

$$x P_n(x) = A_n P_{n-1}(x) + B_n P_n(x) + C_n P_{n+1}(x),$$

where

$$A_n = \frac{2(n+\alpha)(n+\beta)}{(2n+\alpha+\beta+1)(2n+\alpha+\beta)},$$

$$B_n = -\frac{\alpha^2-\beta^2}{(2n+\alpha+\beta+2)(2n+\alpha+\beta)},$$

$$C_n = \frac{2(n+1)(n+\alpha+\beta+1)}{(2n+\alpha+\beta+2)(2n+\alpha+\beta+1)},$$

we can say,

$$\begin{aligned}\int_{-1}^{1} xw(x)u_N R_N dx = \int_{-1}^{1} w(x)\sum_{n=0}^{N} a_n(t)\big[A_n\big(P_{n-1}(x)-P_{n-1}(1)\big) \\ + B_n\big(P_n(x)-P_n(1)\big) \\ + C_n\big(P_{n+1}(x)-P_{n+1}(1)\big)\big]R_N dx.\end{aligned}$$

Since R_N is orthogonal to all $(P_n(x) - P_n(1))$ in this sum except for $n = N+1$, we are left with,

$$\int_{-1}^{1} xw(x)u_N R_N dx = \int_{-1}^{1} w(x)a_N(t)C_N\big(P_{N+1}(x)-P_{N+1}(1)\big)R_N dx.$$

We can write

$$R_N = \sum_{k=0}^{N} r_k P_k(x),$$

and so

$$\begin{aligned}RHS &= a_N(t)C_N\int_{-1}^{1} w(x)\big(P_{N+1}(x)-P_{N+1}(1)\big)R_N dx \\ &= a_N(t)C_N\tau(t)\int_{-1}^{1} w(x)\big(P_{N+1}(x)-P_{N+1}(1)\big)\sum_{k=0}^{N} r_k(t)P_k(x)dx \\ &= -a_N(t)C_N\tau(t)r_0(t)\int_{-1}^{1} w(x)P_{N+1}(1)P_0(x)dx.\end{aligned}$$

To observe that the RHS is negative, we must look at each term. $P_0(x) = 1 \geq 0$ and

$$P_{N+1}(1) = \frac{\Gamma(n+\alpha+1)}{n!\Gamma(\alpha+1)} \geq 0,$$

due to the way the Jacobi polynomials are normalized. As we saw above, $C_N \geq 0$ as long as $N+\alpha+\beta+1 \geq 0$. Now we are left with $a_N(t)\tau(t)r_0(t)$.

To determine the signs of these we look at the definition of the residual,

$$\frac{\partial u_N}{\partial t} = \frac{\partial u_N}{\partial x} + \tau(t) \sum_{k=0}^{N} r_k(t) P_k(x).$$

Since the term $\partial u_N / \partial x$ is a polynomial of degree $N - 1$, we know that the highest degree polynomial is present only in the first and third terms. Equating these terms, we have

$$\frac{da_N}{dt} = \tau r_N.$$

Furthermore, since R_N is orthogonal to the space of u_N,

$$\begin{aligned} 0 &= \left(\sum_{k=0}^{N} r_k(t) P_k(x), \; P_n(x) - P_n(1) \right)_w \\ &= r_n \left(P_n, P_n\right)_w - r_0 \left(P_0, P_n(1)\right)_w \end{aligned}$$

so that r_n is the same sign as r_0 for all $n = 1, \ldots, N$. Thus, we have

$$\frac{d}{dt} \|u_N\|^2_{\tilde{w}(x)} + \int_{-1}^{1} \tilde{w}'(x) \left(u_N\right)^2 dx = -K \frac{d}{dt} \left(a_N^2\right),$$

where

$$K = \frac{r_0}{r_N} C_N \int_{-1}^{1} w(x) P_{N+1}(1) P_0(x) dx \geq 0.$$

Thus, we have

$$\frac{d}{dt} \left(\|u_N\|^2_{\tilde{w}(x)} + K a_N^2 \right) \leq 0,$$

and so the method is stable

$$\|u_N(t)\|^2_{\tilde{w}(x)} + K a_N^2(t) \leq \|u_N(0)\|^2_{\tilde{w}(x)} + K a_N^2(0).$$

Since $\tilde{w}(x) \leq 2w(x)$, the method is also consistent in the new norm defined by $\tilde{w}(x)$. Taken with stability, this yields convergence.

8.2 The collocation approach

To analyze the stability for collocation methods, we use the fact that the collocation points are related to Gauss-type integration formulas. The properties of these quadrature formulas allow us to pass from summation to integration.

Example 8.8 Consider the Chebyshev collocation method for the equation

$$u_t = \sigma(x) u_{xx} \quad \sigma(x) > 0,$$

with boundary conditions

$$u(\pm 1, t) = 0.$$

Recall from Example 8.2 that this problem is wellposed in the Chebyshev norm. In the collocation method we seek solutions $u_N \in \mathsf{B}_N$ such that the equation

$$\frac{\partial u_N}{\partial t} - \sigma(x)\frac{\partial^2 u_N}{\partial x^2}$$

is satisfied at the points

$$x_j = -\cos\left(\frac{\pi j}{N}\right) \quad j = 1, \ldots, N-1,$$

and $u_N(\pm 1) = 0$. Multiplying by $u_N(x_j, t)$ and the weights w_j, and summing, we have

$$\sum_{j=0}^{N} \frac{1}{\sigma(x_j)} u_N(x_j) \frac{\partial u_N(x_j)}{\partial t} w_j = \sum_{j=0}^{N} u_N(x_j) \frac{\partial^2 u_N(x_j)}{\partial x^2} w_j.$$

Let w_j be the Chebyshev Gauss–Lobatto weights; then since $u_N(\partial^2 u_N/\partial x^2)$ is a polynomial of degree $2N-2$, the quadrature formula is exact, and the right hand side

$$\sum_{j=0}^{N} u_N(x_j) \frac{\partial^2 u_N(x_j)}{\partial x^2} w_j = \int_{-1}^{1} \frac{1}{\sqrt{1-x^2}} u_N \frac{\partial^2 u_N}{\partial x^2} dx \leq 0.$$

Now, the left hand side

$$\begin{aligned}\sum_{j=0}^{N} \frac{1}{\sigma(x_j)} u_N(x_j) \frac{\partial u_N(x_j)}{\partial t} w_j &= \frac{1}{2} \sum_{j=0}^{N} \frac{1}{\sigma(x_j)} \frac{\partial u_N^2}{\partial t}(x_j) w_j \\ &= \frac{1}{2}\frac{d}{dt} \sum_{j=0}^{N} \frac{1}{\sigma(x_j)} u_N^2(x_j) w_j,\end{aligned}$$

and so we can conclude that

$$\frac{d}{dt} \sum_{j=0}^{N} \frac{1}{\sigma(x_j)} u_N^2(x_j) w_j \leq 0.$$

This implies that

$$\sum_{j=0}^{N} \frac{1}{\sigma(x_j)} u_N^2(x_j, t) w_j \leq \sum_{j=0}^{N} \frac{1}{\sigma(x_j)} u_N^2(x_j, 0) w_j$$

and since

$$\frac{1}{\max \sigma(x_j)} \sum_{j=0}^{N} u_N^2(x_j, t) w_j \leq \sum_{j=0}^{N} \frac{1}{\sigma(x_j)} u_N^2(x_j, t) w_j$$
$$\leq \frac{1}{\min \sigma(x_j)} \sum_{j=0}^{N} u_N^2(x_j, t) w_j$$

we have

$$\sum_{j=0}^{N} u_N^2(x_j, t) w_j \leq \frac{\max \sigma(x_j)}{\min \sigma(x_j)} \sum_{j=0}^{N} u_N^2(x_j, 0) w_j.$$

Thus, the method is stable in L_w^2.

Next, we consider the Chebyshev collocation method based on the inner and the left boundary Gauss–Lobatto points for the linear hyperbolic equation:

Example 8.9 To approximate the solution to the equation

$$\frac{\partial u}{\partial t} = \frac{\partial u}{\partial x}$$
$$u(1, t) = 0$$

we seek a polynomial of degree $(N-1)$, $u_N(x, t)$, which satisfies the equation

$$\left.\frac{\partial u_N}{\partial t}(x)\right|_{x=x_j} = \left.\frac{\partial u_N}{\partial x}(x)\right|_{x=x_j}$$

at the points

$$x_j = -\cos\left(\frac{\pi j}{N}\right) \quad j = 1, \ldots, N.$$

If we multiply this equation by $(1 + x_j)w_j$ and sum over all x_j, we obtain

$$\sum_{j=0}^{N} (1 + x_j) u_N \frac{\partial u_N}{\partial t} w_j = \sum_{j=0}^{N} (1 + x_j) u_N \frac{\partial u_N}{\partial x} w_j.$$

Since $u_N(\partial u_N / \partial x)$ is a polynomial of degree $2N - 3$, the quadrature is exact and we have

$$\sum_{j=0}^{N} (1 + x_j) u_N \frac{\partial u_N}{\partial t} w_j = \sum_{j=0}^{N} (1 + x_j) u_N \frac{\partial u_N}{\partial x} w_j$$
$$= \int_{-1}^{1} \frac{1 + x}{\sqrt{1 - x^2}} u_N \frac{\partial u_N}{\partial x} dx \leq 0.$$

The stability of the Chebyshev method based on Gauss–Lobatto points has also been proved, but this proof is more complicated and we will not present it here.

8.3 Stability of penalty methods

We turn to the stability of the penalty methods introduced in Chapter 7.

Example 8.10 Consider the wave equation

$$u_t = u_x,$$

with some specified initial condition, and a boundary condition

$$u(1, t) = g(t).$$

The Legendre penalty method is

$$\frac{\partial u_N}{\partial t} = \frac{\partial u_N}{\partial x}$$

for $x = x_j$, the zeroes of $(1 - x)P_N'(x)$, where $P_N(x)$ is the Legendre polynomial, and

$$\frac{\partial u_N}{\partial t} = \frac{\partial u_N}{\partial x} - \tau(u_N(1) - g(t)),$$

at $x = 1$.

This scheme is consistent, because the exact solution satisfies the boundary condition. For stability we consider the case $g(t) = 0$,

$$\frac{\partial u_N}{\partial t} = \frac{\partial u_N}{\partial x} \quad \text{for } -1 \leq x < 1$$

and

$$\frac{\partial u_N}{\partial t} = \frac{\partial u_N}{\partial x} - \tau u_N(1) \quad \text{for } x = 1.$$

Lemma 8.11 *Let $\tau > 1/2w_0$, where $w_0 = \frac{2}{N(N+1)}$ is the weight on the Legendre Gauss–Lobatto formula at the endpoints. Then,*

$$\sum_{j=0}^{N} u_N^2(x_j, t)w_j \leq \sum_{j=0}^{N} u_N^2(x_j, 0)\, w_j.$$

Proof:

$$
\begin{aligned}
\frac{1}{2}\frac{d}{dt}\sum_{j=0}^{N} u_N^2(x_j,t)w_j &= \sum_{j=0}^{N} u_N \frac{\partial u_N}{\partial t} w_j - \tau u_N^2(1)w_0 \\
&= \frac{1}{2}\int_{-1}^{1} \left(u_N^2\right)_x dx - \tau u_N^2(1)w_0 \\
&= \frac{1}{2}u_N^2(1) - \frac{1}{2}u_N^2(-1) - \tau u_N^2(1)w_0 \\
&\leq 0.
\end{aligned}
$$

QED

In the next example, we prove the stability of the Chebyshev penalty method for the linear scalar hyperbolic problem.

Example 8.12 Again we consider the wave equation

$$
\frac{\partial u}{\partial t} = \frac{\partial u}{\partial x},
$$

with boundary condition given by

$$
u(1,t) = g(t).
$$

Let $v = u_{N-1}$, so that the Chebyshev penalty method is

$$
\frac{\partial v}{\partial t} = \frac{\partial v}{\partial x} \quad \text{at } x_j = -\cos\left(\tfrac{\pi j}{N}\right) \quad \forall\, j = 1, \ldots, N
$$

and

$$
\frac{\partial v}{\partial t} = \frac{\partial v}{\partial x} - \tau(v(1,t) - g(t)),
$$

at $x = x_0 = 1$.

As before, this scheme is consistent because the exact solution satisfies the boundary condition. For stability we consider the case $g(t) = 0$:

Lemma 8.13 *There exists some constant β, so that the method is stable for $\tau > \beta N^2$.*

Proof: Let w_j be the Chebyshev Gauss–Lobatto weights. Noting that $x_0 = -1$, and using the definition of the method, we have

$$
\begin{aligned}
\sum_{j=1}^{N} v\frac{\partial v}{\partial t}(1+x_j)w_j &= \sum_{j=0}^{N} v\frac{\partial v}{\partial t}(1+x_j)w_j \\
&= \sum_{j=0}^{N} v\frac{\partial v}{\partial x}(1+x_j)w_j - 2\tau v^2(1,t)w_N.
\end{aligned}
$$

If we look at the term $v(\partial v/\partial t)(1 + x_j)$ we see that it is a $(2N - 1)$-order polynomial, so the Gauss–Lobatto quadrature is exact and the left-hand side becomes

$$\sum_{j=1}^{N} v\frac{\partial v}{\partial t}(1 + x_j)w_j^N = \int_{-1}^{1} \frac{(1+x)}{\sqrt{1-x^2}} v\frac{\partial v}{\partial t}dx = \frac{1}{2}\frac{d}{dt}\int_{-1}^{1} \frac{(1+x)}{\sqrt{1-x^2}} v^2 dx.$$

Putting this all together,

$$\begin{aligned}\frac{1}{2}\frac{d}{dt}\int_{-1}^{1} \frac{(1+x)}{\sqrt{1-x^2}} v^2 dx &= \sum_{j=0}^{N} v\frac{\partial v}{\partial x}(1 + x_j)w_j - 2\tau v^2(1,t)w_N \\ &= \frac{1}{2}\sum_{j=0}^{N} \frac{\partial v^2}{\partial x}(1 + x_j)w_j - 2\tau v^2(1,t)w_N.\end{aligned}$$

Now recall that $T_N' = 0$ for the inner Gauss–Lobatto points, so we add and subtract these as desired,

$$\begin{aligned}\frac{1}{2}\frac{\partial}{\partial t}\int_{-1}^{1} \frac{(1+x)}{\sqrt{1-x^2}} v^2 dx &= \frac{1}{2}\sum_{j=0}^{N}(1 + x_j)\left[\frac{\partial v^2}{\partial x} - v^2(1)T_N'\right] w_j dx \\ &\quad + N^2 w_N v^2(1,t) - 2\tau v^2(1,t)w_N \\ &= -\frac{1}{2}\int_{-1}^{1} \frac{1}{(1-x)\sqrt{1-x^2}}(v^? - v^2(1)T_N)dx \\ &\quad + N^2 w_N v^2(1) - \tau v^2 w_N \\ &\le \frac{1}{2}\int_{-1}^{1} \frac{1}{(1-x)\sqrt{1-x^2}} v^2(1)T_N dx \\ &\quad + N^2 w_N v^2(1) - \tau v^2 w_N.\end{aligned}$$

Now if we let z_j be the Gauss points of the $(N + 1)$-order polynomial, with the corresponding weights v_j the integral can be converted back into the sum by the Gauss quadrature,

$$\begin{aligned}\frac{1}{2}\frac{\partial}{\partial t}\int_{-1}^{1} \frac{(1+x)}{\sqrt{1-x^2}} v^2 dx &\le \frac{1}{2}v^2(1,t)\sum_{j=1}^{N} \frac{1}{1-y_j}T_N(y_j)w_j \\ &\quad + N^2 w_N v^2(1) - \tau v^2 w_N.\end{aligned}$$

Since the first two terms are of order N^2, and there exists some β so that $\tau > \beta N^2$, we can conclude that this term is bounded.

QED

The next result is on the stability of the Legendre collocation penalty method for a constant coefficient hyperbolic system.

Example 8.14 Consider the system

$$\frac{\partial u}{\partial t} = \mathrm{A}\frac{\partial u}{\partial x}$$

with a set of initial and boundary conditions. If we diagonalize the matrix A, we obtain the diagonal system

$$\frac{\partial \boldsymbol{u}^I}{\partial t} = \Lambda^I \frac{\partial \boldsymbol{u}^I}{\partial x}$$

$$\boldsymbol{u}^I(1,t) = \mathrm{R}\boldsymbol{u}^{II}(1,t),$$

$$\frac{\partial \boldsymbol{u}^{II}}{\partial t} = -\Lambda^{II} \frac{\partial \boldsymbol{u}^{II}}{\partial x}$$

$$\boldsymbol{u}^{II}(-1,t) = \mathrm{L}\boldsymbol{u}^I(-1,t),$$

where the vector $\boldsymbol{u}^I = (u_1^I, u_2^I, \ldots, u_p^I)$ is comprised of all the left-entering characteristic variables, and $\boldsymbol{u}^{II} = (u_1^{II}, u_2^{II}, \ldots, u_q^{II})$ of the right-entering characteristic variables. The constant matrices $\mathrm{R}_{p\times q}$ and $\mathrm{L}_{q\times p}$ represent the coupling of the system through the boundary conditions. The spectral norms of these matrices $|R| = r$ and $|L| = l$, and we require that $rl \leq 1$, so that the energy of the system is non-increasing. The matrices

$$\Lambda^I = diag\left(\lambda_1^I, \lambda_2^I, \ldots, \lambda_p^I\right),$$

and

$$\Lambda^{II} = diag\left(\lambda_1^{II}, \lambda_2^{II}, \ldots, \lambda_q^{II}\right),$$

are the diagonal matrices of eigenvalues.

In the Legendre collocation penalty method we seek solution vectors, $\boldsymbol{v}^I$ of polynomials $(v_1^I(x), \ldots, v_p^I(x))$, and $\boldsymbol{v}^{II} = (v_1^{II}(x), \ldots, v_q^{II}(x))$, which satisfy

$$\frac{\partial \boldsymbol{v}^I}{\partial t} = \Lambda^I \frac{\partial \boldsymbol{v}^I}{\partial x} - \Lambda^I\left(\boldsymbol{v}^I(1,t) - R\boldsymbol{v}^{II}(1,t)\right) Q^+(x)$$

$$\frac{\partial \boldsymbol{v}^{II}}{\partial t} = -\Lambda^{II} \frac{\partial \boldsymbol{v}^{II}}{\partial x} - \Lambda^{II}\left(\boldsymbol{v}^{II}(-1,t) - L\boldsymbol{v}^I(-1,t)\right) Q^-(x).$$

Where Q^- (or Q^+) is a vector of polynomials which vanishes on all the grid points except the left (or right, respectively) boundary, and has the value $1/\omega_0$ at the boundary, where ω_0 is the Legendre Gauss–Lobatto weight associated with the boundary points.

We assume that each polynomial $v_j^I(x,t)$ (where $1 \leq j \leq p$) and $v_k^{II}(x,t)$ (where $1 \leq k \leq q$) is a good approximation to the corresponding u_j^I and u_k^{II}. To see this, we recall that consistency is assured by the penalty method formulation, and we proceed to prove stability.

Let's look at the jth element of the $\boldsymbol{v}^I$ vector. It satisfies

$$\frac{1}{\lambda_j^I}\frac{\partial v_j^I}{\partial t} = \frac{\partial v_j^I}{\partial x} - \left[\boldsymbol{v}^I(1,t) - R\boldsymbol{v}^{II}(1,t)\right]_j Q^+ \ \forall x.$$

Let's multiply this expression by $v_j^I(x_l)\omega_l$ and sum over the Legendre Gauss–Lobatto points x_l, to obtain

$$\frac{1}{\lambda_j^I}\frac{1}{2}\frac{d}{dt}\sum_{l=0}^{N}\left(v_j^I(x_l)\right)^2 w_j = \sum_{l=0}^{N}\omega_l v_j^I(x_l)\left.\frac{\partial v_j^I}{\partial x}\right|_{x=x_l} - [\boldsymbol{v}^I(1,t) - R\boldsymbol{v}^{II}(1,t)]_j v_j^I(1).$$

The first term on the right-hand side,

$$\sum_{l=0}^{N}\omega_l v_j^I(x_l)\left.\frac{\partial v_j^I}{\partial x}\right|_{x=x_l} = \frac{1}{2}\int_{-1}^{1}\frac{\partial\left(v_j^I(x)\right)^2}{\partial x}dx = \frac{1}{2}\left(v_j^I(1)\right)^2 - \frac{1}{2}\left(v_j^I(-1)\right)^2.$$

Let

$$(u, v)_w = \sum_{l=0}^{N} u(x_l)v(x_l)w_l,$$

and

$$<\boldsymbol{u}^I, \boldsymbol{v}^I>_p = \sum_{j=0}^{p}(\boldsymbol{u}^I, \boldsymbol{v}^I)_w, \quad \text{and} \quad <\boldsymbol{u}^{II}, \boldsymbol{v}^{II}>_q = \sum_{j=0}^{q}(\boldsymbol{u}^{II}, \boldsymbol{v}^{II})_w.$$

Summing over all points, we obtain

$$\begin{aligned}
\frac{1}{2}l(\Lambda^I)^{-1}\frac{d}{dt} <\boldsymbol{v}^I, \boldsymbol{v}^I>_p &= \frac{l}{2}(\boldsymbol{v}^I(1))^2 - \frac{l}{2}(\boldsymbol{v}^I(-1))^2 - l(\boldsymbol{v}^I(1))^2 \\
&\quad - l <\boldsymbol{v}^I, R\boldsymbol{v}^{II}>_p (1) \\
\frac{1}{2}r(\Lambda^{II})^{-1}\frac{d}{dt} <\boldsymbol{v}^{II}, \boldsymbol{v}^{II}>_q &= \frac{r}{2}(\boldsymbol{v}^{II}(-1))^2 - \frac{r}{2}(\boldsymbol{v}^{II}(1))^2 - r(\boldsymbol{v}^{II}(-1))^2 \\
&\quad + r <\boldsymbol{v}^{II}, L\,\boldsymbol{v}^I>_q (-1).
\end{aligned}$$

Adding these, we have

$$\begin{aligned}
&\frac{1}{2}l(\Lambda^I)^{-1}\frac{d}{dt} <\boldsymbol{v}^I, \boldsymbol{v}^I>_p + \frac{1}{2}r(\Lambda^{II})^{-1}\frac{d}{dt} <\boldsymbol{v}^{II}, \boldsymbol{v}^{II}>_q \\
&\le -\frac{1}{2}\left(l(\boldsymbol{v}^I(1))^2 - 2lr|\boldsymbol{v}^I(1)||\boldsymbol{v}^{II}(1)| + r(\boldsymbol{v}^{II}(1))^2\right) \\
&\quad -\frac{1}{2}\left(l(\boldsymbol{v}^I(-1))^2 + 2lr|\boldsymbol{v}^I(-1)||\boldsymbol{v}^{II}(-1)| - r(\boldsymbol{v}^{II}(-1))^2\right) \\
&\le 0,
\end{aligned}$$

since each of the terms in brackets is positive as long as $(lr)^2 \leq lr$, which is correct since $rl \leq 1$.

Remark A polynomial Galerkin penalty method consists of finding a polynomial solution which has the property that the residual is orthogonal to the polynomial space, and the boundary conditions are enforced using a penalty method. This is exactly the same process employed in each subdomain of the discontinuous Galerkin method. Thus, the multi-domain spectral penalty method bears a striking similarity to the discontinuous Galerkin method. This topic will be further discussed in Chapter 12.

8.4 Stability theory for nonlinear equations

Consider the nonlinear hyperbolic system

$$\frac{\partial U}{\partial t} + \frac{\partial f(U)}{\partial x} = 0 \tag{8.2}$$

Polynomial spectral methods for this equation are unstable in general but can be stabilized by the Super Spectral Viscosity (SSV) method.

$$\frac{\partial u_N}{\partial t} + \frac{\partial \mathcal{P}_N f(u_N)}{\partial x} = \epsilon_N (-1)^{s-1} \left(\sqrt{1-x^2} \frac{\partial}{\partial x} \right)^{2s} u_N, \tag{8.3}$$

for the Chebyshev polynomial basis and

$$\frac{\partial u_N}{\partial t} + \frac{\partial \mathcal{P}_N f(u_N)}{\partial x} = \epsilon_N (-1)^{s-1} \left(\frac{\partial}{\partial x} (1-x^2) \frac{\partial}{\partial x} \right)^{s} u_N, \tag{8.4}$$

for the Legendre polynomial basis.

Note that for spectral accuracy the order, s, must be proportional to N, the number of polynomials (or grid points) in the approximation. Thus viscosity changes with mesh refinement.

Maday and Tadmor (1989) showed that the SSV methods converge to the correct entropy solution for Legendre approximations to scalar nonlinear hyperbolic equations. Furthermore, it has been proven that even for systems, if the solution converges it converges to the correct entropy solution.

The straightforward use of the SSV method requires extra derivative computations thus effectively doubling the computational cost. For this reason, most of the large scale spectral codes for high Mach number flows use filtering to stabilize the code. In fact, as will be explained in this section, filtering can be seen as an efficient way of applying the SSV method.

To understand the relationship between SSV and filtering let

$$u_N(x,t) = \sum_{k=0}^{N} a_k(t)\phi_k(x)$$

where ϕ_k are the basis function used (Chebyshev or Legendre). Also, let $b_k(a_0, \ldots, a_N)$ be the coefficients in the expansion

$$\mathcal{P}_N f(u_N) = \sum_{k=0}^{N} b_k(t)\phi_k(x).$$

Because we include the terms $(\sqrt{1-x^2})$ in the Chebyshev, the right-hand side is

$$\left(\sqrt{1-x^2}\frac{d}{dx}\right)^{2s} T_k = k^{2s} T_k,$$

by the Sturm–Liouville equation. And so Equation (8.3) becomes

$$\frac{\partial a_k}{\partial t} = b_k - c\epsilon_N k^{2s} a_k. \tag{8.5}$$

Similarly, for the Legendre case

$$\frac{d}{dx}\left((\sqrt{1-x^2})\frac{d}{dx}\right)^{s} P_k = k^s(1+k)^s P_k,$$

and so Equation (8.4) becomes

$$\frac{\partial a_k}{\partial t} = b_k - c\epsilon_N k^s(1+k)^s a_k. \tag{8.6}$$

Thus a typical step in a Runge–Kutta method for the Chebyshev SSV method is

$$a_k^{n+1} = a_k^n + \Delta t b_k^n - \Delta t c\epsilon_N k^{2s} a_k^{n+1}.$$

Note that the stabilizing term is implicit, to prevent further limitation on the time step. In the filtering method we change slightly the last term,

$$a_k^{n+1} = a_k^n + \Delta t b_k^n + \left(1 - e^{\Delta t c\epsilon_N k^{2s}}\right) a_k^{n+1}$$

yielding

$$a_k^{n+1} = e^{-\Delta t c\epsilon_N k^{2s}}\left(a_k^n + \Delta t b_k^n\right).$$

This is an exponential filter. The benefit of this method is that it does not require any extra derivative computation and therefore does not increase the computational cost. Similar results can be obtained for the Legendre method.

We note here that the parameter s can be a function of x. This means that in different regions one can use viscosity terms of different orders. In the presence of local sharp gradients one should reduce the order of the filter. To maintain spectral accuracy, though, s should be an increasing function of N.

Thus the local adaptive filter is defined by

$$u_N^\sigma = \sum_{k=0}^{N} \sigma\left(\frac{k}{N}\right) a_k(t)\phi(x), \tag{8.7}$$

where

$$\sigma(\omega) = e^{-\alpha\omega^{2\gamma s}},$$

and $\gamma = \gamma(x)$ can change within the domain.

8.5 Further reading

The stability theory for the polynomial case has largely been developed by Gottlieb and collaborators (1981, 1983, 1987) with contributions also by Canuto and Quarteroni (1981, 1986), Bressan and Quarteroni (1986). Further results and overviews can be found in the review by Gottlieb and Hesthaven (2001) and in the texts by Canuto et al (1986, 2006) and by Guo (1998). The penalty methods were proposed for hyperbolic problems by Funaro and Gottlieb (1988, 1991). For the stability and convergence for nonlinear problems we refer to the papers by Maday and Tadmor (1989), Maday et al (1993), Tadmor (1997), and Ma (1998). More details on stabilization using filters can be found in the review by Gottlieb and Hesthaven (2001) and the papers by Don and collaborators (1994, 1995, 1998, 2003).

9

Spectral methods for nonsmooth problems

The error estimates for spectral approximations of solutions of PDEs rely heavily on the fact that the solution is smooth everywhere in the interval of interest. If the function has a discontinuity, even at one point, the error estimates fail. Moreover, the rate of convergence deteriorates to first order. It would seem that spectral methods are not designed for problems with discontinuities. However, in this chapter we will show that spectral methods retain high resolution information even for discontinuous solutions, and that high-order approximations can be obtained by appropriate postprocessing in smooth regions.

Let's consider a simple example.

Example 9.1 We approximate the solution of the scalar hyperbolic equation

$$\frac{\partial u}{\partial t} = -2\pi \frac{\partial u}{\partial x}, \tag{9.1}$$
$$u(0, t) = u(2\pi, t),$$

with initial conditions

$$u(x, 0) = \begin{cases} x & 0 \leq x \leq \pi, \\ x - 2\pi & \pi < x \leq 2\pi, \end{cases}$$

and periodic boundary conditions, using a Fourier collocation method with $N = 64$ grid points in space and a fourth-order Runge–Kutta method in time. The initial profile, a saw-tooth function which is linear in each of the regions $0 \leq x < \pi$ and $\pi < x \leq 2\pi$ with a discontinuity at $x = \pi$, is convected around the domain.

In Figure 9.1 we plot the computed approximation at $t = 0$ and after 100 periodic revolutions. This simple experiment demonstrates many of the issues which arise in the spectral solution of discontinuous problems. The initial approximation ($t = 0$), features strong oscillations, particularly near the shock. Although the steep gradient characterizing the discontinuity is fairly clear, the

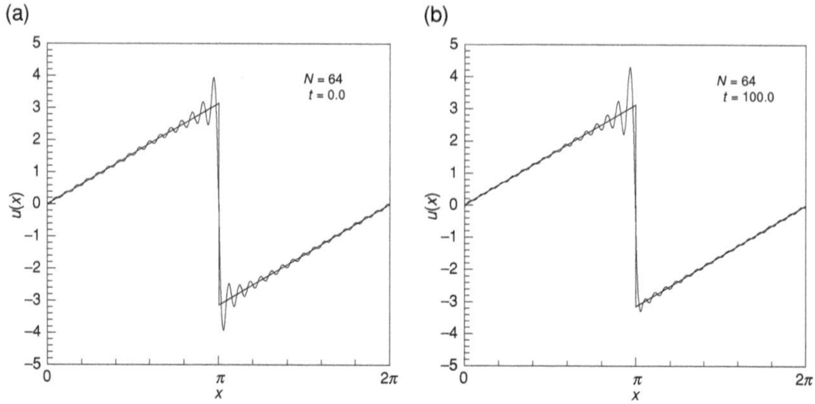

Figure 9.1 (a) The initial saw-tooth function and the associated discrete Fourier series approximation. (b) The saw-tooth function and the convected Fourier approximation at $t = 100$ obtained by solving Equation (9.1).

approximation of the solution is not very good in the neighborhood of the discontinuity. However, the discontinuity remains well approximated even after a very long time ($t = 100$) and only little smearing of the initial front can be observed. All in all, although the approximation itself is oscillatory, the initial profile is advected accurately.

While the initial spectral approximation of nonsmooth functions introduces very significant oscillations, the superior phase properties of the Fourier method remain intact. In the first section of this chapter, we will discuss in detail the appearance of the oscillations, known as the Gibbs phenomenon, and the nonuniform convergence which characterizes this phenomenon. The second part of this chapter is devoted to techniques, such as filtering and postprocessing, that will allow us to overcome or at least decrease the effect of the Gibbs phenomenon when solving general partial differential equations. This general theory shows that we can recover exponentially convergent series for piecewise analytic problems, and completely overcome the Gibbs phenomenon. Finally, we show that this methodology can be applied in the context of the numerical solution of PDEs.

9.1 The Gibbs phenomenon

Let us begin by considering the behavior of trigonometric expansions of piecewise continuous functions. To understand the problems which we saw in Example 9.1, we focus on the approximation of the initial profile.

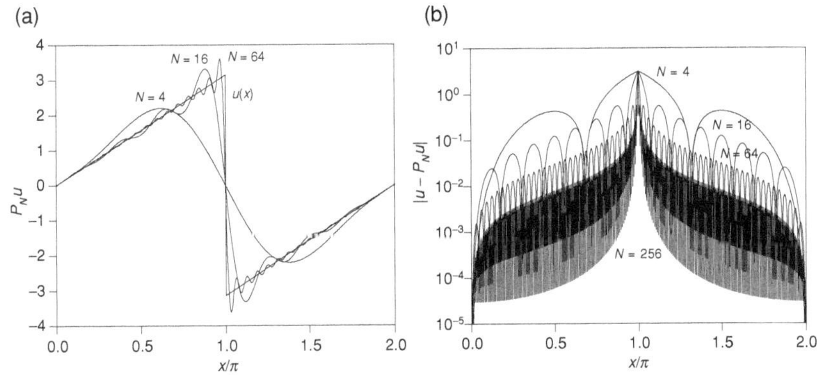

Figure 9.2 On the left we show the continuous Fourier series approximation of a discontinuous function illustrating the appearance of the Gibbs phenomenon. The figure on the right displays the pointwise error of the continuous Fourier series approximation for increasing resolution. Note that the value of the pointwise error at the discontinuity corresponds exactly to $(u(\pi+) + u(\pi-))/2$.

Example 9.2 Consider the function

$$u(x) = \begin{cases} x & 0 \leq x \leq \pi, \\ x - 2\pi & \pi < x \leq 2\pi, \end{cases}$$

and assume that it is periodically extended. The continuous Fourier series expansion coefficients are

$$\hat{u}_n = \begin{cases} i(-1)^{|n|}/n & n \neq 0, \\ 0 & n = 0. \end{cases}$$

In Figure 9.2 we show the Fourier series approximation to $u(x)$. Observe that the strong oscillations around the discontinuity do not decay as we increase the number of points, but that the approximation converges, albeit slowly, away from the discontinuity. The convergence of the approximation away from the discontinuity can be seen in Figure 9.2 (right) where we plot the pointwise error. The convergence is, at most, linear, corresponding to the decay of the expansion coefficients.

As Figure 9.2 shows, the error at the discontinuity does not disappear as we increase the resolution. Thus, although the approximation converges in the mean, the approximation is not uniformly convergent. To see that the approximation converges in the mean, we can estimate the L^2-error using Parseval's identity,

$$\|u - \mathcal{P}_N u\|_{L^2[0,2\pi]} = 2\pi \left(\sum_{|n|>N} \frac{1}{n^2} \right)^{1/2} \simeq \frac{1}{\sqrt{N}}.$$

Although the function is smooth and periodic away from the discontinuity, the global rate of convergence is dominated by the presence of the discontinuity. Thus, we see two aspects of the problem: the slow convergence away from the discontinuity, and the non-uniform convergence near the discontinuity. This characteristic behavior is known as the Gibbs phenomenon.

To study the Gibbs phenomenon in more detail, we look at the truncated Fourier series sum

$$\mathcal{P}_N u(x) = \sum_{|n| \le N/2} \hat{u}_n e^{inx},$$

where the continuous expansion coefficients are

$$\hat{u}_n = \frac{1}{2\pi} \int_0^{2\pi} u(x) e^{-inx} \, dx.$$

Theorem 9.3 *Every piecewise continuous function, $u(x) \in L^2[0, 2\pi]$, has a converging Fourier series*

$$\mathcal{P}_N u(x) \to \frac{u(x^+) + u(x^-)}{2} \quad \textit{as} \quad N \to \infty.$$

Proof: We begin by rewriting the truncated series

$$\begin{aligned}
\mathcal{P}_N u(x) &= \frac{1}{2\pi} \sum_{|n| \le N/2} \left(\int_0^{2\pi} u(t) e^{-int} \, dt \right) e^{inx} \\
&= \frac{1}{2\pi} \int_0^{2\pi} u(t) \left(\sum_{|n| \le N/2} e^{in(x-t)} \right) dt \\
&= \frac{1}{2\pi} \int_{x-2\pi}^{x} D_N(t) u(x-t) \, dt,
\end{aligned}$$

where

$$D_N(y) = \sum_{|n| \le N/2} e^{iny} = \frac{\sin((N+1)t/2)}{\sin(t/2)}, \tag{9.2}$$

is the Dirichlet kernel. The Dirichlet kernel can be viewed as the projection of a delta function onto the space spanned by the Fourier basis and is an even function in y, which oscillates while changing signs at $t_j = 2\pi j/(N+1)$. The kernel is shown for increasing values of N in Figure 9.3.

Since the main contribution of the integral originates from a narrow region around zero, we have

$$\mathcal{P}_N u(x) \simeq \frac{1}{2\pi} \int_{-\varepsilon}^{\varepsilon} D_N(t) u(x-t) \, dt,$$

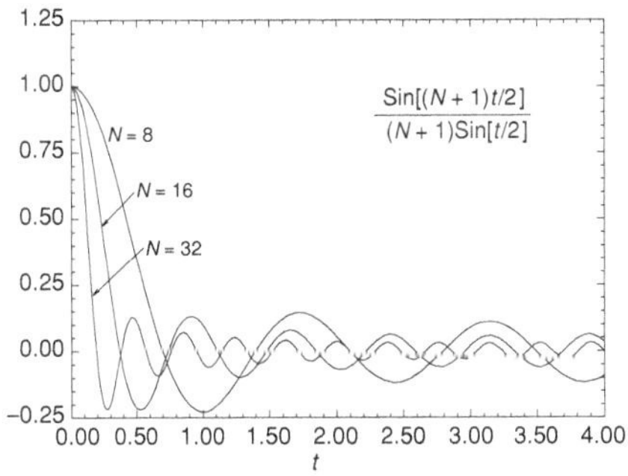

Figure 9.3 The normalized Dirichlet kernel for various values of N.

where $\varepsilon \ll 1$. Because $u(x)$ is at least piecewise continuous we may assume that $u(x-t) \simeq u(x^-)$ for $0 \leq t \leq \varepsilon$ and $u(x-t) \simeq u(x^+)$ for $0 \geq t \geq -\varepsilon$. Since t is small, we can also assume $\sin(t/2) \simeq t/2$, and therefore

$$\mathcal{P}_N u(x) \simeq \frac{1}{\pi}(u(x^+) + u(x^-)) \int_0^{\varepsilon} \frac{\sin[(N+1)t/2]}{t} dt.$$

This integral goes to $1/2$ as $N \to \infty$,

$$\begin{aligned} \frac{1}{\pi}\int_0^{\varepsilon} \frac{\sin[(N+1)t/2]}{t} dt &= \frac{1}{\pi}\int_0^{(N+1)\varepsilon/2} \frac{\sin s}{s} ds \\ &\simeq \frac{1}{\pi}\int_0^{\infty} \frac{\sin s}{s} ds = \frac{1}{2} \quad \text{as} \quad N \to \infty, \end{aligned}$$

which yields

$$\mathcal{P}_N u(x) \to \frac{u(x^+) + u(x^-)}{2} \quad \text{as} \quad N \to \infty.$$

QED

This theorem confirms that the trigonometric polynomial approximation converges pointwise for each point of continuity, while at the discontinuity it converges to the average. However, as we saw in the example, the convergence is non-uniform close to a discontinuity. To explain why this happens we look at the approximation at $x_0 + 2z/(N+1)$ (where z is a parameter) in the neighborhood of the point of discontinuity x_0. We will need to use the assymptotic behavior mentioned in the proof above, as well as the Sine integral function (plotted in Figure 9.4),

$$\text{Si}(z) = \int_0^z \frac{\sin s}{s} ds$$

and its properties

$$\lim_{z \to \infty} \text{Si}(z) = \frac{\pi}{2}, \qquad \text{Si}(-z) = -\text{Si}(z).$$

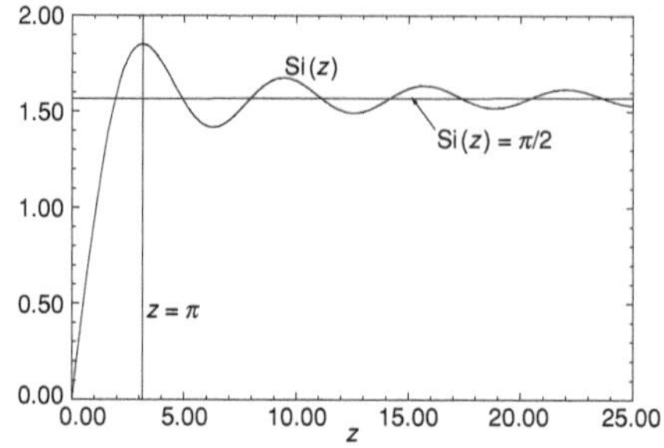

Figure 9.4 The Sine integral function, Si(z).

With these facts in place, the truncated sum

$$\mathcal{P}_N u\left(x_0 + \frac{2z}{N+1}\right) \simeq \frac{1}{2\pi}\int_{-\varepsilon}^{\varepsilon} D_N(t) u\left(x_0 + \frac{2z}{N+1} - t\right) dt,$$

becomes

$$\begin{aligned}\mathcal{P}_N u\left(x_0 + \frac{2z}{N+1}\right) &\simeq \frac{u(x_0^+)}{\pi}\int_{-\infty}^{z} \frac{\sin s}{s} ds + \frac{u(x_0^-)}{\pi}\int_{z}^{\infty} \frac{\sin s}{s} ds \\ &\simeq \frac{1}{2}(u(x_0^+) + u(x_0^-)) + \frac{1}{\pi}(u(x_0^+) - u(x_0^-))\mathrm{Si}(z).\end{aligned}$$

If $u(x)$ were continuous at x_0, the second part would vanish, and we would recover the pointwise convergence result. However, in the case of a discontinuity, this second term will cause a problem. For each N, if we look at a point x such that $x - x_0 = \mathcal{O}(1/N)$, the approximation

$$\mathcal{P}_N u(x) - \frac{1}{2}(u(x_0^+) + u(x_0^-)) = \mathcal{O}(1).$$

This means that as the point x approaches x_0 (i.e., as $N \to \infty$), the truncated series does not converge to the function. Thus, even though we have pointwise convergence at any point of continuity, the convergence is not uniform. We can see this non-uniform convergence in Figure 9.2, where the overshoot comes closer to the discontinuity as N is increased, but it does not decay.

The maximum size of the overshoot occurs at $z = \pi$, which is where the Sine integral obtains its maximum,

$$\frac{1}{\pi}\mathrm{Si}(\pi) = 0.58949.$$

Thus, the maximum overshoot or undershoot at a point of discontinuity asymptotically approaches

$$\mathcal{P}_N u(x) - u(x_0^-) \simeq 1.08949(u(x_0^+) - u(x_0^-)).$$

Hence, it is approximately 9% of the magnitude of the jump, and happens for $x \approx x_0 \pm 2\pi/(N+1)$. The following examples illustrate this aspect of the Gibbs phenomenon.

Example 9.4 Once again, we look at the function

$$u(x) = \begin{cases} x & 0 \le x \le \pi, \\ x - 2\pi & \pi < x \le 2\pi, \end{cases}$$

and assume that it is periodically extended. Using the theory above we expect the maximum overshoot around $x_0 = \pi$ to be

$$|\mathcal{P}_N u(x) - u(\pi^{\pm})| \simeq 0.08949|u(\pi^+) - u(\pi^-)| = 0.08949\, 2\pi \simeq 0.5623,$$

which agrees well with the results displayed in Figure 9.2.

Example 9.5 [Fornberg (1996)] Let's consider the periodically extended unit step function

$$u(x) = \begin{cases} -\frac{1}{2} & -\pi < x < 0, \\ \frac{1}{2} & 0 < x < \pi. \end{cases}$$

The continuous Fourier series expansion is

$$u(x) = \frac{2}{\pi} \sum_{k=0}^{\infty} \frac{\sin(2k+1)x}{2k+1}.$$

The extremum of the truncated expansion

$$\mathcal{P}_N u(x) = \frac{2}{\pi} \sum_{k=0}^{N} \frac{\sin(2k+1)x}{2k+1},$$

can be found by taking its derivative

$$(\mathcal{P}_N u(x))' = \frac{2}{\pi} \sum_{k=0}^{N} \cos(2k+1)x = \frac{\sin 2(N+1)x}{\pi \sin x},$$

and observing that this vanishes for $x = \frac{\pi}{2(N+1)}$. If we insert this into the truncated expansion and identify this with a Riemann sum, we obtain the asymptotic result

$$\begin{aligned} \mathcal{P}_N u\big(\pi/(2(N+1))\big) &= \frac{1}{\pi(N+1)} \sum_{k=0}^{N} \frac{\sin \frac{(2k+1)\pi}{2(N+1)}}{\frac{2k+1}{2(N+1)}} \\ &\simeq \frac{1}{\pi} \int_0^1 \frac{\sin \pi t}{t}\, dt = \frac{1}{\pi} \mathrm{Si}(\pi) = 0.58949. \end{aligned}$$

The above discussion concerned continuous expansions of a discontinuous function. The Gibbs phenomenon also appears in the same way in discrete expansions, and so we will not discuss this topic explicitly. The Gibbs phenomenon also appears in polynomial approximations and Bessel function expansions. For example, Chebyshev expansions do not exhibit the Gibbs phenomenon at the boundaries $x = \pm 1$, but they do exhibit the phenomenon at interior discontinuities of the function.

9.2 Filters

One of the manifestations of the Gibbs phenomenon is that the expansion coefficients decay slowly. This is due to the *global* nature of the approximation: the expansion coefficients are obtained by integration or summation over the entire domain, including the point(s) of discontinuity. The idea of filtering is to alter the expansion coefficients so that they decay faster. While, as we shall see, this will not impact the non-uniform convergence near the shock, a well chosen filter will speed convergence away from the discontinuity.

Before we can discuss the design or effect of a filter, we need to define it.

Definition 9.6 *A filter of order q is a real and even function $\sigma(\eta) \in C^{q-1}[-\infty, \infty]$ with the following properties:*

(a) $\sigma(\eta) = 0 \quad for \quad |\eta| > 1.$

(b) $\sigma(0) = 1 \quad and \quad \sigma(1) = 0.$

(c) $\sigma^{(m)}(0) = \sigma^{(m)}(1) = 0 \quad \forall m \in [1, \ldots, q-1].$

A filtered truncated continuous polynomial approximation can be written as

$$\mathcal{P}_N u(x) = u_N^\sigma(x) = \sum_{n=0}^{N} \sigma\left(\frac{n}{N}\right) \hat{u}_n \phi_n(x), \tag{9.3}$$

where $\hat{u}_n$ are the continuous expansion coefficients and $\sigma(n/N)$ is the filter. Note that the filter is a continuous function which does not affect the low modes, only the high modes. Cutting the high modes abruptly (corresponding to a step function filter) does not enhance the convergence rate even in smooth regions.

The filter can also be defined in physical space. For a Fourier space filter $\sigma(n/N)$, the physical filter function

$$S(x, y) = \sum_{n=0}^{N} \sigma\left(\frac{n}{N}\right) \frac{1}{\gamma_n} \phi_n(x) \phi_n(y), \tag{9.4}$$

enters into the filtered approximation

$$u_N^\sigma(x) = \int_D u(y) w(y) S(x, y)\, dy.$$

Note that if we let $\sigma(\eta) = 1$, the filter function is just the polynomial Dirichlet kernel. Since the highly oscillatory behavior of the Dirichlet kernel is a representation of the global nature of the approximation, one could attempt to specify $\sigma(\eta)$ with the aim of localizing the influence of $S(x, y)$. As it turns out, proper localization of the Dirichlet kernel is at least as complex as that of increasing the decay rate and it is indeed the latter procedure that has received most of the attention in the past.

9.2.1 A first look at filters and their use

Let's look at some examples of filters, and see how the filtered approximations behave. For the sake of simplicity we restrict our examples to continuous Fourier series. Similar results apply when filtering polynomial expansions.

Example 9.7 (Cesáro filter) The filter

$$\sigma(\eta) = 1 - \eta,$$

results from an arithmetic mean of the truncated series, and is only linear ($q = 1$). In Figure 9.5 we have plotted the Cesáro filtered Fourier series approximation of the discontinuous saw-tooth function and compared it with the un-smoothed approximation. We observe that the Cesáro filter eliminates the oscillations and overshoot characteristic of the Gibbs phenomenon, however it also produces a heavily smeared approximation to the original function. Although the Cesáro Filter allows us to recover uniform convergence, nothing is gained in terms of accuracy away from the discontinuity, since the filter is only first order. In Figure 9.6 we plot the pointwise error of the filtered approximation as compared to the un-filtered approximation.

The effect of the Cesáro filter in physical space can be understood by considering the modified Dirichlet kernel

$$S(x) = \frac{2}{N+2} \begin{cases} \frac{\sin^2((N/2+1)x/2)}{\sin^2(x/2)} & x \neq 2\pi p \\ 1 & x = 2\pi p \end{cases} \qquad p = 0, \pm 1, \pm 2, \ldots$$

First, we observe that $S(x) \geq 0$, i.e., the Cesáro filtered Fourier series approximation can be expected to be non-oscillatory in physical space as observed in Figure 9.5. Next, we notice that the first zero of the modified kernel appears at $x = 4\pi/(N+2)$ while the original Dirichlet kernel has it first zero at $x = 2\pi/(N+1)$. As a consequence, we expect the significant smearing of the discontinuity observed in Figure 9.5. Indeed, the smearing is so severe that we lose the ability to accurately identify the location of the discontinuity, in reality reducing the Cesáro filter to an analytical tool, rather than a practical one.

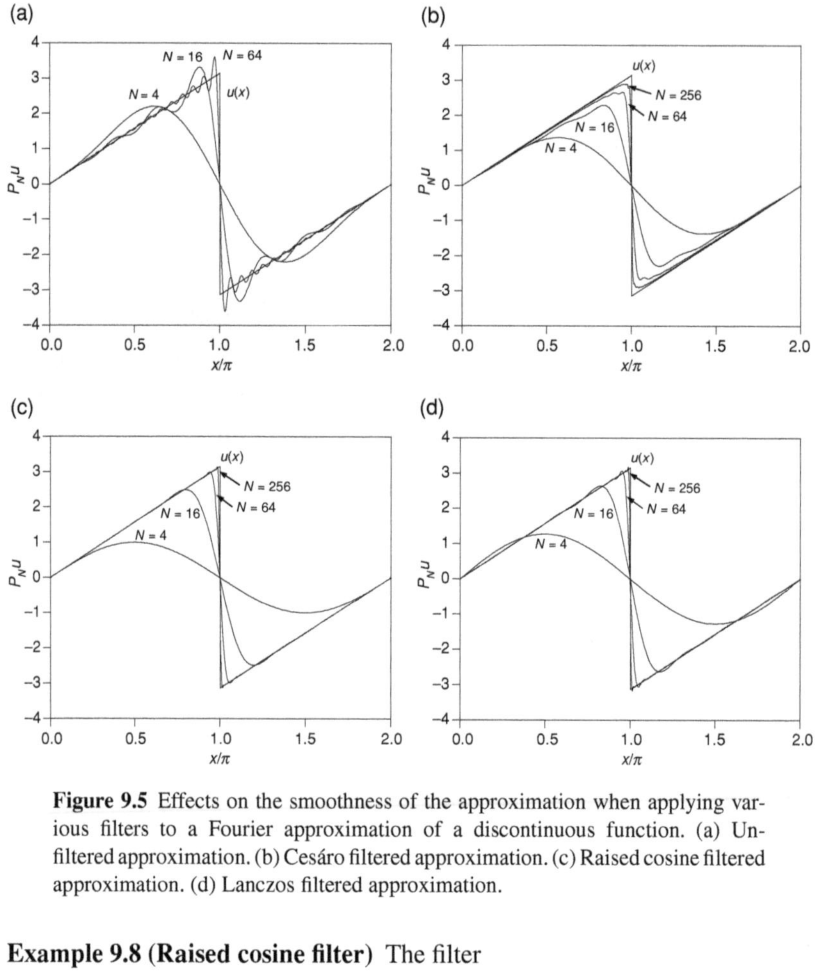

Figure 9.5 Effects on the smoothness of the approximation when applying various filters to a Fourier approximation of a discontinuous function. (a) Unfiltered approximation. (b) Cesáro filtered approximation. (c) Raised cosine filtered approximation. (d) Lanczos filtered approximation.

Example 9.8 (Raised cosine filter) The filter

$$\sigma(\eta) = \frac{1}{2}(1 + \cos(\pi\eta)),$$

is formally of second order. In fact the application of this filter is equivalent to taking the spatial average

$$u_j \simeq \frac{u_{j+1} + 2u_j + u_{j-1}}{4},$$

and has the modified kernel

$$S(x) = \frac{1}{4}\left[D_N\left(x - \frac{2\pi}{N}\right) + 2D_N(x) + D_N\left(x + \frac{2\pi}{N}\right)\right],$$

where $D_N(x)$ is the Dirichlet kernel discussed in the previous section.

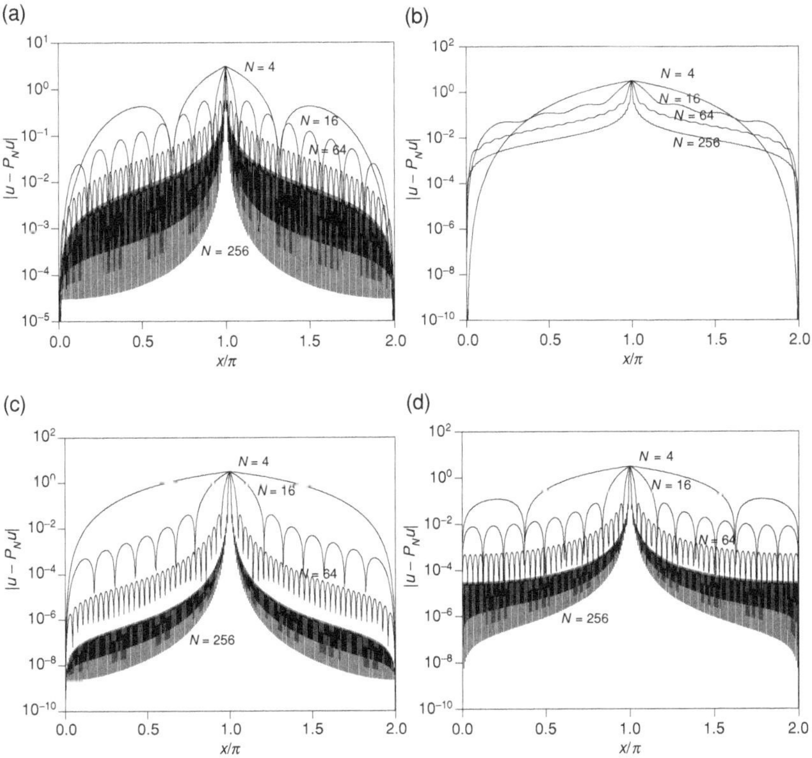

Figure 9.6 Pointwise errors of the approximation when applying various filters to a Fourier approximation of a discontinuous function. (a) Un-filtered approximation (Note different scale). (b) Cesáro filtered approximation. (c) Raised cosine filtered approximation. (d) Lanczos filtered approximation.

From Figure 9.5 we observe that the raised cosine filter reduces the oscillations present in the un-filtered approximation, but does not eliminate the overshoot near the discontinuity. Figure 9.6 demonstrates that away from the discontinuity the approximation error is clearly reduced by applying the filter.

Example 9.9 (Lanczos filter) The Lanczos filter

$$\sigma(\eta) = \frac{\sin \pi \eta}{\pi \eta},$$

is also second order. As is evident from Figure 9.5, the Lanczos filter eliminates the oscillations away from the discontinuity, but not the overshoot at the discontinuity. Figure 9.6 shows that the pointwise error indeed decreases as compared to the un-filtered approximation. However, in this case the error is slightly larger than that obtained by using the raised cosine filter.

The classical filters considered so far lead to a significant reduction of the Gibbs phenomenon away from the discontinuity. However, the pointwise convergence remains first or second-order in N. Is it possible to recover higher-order accuracy away from the discontinuity?

Example 9.10 (Exponential filters) We have the family of exponential filters

$$\sigma(\eta) = \begin{cases} 1 & |\eta| \leq \eta_c, \\ e^{(-\alpha(\frac{\eta-\eta_c}{1-\eta_c})^p)} & \eta > \eta_c, \end{cases}$$

where α is a measure of how strongly the modes should be filtered and p is the order of the filter. Note that the exponential filter does not conform with the definition of a filter in Definition 9.6 because $\sigma(1) = e^{-\alpha}$. However, in practice we choose α such that $\sigma(1) \simeq \mathcal{O}(\varepsilon_M)$ where ε_M is the machine accuracy, so that the $\pm N/2$ modes are completely removed.

In Figure 9.7 and Figure 9.8 we illustrate the effect of applying exponential filters of various orders to the saw-tooth function. We observe that even though the Gibbs phenomenon remains, as evidenced by the overshoot near the discontinuity, we recover p-order convergence in N away from the discontinuity. Choosing p as a linear function of N yields exponential accuracy away from the discontinuity. The recovery of convergence of any desired order has made the exponential filter a popular choice when performing simulations of nonlinear partial differential equations, where the Gibbs phenomenon may cause a stable scheme to become unstable due to amplification of oscillations. As we have seen in Chapter 3, filtering may also serve to stabilize a scheme.

9.2.2 Filtering Fourier spectral methods

Recall that, in Fourier spectral methods, the use of continuous and discrete expansion coefficients leads to different methods and different implementations when computing derivatives. This is seen in filtering, as well.

Filtering the continuous expansion Filtering the continuous expansion is straightforward as it only involves summing the filtered series

$$u_N^\sigma = \sum_{|n| \leq N/2} \sigma\left(\frac{|n|}{N/2}\right) \hat{u}_n e^{inx},$$

and the filtered differentiation operators

$$\frac{d^q}{dx^q} u_N^\sigma = \sum_{|n| \leq N/2} \sigma\left(\frac{|n|}{N/2}\right) \hat{u}_n^{(q)} e^{inx}.$$

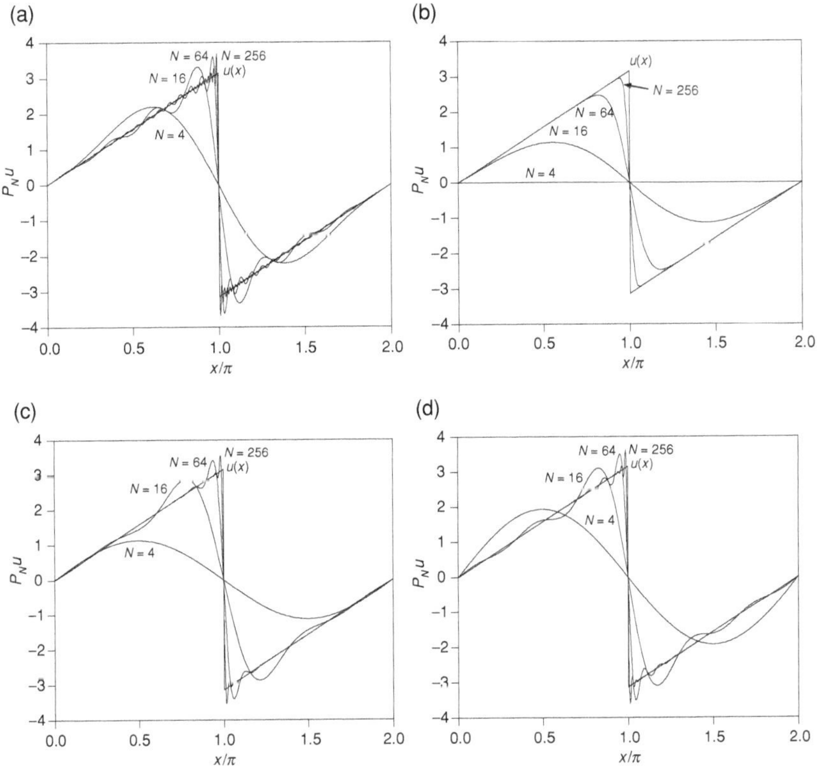

Figure 9.7 Effects on the smoothness of the approximation when applying exponential filters of increasing order to a Fourier approximation of a discontinuous function. In all cases we used $\alpha = -\log \varepsilon_M \sim 35$. (a) Un-filtered approximation. (b) Exponential filter of order 2. (c) Exponential filter of order 6. (d) Exponential filter of order 10.

In both these contexts, filtering consists of multiplying the expansion coefficients by the filter function, and then summing as usual. There is no significant increase in the computational work involved.

Filtering the discrete expansion As we have seen previously, there are two mathematically equivalent but computationally different methods of expressing the discrete expansions, leading to two different ways of filtering the expansions.

The first method follows the approach used for the continuous expansion: we multiply the expansion coefficients by the filter and proceed as usual. The only difference is that we use the discrete expansion coefficients rather than the continuous ones.

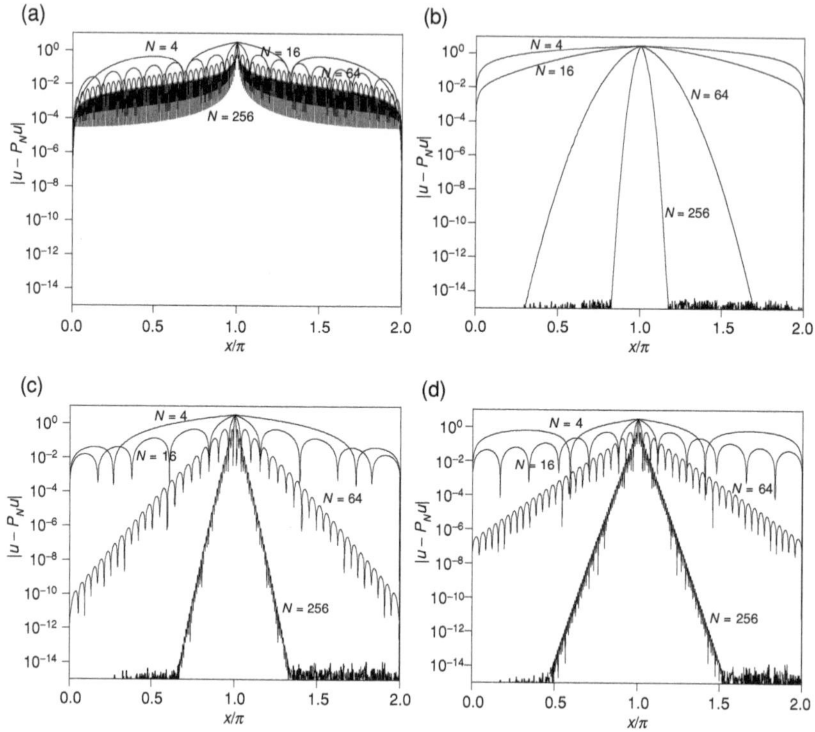

Figure 9.8 Pointwise error of the approximation when applying exponential filters of increasing order to a Fourier approximation of a discontinuous function. In all cases we used $\alpha = -\log \varepsilon_M \sim 35$. (a) Un-filtered approximation. (b) Exponential filter of order 2. (c) Exponential filter of order 6. (d) Exponential filter of order 10.

The second method involves matrices rather than summation of series. First, consider the case of an even number of points

$$x_j = \frac{2\pi}{N} j, \quad j \in [0, \ldots, N-1],$$

with the corresponding approximation

$$\mathcal{I}_N u(x) = \sum_{j=0}^{N-1} u(x_j) g_j(x),$$

where $g_j(x)$ is the interpolating Lagrange polynomial. To obtain the filtered approximation

$$u_N^\sigma(x) = \sum_{j=0}^{N-1} u(x_j) g_j^\sigma(x),$$

we need the filtered version of the interpolating Lagrange polynomial, $g_j^\sigma(x)$. Since the filter $\sigma(\eta)$ is even, we have

$$g_j^\sigma(x) = \frac{2}{N}\sum_{n=0}^{N/2}\frac{1}{c_n^\sigma}\sigma\left(\frac{n}{N/2}\right)\cos[n(x-x_j)],$$

where $c_0^\sigma = c_{N/2}^\sigma = 2$ and $c_n^\sigma = 1$ otherwise.

This allows us to express filtering as a matrix operation

$$u_N^\sigma(x_l) = \sum_{j=0}^{N-1} u_N(x_j) g_j^\sigma(x_l).$$

Note that the filter matrix is a symmetric, Toeplitz matrix. Similarly, we can obtain matrix forms for the combination of filtering and differentiation

$$\mathrm{D}_{lj}^{(q),\sigma} = \frac{2}{N}\sum_{n=0}^{N/2}\frac{1}{c_n^\sigma}\sigma\left(\frac{n}{N/2}\right)\begin{cases}(in)^q\cos[n(x_l-x_j)] & q \text{ even},\\ i(in)^q\sin[n(x_l-x_j)] & q \text{ odd},\end{cases},$$

so that the filtered and differentiated approximation is

$$\frac{d^q}{dx^q}u_N^\sigma(x_l) = \sum_{j=0}^{N-1}\mathrm{D}_{lj}^{(q),\sigma} u_N(x_j).$$

Note that $\mathrm{D}^{(q),\sigma}$ has the same properties as $\mathrm{D}^{(q)}$, i.e., it is a circulant Toeplitz matrix that is symmetric when q is even and skew-symmetric when q is odd.

The filtered versions of the Lagrange interpolation polynomials and the differentiation matrices for an odd number of collocation points

$$y_j = \frac{2\pi}{N+1}j, \quad j \in [0, \ldots, N],$$

can be obtained for the above results by setting $c_n^\sigma = 1$ for all values of n.

9.2.3 The use of filters in polynomial methods

Now we turn our attention to the use of filters in continuous and discrete polynomial expansions.

Filtering of the continuous expansion Once again, the filtering of the continuous expansions is a multiplication of the coefficients by the filter,

$$u_N^\sigma = \sum_{n=0}^{N}\sigma\left(\frac{n}{N}\right)\hat{u}_n\phi_n(x),$$

and likewise, the filtered differentiation operators

$$\frac{d^q}{dx^q}u_N^\sigma = \sum_{n=0}^{N} \sigma\left(\frac{n}{N}\right) \hat{u}_n^{(q)} \phi_n(x).$$

Filtering of the discrete expansion As before, the two equivalent formulations of the discrete polynomial approximation result in two equivalent ways in which to apply a filter. The first approach is the same as the one we use for filtering the continuous expansion, a straightforward multiplication of the expansion coefficients (in this case, the discrete ones) by the filter, and the usual summation. However, if we wish to employ the Lagrange polynomial formulation, we have

$$\mathcal{I}_N u(x) = \sum_{j=0}^{N} u(x_j) l_j(x),$$

where x_j is some chosen set of grid points and $l_j(x)$ are the associated Lagrange polynomials. For simplicity, we'll use the Gauss–Lobatto quadrature points associated with the ultraspherical polynomials, $P_n^{(\alpha)}(x)$, in which case the Lagrange polynomials are

$$l_j(x) = w_j \sum_{n=0}^{N} \frac{P_n^{(\alpha)}(x) P_n^{(\alpha)}(x_j)}{\tilde{\gamma}_n},$$

with the Gauss–Lobatto quadrature weights w_j.

The filtered approximation

$$u_N^\sigma(x) = \sum_{j=0}^{N} u(x_j) l_j^\sigma(x),$$

is obtained by filtering the Lagrange polynomials

$$l_j^\sigma(x) = w_j \sum_{n=0}^{N} \sigma\left(\frac{n}{N}\right) \frac{P_n^{(\alpha)}(x) P_n^{(\alpha)}(x_j)}{\tilde{\gamma}_n},$$

which we can not, in general, express in a simple form. However, we can express the action of the filter at the grid points by a filter matrix $\mathrm{F}_{ij} = l_j^\sigma(x_i)$, so that the filtered expansion is

$$u_N^\sigma(x_i) = \sum_{j=0}^{N} \mathrm{F}_{ij} u(x_j).$$

Note that F_{ij} is centro-symmetric as a consequence of the symmetry of the quadrature points. A similar approach can be taken if the Gauss quadrature points are chosen.

Likewise, we obtain matrix forms for the combination of filtering and differentiation,

$$\mathrm{D}_{ij}^{(q),\sigma} = w_j \sum_{n=0}^{N} \sigma\left(\frac{n}{N}\right) \frac{P_n^{(\alpha)}(x_j)}{\tilde{\gamma}_n} \left.\frac{d^q P_n^{(\alpha)}}{dx^q}\right|_{x_i},$$

so that the filtered and differentiated approximation is

$$\frac{d^q}{dx^q} u_N^\sigma(x_i) = \sum_{j=0}^{N} \mathrm{D}_{ij}^{(q),\sigma} u(x_j).$$

The filtered matrices, $\mathrm{D}_{ij}^{(q),\sigma}$, share the properties of the unfiltered versions, i.e., they are centro-antisymmetric when q is odd and centro-symmetric when q is even.

9.2.4 Approximation theory for filters

In the previous section we saw that filtering improves convergence away from the discontinuity. In this section we shall prove this result in a more rigorous sense. The theory is more complete for filtering of trigonometric expansions than for polynomial expansions, and we review it here.

We assume that we know the first $(2N+1)$ expansion coefficients, $\hat{u}_n$, of a piecewise analytic function, u, and that the function is known to have a discontinuity at $x = \xi$. The aim is to recover the value of u at any point in the interval $[0, 2\pi]$. For this purpose we introduce a filter $\sigma(|n|/N)$ with the hope that the modified approximation

$$u_N^\sigma(x) = \sum_{n \leq N} \sigma\left(\frac{|n|}{N}\right) \hat{u}_n e^{inx},$$

converges faster than the original series

$$u_N(x) = \sum_{n \leq N} \hat{u}_n e^{inx}.$$

The filter σ has the property that $\sigma(\eta) = 0$ for $\eta \geq 1$, so that the modified approximation can be written as

$$u_N^\sigma(x) = \sum_{|n|<\infty} \sigma\left(\frac{|n|}{N}\right) \hat{u}_n e^{inx}.$$

If we use the definition of $\hat{u}_n$ and rearrange the terms, the filtered approximation can be expressed in terms of the filter function

$$S(z) = \sum_{|n|<\infty} \sigma(\eta) e^{inz}. \tag{9.5}$$

and the function u,

$$u_N^\sigma(x) = \frac{1}{2\pi}\int_0^{2\pi} S(x-y)u(y)\,dy. \tag{9.6}$$

On our way to an understanding of the properties of filters, we need to introduce the concept of a filter function *family*, which will play an important role.

Definition 9.11 *A filter function $S(z)$ has an associated* family of functions, *$S_l(z)$, defined recursively,*

$$\begin{aligned} S_0(z) &= S(z) \\ S_l'(z) &= S_{l-1}(z) \\ \int_0^{2\pi} S_l(z)\,dz &= 0, \quad l \geq 1. \end{aligned}$$

The filter function defined in Equation (9.5) leads to several equivalent representations of the corresponding filter family, as stated in the following lemma.

Lemma 9.12 *Assume that $\sigma(\eta)$ is a filter of order p, as defined in Definition 9.6, with the associated filter function*

$$S(z) = S_0(z) = \sum_{|n|<\infty} \sigma(\eta)e^{inz},$$

where $\eta = |n|/N$. The corresponding filter family, $S_l(z)$, defined in Definition 9.11, has the following equivalent representations ($1 \leq l \leq p$);

(a)

$$S_l(z) = \frac{1}{N^l}\sum_{|n|<\infty} G_l(\eta)i^l e^{inz},$$

where

$$G_l(\eta) = \frac{\sigma(\eta)-1}{\eta^l}.$$

(b)

$$S_l(z) = \frac{1}{N^{l-1}}\sum_{|m|<\infty}\int_{-\infty}^{\infty} e^{iN(z+2\pi m)\eta}G_l(\eta)i^l\,dz.$$

(c)

$$l = 1:\ S_1(z) = z - \pi + \sum_{\substack{|n|<\infty\\ n\neq 0}} \sigma(\eta)(in)^{-1}e^{inx},$$

$$l \geq 2:\ S_l(z) = B_l(z) + \sum_{\substack{|n|<\infty\\ n\neq 0}} \sigma(\eta)(in)^{-l}e^{inx}.$$

The polynomials $B_l(z)$ are the Bernoulli polynomial of order l.

The filter family and its integrals will be important for the following theorem. This theorem shows the difference between the filtered approximation and the function $u(x)$ in terms of the filter function.

Theorem 9.13 *Let $u(x)$ be a piecewise $C^p[0, 2\pi]$ function with a single point of discontinuity at $x = \xi$. Then*

$$
\begin{aligned}
u_N^\sigma(x) - u(x) = &\frac{1}{2\pi} \sum_{l=0}^{p-1} S_{l+1}(c) \left(u^{(l)}(\xi^+) - u^{(l)}(\xi^-)\right) \\
&+ \frac{1}{2\pi} \int_0^{2\pi} S_p(x-y) u^{(p)}(y)\, dy
\end{aligned}
$$

where $c = x - \xi$ for $x > \xi$ and $c = 2\pi + x - \xi$ for $x < \xi$.

Proof: First, we note that by using expression (c) in Lemma 9.12 for $S_1(z)$ and the integral condition on $S_l(z)$ for $l > 1$,

$$
\begin{aligned}
&S_1(2\pi) - S_1(0) = 2\pi \\
&S_l(2\pi) - S_l(0) = 0, \quad l \geq 2.
\end{aligned}
\tag{9.7}
$$

Consider the case $x > \xi$. Integrating Equation (9.6) by parts p times, taking care not to integrate over the point of discontinuity, yields

$$
\begin{aligned}
2\pi u_N^\sigma(x) &= \int_0^{\xi^-} S(x-y)u(y)\,dy + \int_{\xi^+}^{2\pi} S(x-y)u(y)\,dy \\
&= \sum_{l=0}^{p-1} \left(S_{l+1}(2\pi) - S_{l+1}(0)\right) u^{(l)}(x) \\
&\quad + \sum_{l=0}^{p-1} S_{l+1}(x-\xi)\left(u^{(l)}(\xi^+) - u^{(l)}(\xi^-)\right) \\
&\quad + \int_0^{2\pi} S_p(x-y)u^{(p)}(y)\,dy \\
&= 2\pi u(x) + \sum_{l=0}^{p-1} S_{l+1}(x-\xi)\left(u^{(l)}(\xi^+) - u^{(l)}(\xi^-)\right) \\
&\quad + \int_0^{2\pi} S_p(x-y)u^{(p)}(y)\,dy,
\end{aligned}
$$

where the last reduction follows from Equation (9.7), and we used the fact that u as well as S_l are periodically extended.

Likewise, for $x < \xi$ we obtain the result

$$
\begin{aligned}
2\pi u_N^\sigma(x) &= \int_0^{\xi^-} S(x-y)u(y)\,dy + \int_{\xi^+}^{2\pi} S(x-y)u(y)\,dy \\
&= \sum_{l=0}^{p-1} (S_{l+1}(2\pi) - S_{l+1}(0))\, u^{(l)}(x) \\
&\quad + \sum_{l=0}^{p-1} S_{l+1}(2\pi + x - \xi)\left(u^{(l)}(\xi^+) - u^{(l)}(\xi^-)\right) \\
&\quad + \int_0^{2\pi} S_p(x-y)u^{(p)}(y)\,dy \\
&= 2\pi u(x) + \sum_{l=0}^{p-1} S_{l+1}(2\pi + x - \xi)\left(u^{(l)}(\xi^+) - u^{(l)}(\xi^-)\right) \\
&\quad + \int_0^{2\pi} S_p(x-y)u^{(p)}(y)\,dy.
\end{aligned}
$$

QED

This result provides a precise expression of the error between the unknown point value of $u(x)$ and its filtered and truncated approximation, $u_N^\sigma(x)$. Now we would like to estimate the two terms on the right hand side of the expression in Theorem 9.13, to get a clearer idea of how well the filtered series approximates the function.

If $u(x) \in C^{p-1}[0, 2\pi]$, then the first term of Theorem 9.13 vanishes and we are left with the last term. Thus, this last term is really the error term for smooth functions. We estimate this last term in the following lemma.

Lemma 9.14 *Let $S_l(x)$ be defined as in Definition 9.11, then*

$$
\frac{1}{2\pi}\left|\int_0^{2\pi} S_p(x-y)u^{(p)}(y)\,dy\right| \leq C\frac{\sqrt{N}}{N^p}\left(\int_0^{2\pi} \left|u^{(p)}\right|^2 dx\right)^{1/2},
$$

where C is independent of N and u.

Proof: The Cauchy–Schwarz inequality yields

$$
\begin{aligned}
&\frac{1}{2\pi}\left|\int_0^{2\pi} S_p(x-y)u^{(p)}(y)\,dy\right| \\
&\quad \leq \frac{1}{2\pi}\left(\int_0^{2\pi} S_p^2(x)\,dx\right)^{1/2}\left(\int_0^{2\pi} \left|u^{(p)}(x)\right|^2 dx\right)^{1/2}
\end{aligned}
$$

because S_p is periodic. To estimate the first term we express S_p using (a) of

Lemma 9.12

$$
\begin{aligned}
\frac{1}{2\pi}\int_0^{2\pi} S_p^2(x)\,dx &= \frac{1}{2\pi}\int_0^{2\pi} \sum_{|n|<\infty}\left(\frac{1}{N^p}\frac{\sigma(\eta)-1}{\eta^p} i^p e^{inx}\right)^2 \\
&= \sum_{|n|<\infty} \frac{1}{N^{2p}}\left(\frac{\sigma(\eta)-1}{\eta^p}\right)^2 \\
&= \frac{1}{N^{2p-1}}\sum_{|n|\le N}\frac{2}{2N}\left(\frac{\sigma(\eta)-1}{\eta^p}\right)^2 + \sum_{|n|>N}\frac{1}{n^{2p}} \\
&\le \frac{1}{N^{2p-1}}\int_{-1}^{1}\left(\frac{\sigma(\eta)-1}{\eta^p}\right)^2 d\eta + \frac{1}{N^{2p-1}} \\
&\le C\frac{N}{N^{2p}},
\end{aligned}
$$

where we have used the orthogonality of exponential functions and bounded the Riemann sum by its integral, which is convergent under condition (c) of Definition 9.6.

QED

If the function is not smooth, we must also consider the first term on the right-hand side of Theorem 9.13. The conditions given in Lemma 9.12 will be enough to show that this term diminishes away from the discontinuity.

Lemma 9.15 *Let $S_l(x)$ be defined by Definition 9.11 for $l \ge 1$, with*

$$
S_0(x) = \sum_{|n|\le N} \sigma(\eta) e^{inx},
$$

then

$$
|S_l(x)| \le C\frac{1}{N^{p-1}}\left(\frac{1}{|x|^{p-l}} + \frac{1}{|2\pi - x|^{p-l}}\right)\int_{-\infty}^{\infty}\left|G_l^{(p-l)}(\eta)\right| d\eta,
$$

where $G_l(\eta)$ is defined in Lemma 9.12.

Proof: Consider $S_l(x)$ (in part (b) of Lemma 9.12); since $G_l(x)$ is $p-1$ times differentiable and the filter is defined such that $\sigma^{(l)}(0) = \sigma^{(l)}(1) = 0$, we obtain by integrating by parts $p - l$ times,

$$
|S_l(x)| = \frac{1}{N^{l-1}}\left|\sum_{|m|<\infty}\int_{-\infty}^{\infty}\frac{e^{iN(x+2\pi m)\eta}}{N^{p-l}(x+2\pi m)^{p-l}} G_l^{(p-l)}(\eta)\,d\eta\right|.
$$

Since $x \in [0, 2\pi]$, the dominating terms are found for $m = 0$ and $m = -1$. Using the triangle inequality and taking these two contributions outside the integral we obtain the result.

QED

Finally, we are ready to estimate the error between the filtered approximation and the function itself.

Theorem 9.16 *Let $u(x)$ be a piecewise $C^p[0, 2\pi]$ function with only one point of discontinuity at $x = x_d$. Let $\sigma(\eta)$ be a filter function according to Definition 9.6. Let $d(x) = |x - x_d|$ be the distance between a point $x \in [0, 2\pi]$ and the discontinuity, and*

$$u_N^\sigma(x) = \sum_{n=-N}^{N} \sigma(\eta)\hat{u}_N e^{inx},$$

is the filtered truncated series approximation to $u(x)$.

Then the pointwise difference between $u(x)$ and $u_N^\sigma(x)$ at all $x \neq x_d$ is bounded by

$$\left|u(x) - u_N^\sigma(x)\right| \leq C_1 \frac{1}{N^{p-1}} \frac{1}{d(x)^{p-1}} K(u) + C_2 \frac{\sqrt{N}}{N^p} \left(\int_0^{2\pi} \left|u^{(p)}\right|^2 dx \right)^{1/2},$$

where

$$K(u) = \sum_{l=0}^{p-1} d(x)^l \left|u^{(l)}(x_d^+) - u^{(l)}(x_d^-)\right| \int_{-\infty}^{\infty} \left|G_l^{(p-l)}(\eta)\right| d\eta.$$

Proof: The second part of the estimate follows from Lemma 9.14. From Lemma 9.15 we know that the first part of the expression in Theorem 9.13 is bounded. Since $c = x - x_d$ for $x > x_d$ and $c = 2\pi + x - x_d$ for $x < x_d$ in Theorem 9.13 we have that

$$\frac{1}{|c|^{p-l}} + \frac{1}{|2\pi - c|^{p-l}} \leq \frac{2}{d(x)^{p-l}}.$$

Combining this with the estimate of Lemma 9.15 yields the result.

QED

This theorem proves that filtering raises the rate of convergence away from the discontinuity ($d(x) > 0$), as all terms can be bounded by $\mathcal{O}(N^{1-p})$ depending only on the regularity of the piecewise continuous function and the order of the filter. In the case where $u(x)$ is piecewise analytic, one can take the order of the filter $p \propto N$ to yield exponential accuracy away from the discontinuity. The extension to multiple points of discontinuity is straightforward.

9.3 The resolution of the Gibbs phenomenon

While filtering may remove the Gibbs Phenomenon away from the discontinuity, the order of accuracy decreases with proximity to the discontinuity.

However, the Gibbs phenomenon may be completely removed for all expansions commonly used in spectral methods, by re-expanding a slowly converging global expansion using a different set of basis functions.

We begin with an $L^2[-1, 1]$ function, $f(x)$, which is analytic on some subinterval $[a, b] \subset [-1, 1]$. Let $\{\Psi_k(x)\}$ be an orthonormal family under some inner product $(\cdot, \cdot)$. We focus on the Fourier basis or the Jacobi polynomial basis, since they are commonly used in spectral methods. Denote the finite continuous expansion in this basis by

$$f_N(x) = \sum_{k=0}^{N} (f, \Psi_k)\Psi_k(x) \tag{9.8}$$

and assume that

$$|(f, \Psi_k)| \leq C \tag{9.9}$$

for C independent of k, and that

$$\lim_{N\to\infty} |f(x) - f_N(x)| = 0 \tag{9.10}$$

almost everywhere in $x \in [-1, 1]$. This expansion may converge slowly in the presence of a discontinuity outside of the interval of analyticity $[a, b]$. This is due to the global nature of the expansion: since the coefficients are computed over the entire domain $[-1, 1]$, any discontinuity in that domain will contaminate the expansion even in the interval of analyticity $[a, b]$. However, under certain conditions, $f_N(x)$ retains sufficient information to enable us to recover a high order approximation to $f(x)$ in the interval $[a, b]$.

First, let's define the local variable $\xi \in [-1, 1]$ by the transformation

$$\xi = -1 + 2\left(\frac{x-a}{b-a}\right) \tag{9.11}$$

such that if $a \leq x \leq b$ then $-1 \leq \xi \leq 1$.

Now we re-expand $f_N(x)$ (an expansion in the basis Ψ_k), by an orthonormal family of functions $\{\Phi_k^\lambda\}$ defined in an interval of analyticity $[a, b]$,

$$f_N^\lambda(x) = \sum_{k=0}^{N} < f_N, \Phi_k^\lambda >_\lambda \Phi_k^\lambda(\xi(x)) \,. \tag{9.12}$$

For each fixed λ, the family $\{\Phi_k^\lambda\}$ is orthonormal under some inner product $< \cdot, \cdot >_\lambda$.

This new expansion converges exponentially fast in the interval $[a, b]$ if the family $\{\Phi_k^\lambda\}$ is **Gibbs Complementary**.

Definition 9.17 *The family $\{\Phi_k^\lambda(\xi)\}$ is **Gibbs complementary** to the family $\{\Psi_k(x)\}$ if the following three conditions are satisfied:*

(a) **Orthonormality**

$$< \Phi_k^\lambda(\xi), \Phi_l^\lambda(\xi) >_\lambda = \delta_{kl} \tag{9.13}$$

for any fixed λ.

(b) **Spectral convergence** *The expansion of a function $g(\xi)$ which is analytic in $-1 \le \xi \le 1$ (corresponding to $a \le x \le b$), in the basis $\Phi_k^\lambda(\xi)$, converges exponentially fast with λ, i.e.,*

$$\max_{-1\le\xi\le1} \left| g(\xi) - \sum_{k=0}^{\lambda} < g, \Phi_k^\lambda >_\lambda \Phi_k^\lambda(\xi) \right| \le e^{-q_1\lambda}, \quad q_1 > 0. \tag{9.14}$$

(c) **The Gibbs condition** *There exists a number $\beta < 1$ such that if $\lambda = \beta N$, then*

$$\left| < \Phi_l^\lambda(\xi), \Psi_k(x(\xi)) >_\lambda \right| \max_{-1\le\xi\le1} \left| \Phi_l^\lambda(\xi) \right| \le \left(\frac{\alpha N}{k} \right)^\lambda \tag{9.15}$$

$$k > N, \; l \le \lambda, \quad \alpha < 1. \tag{9.16}$$

This means that the projection of the high modes of the basis $\{\Psi_k\}$ (large k) on the low modes in Φ ($\Phi_l^\lambda(\xi)$ with small l) is exponentially small in the interval $-1 \le \xi \le 1$ for λ proportional to N.

In the following we will show that $f_N^\lambda(x)$, the re-expansion of $f_N(x)$ in a Gibbs complementary basis, approximates the original function $f(x)$ exponentially in N everywhere in the domain $[a, b]$.

Theorem 9.18 *Let $f(x) \in L^2[-1, 1]$ be analytic in $[a, b] \subset [-1, 1]$. Suppose that $\{\Psi_k(x)\}$ is an orthonormal family with the inner product $(\cdot, \cdot)$ that satisfies Equations (9.9) and (9.10), and $\{\Phi_k^\lambda(\xi)\}$ is a Gibbs complementary basis to the family $\{\Psi_k(x)\}$, with $\lambda = \beta N$. Furthermore, assume*

$$< f - f_N, \Phi_l^\lambda >_\lambda = \sum_{k=N+1}^{\infty} (f, \Psi_k) < \Phi_l^\lambda, \Psi_k >_\lambda . \tag{9.17}$$

Then

$$\max_{a\le x\le b} \left| f(x) - \sum_{l=0}^{\lambda} < f_N, \Phi_l^\lambda >_\lambda \Phi_l^\lambda(\xi(x)) \right| \le e^{-qN}, \quad q > 0. \tag{9.18}$$

Proof: The proof is divided into two parts. We first note that since $f(x)$ is analytic in $[a, b]$ then condition (b), Equation (9.14), means that the expansion in the new basis converges exponentially

$$\max_{a \le x \le b} \left| f(x) - \sum_{l=0}^{\lambda} < f, \Phi_l^\lambda >_\lambda \Phi_l^\lambda(\xi(x)) \right| \le e^{-q_2 \lambda}, \quad q_2 > 0. \qquad (9.19)$$

However, we do not have $< f, \Phi_l^\lambda >_\lambda$, but only $< f_N, \Phi_l^\lambda >_\lambda$. We therefore have to estimate

$$\max_{a \le x \le b} \left| \sum_{l=0}^{\lambda} < f - f_N, \Phi_l^\lambda >_\lambda \Phi_l^\lambda(\xi(x)) \right|.$$

Using Equations (9.9), (9.10) and (9.17),

$$\begin{aligned}
&\max_{-1 \le \xi \le 1} \left| \sum_{l=0}^{\lambda} < f - f_N, \Phi_l^\lambda >_\lambda \Phi_l^\lambda(\xi) \right| \\
&\quad \le \max_{-1 \le \xi \le 1} \sum_{l=0}^{\lambda} \sum_{k=N+1}^{\infty} \left| (f, \Psi_k) < \Phi_l^\lambda, \Psi_k >_\lambda \Phi_l^\lambda(\xi) \right| \\
&\quad \le C \sum_{l=0}^{\lambda} \sum_{k=N+1}^{\infty} \left(\frac{\alpha N}{k} \right)^\lambda \le e^{-qN}, \quad q > 0.
\end{aligned}$$

where the second inequality follows from Equation (9.9) and condition (c).

To conclude the proof we realize that

$$\begin{aligned}
&\left| f(x) - \sum_{l=0}^{\lambda} < f_N, \Phi_l^\lambda >_\lambda \Phi_l^\lambda(\xi(x)) \right| \\
&\quad \le \left| f(x) - \sum_{l=0}^{\lambda} < f, \Phi_l^\lambda >_\lambda \Phi_l^\lambda(\xi(x)) \right| + \left| \sum_{l=0}^{\lambda} < f - f_N, \Phi_l^\lambda >_\lambda \Phi_l^\lambda(\xi) \right|
\end{aligned}$$

QED

This Theorem implies that even if we have a slowly converging series

$$f_N(x) = \sum_{k=0}^{N} (f, \Psi_k) \Psi_k(x),$$

it is still possible to get a rapidly converging approximation to $f(x)$, if one can find another basis set that yields a rapidly converging series to f, as long as the projection of the high modes in the basis $\{\Psi\}$ on the low modes in the new basis is exponentially small. The basis $\Psi_k(x)$ does not provide a good approximation, but it contains meaningful information such that a better approximation can be constructed. The fact that the expansion in $\Psi_k(x)$ contains enough information is manifested by the fact that the projection of the high modes in $\Psi_k(x)$ on the low modes in Φ_l^λ is small, thus a finite expansion in $\Psi_k(x)$ neglects modes with low information contents.

This is the way that the Gibbs phenomenon can be resolved: if we can find a Gibbs complementary basis to the expansions that converge slowly and non-uniformly then we can remove the Gibbs phenomenon.

In the following examples we present a Gibbs complementary basis for the most commonly used spectral bases.

Example 9.19 The Fourier basis is

$$\Psi_k(x) = \frac{1}{\sqrt{2}} e^{ik\pi x}, \tag{9.20}$$

where k runs from $-\infty$ to ∞. A Gibbs complementary basis (a complementary basis need not be unique) is

$$\Phi_k^\lambda(\xi) = \frac{1}{\sqrt{\gamma^{(\lambda_k)}}} C_k^{(\lambda)}(\xi) \tag{9.21}$$

where $C_k^{(\lambda)}(\xi)$ is the Gegenbauer polynomial and $\gamma_k^{(\lambda)}$ is the normalization factor. The λ inner product is defined by

$$< f, g >_\lambda = \int_{-1}^{1} (1 - x^2)^{\lambda - \frac{1}{2}} f(\xi) g(\xi)\, d\xi \tag{9.22}$$

To show that this basis is Gibbs complementary, we must check the three conditions.

The orthogonality condition, Equation (9.13), follows from the definition of $C_k^{(\lambda)}(\xi)$ and $\gamma_k^{(\lambda)}$.

To prove **the spectral convergence condition,** Equation (9.14), we begin by assuming that $f(x)$ is an analytic function on $[-1, 1]$, and that there exist constants $\rho \geq 1$ and $C(\rho)$ such that, for every $k \geq 0$,

$$\max_{-1 \leq x \leq 1} \left| \frac{d^k f}{dx^k}(x) \right| \leq C(\rho) \frac{k!}{\rho^k}. \tag{9.23}$$

The spectral convergence condition is proved by estimating the approximation error of using a finite (first $m + 1$ terms) Gegenbauer expansion to approximate an analytic function $f(x)$ in $[-1, 1]$:

$$\max_{-1 \leq x \leq 1} \left| f(x) - \sum_{l=0}^{m} \hat{f}_l^\lambda C_l^{(\lambda)}(x) \right|. \tag{9.24}$$

What makes this different from the usual approximation results, is that we require λ, the exponent in the weight function $(1 - x^2)^{\lambda - \frac{1}{2}}$, to be growing linearly with m, the number of terms retained in the Gegenbauer expansion. Recall that the Gegenbauer coefficients, $\hat{f}_l^\lambda$, are

$$\hat{f}_l^\lambda = \frac{1}{\gamma_l^{(\lambda)}} \int_{-1}^{1} (1-x^2)^{\lambda-1/2} f(x) C_l^{(\lambda)}(x) dx.$$

We use the Rodrigues' formula

$$(1-x^2)^{\lambda-1/2} C_l^{(\lambda)}(x) = \frac{(-1)^l}{2^l l!} G(\lambda, l) \frac{d^l}{dx^l}\left[(1-x^2)^{l+\lambda-1/2}\right]$$

where

$$G(\lambda, l) = \frac{\Gamma\left(\lambda+\frac{1}{2}\right)\Gamma\left(l+2\lambda\right)}{\Gamma\left(2\lambda\right)\Gamma\left(l+\lambda+\frac{1}{2}\right)},$$

to obtain

$$\begin{aligned}
\hat{f}_l^\lambda &= \frac{(-1)^l}{\gamma_l^{(\lambda)} 2^l l!} G(\lambda, l) \int_{-1}^{1} \frac{d^l}{dx^l}\left[(1-x^2)^{l+\lambda-1/2}\right] f(x) dx \\
&= \frac{G(\lambda, l)}{\gamma_l^{(\lambda)} 2^l l!} \int_{-1}^{1} \frac{d^l f(x)}{dx^l} (1-x^2)^{l+\lambda-1/2} dx
\end{aligned}$$

by integrating by parts l times. We can now use the assumption in Equation (9.23) on $f(x)$ to get the bound

$$\left|\hat{f}_l^\lambda\right| \leq \frac{G(\lambda, l) C(\rho)}{\gamma_l^{(\lambda)} 2^l \rho^l !} \int_{-1}^{1} (1-x^2)^{l+\lambda-1/2} dx.$$

Noting that

$$\int_{-1}^{1} (1-x^2)^{l+\lambda-1/2} dx = \sqrt{\pi} \frac{\Gamma\left(l+\lambda+\frac{1}{2}\right)}{(l+\lambda)\Gamma(l+\lambda)},$$

we bound the Gegenbauer coefficients

$$\left|\hat{f}_l^\lambda\right| \leq \sqrt{\pi} \frac{G(\lambda, l) C(\rho) \Gamma\left(l+\lambda+\frac{1}{2}\right)}{\gamma_l^{(\lambda)} (l+\lambda)\Gamma(l+\lambda) 2^l \rho^l},$$

that is,

$$\left|\hat{f}_l^\lambda\right| \leq \sqrt{\pi} \frac{C(\rho)\Gamma\left(\lambda+\frac{1}{2}\right)\Gamma(l+2\lambda)}{\gamma_l^{(\lambda)} (2\rho)^l \Gamma(2\lambda)} \Gamma(l+\lambda+1),$$

Next, we use the bound on the Gegenbauer coefficients and the fact that the Gegenbauer polynomial is bounded

$$\left|C_l^{(\lambda)}(x)\right| \leq C_l^{(\lambda)}(1) \leq A \frac{l+\lambda}{\sqrt{\lambda}} \gamma_l^{(\lambda)}$$

for some constant, A, to obtain

$$|\hat{f}_l^\lambda|\, C_l^{(\lambda)}(1) \leq K \frac{C(\rho)\Gamma\left(\lambda+\frac{1}{2}\right)\Gamma(l+2\lambda)}{\sqrt{\lambda}(2\rho)^l\Gamma(2\lambda)\Gamma(l+\lambda)}.$$

Define

$$B_l = K \frac{C(\rho)\Gamma\left(\lambda+\frac{1}{2}\right)\Gamma(l+2\lambda)}{\sqrt{\lambda}(2\rho)^l\Gamma(2\lambda)\Gamma(l+\lambda)},$$

and so, since $\rho \geq 1$,

$$\frac{B_{l+1}}{B_l} = \frac{l+2\lambda}{2\rho(l+\lambda)} \leq \frac{l+2\lambda}{2(l+\lambda)}.$$

Finally, we look at the approximation error

$$\begin{aligned}\max_{-1\leq x\leq 1}\left|f(x) - \sum_{l=0}^{m} \hat{f}_l^\lambda C_l^{(\lambda)}(x)\right| &\leq \sum_{l=m+1}^{\infty} |\hat{f}_l^\lambda|\, C_l^{(\lambda)}(1) \\ &\leq \sum_{l=m+1}^{\infty} B_l \leq B_{m+1} \sum_{q=0}^{\infty}\left(\frac{m+2\lambda}{2(m+\lambda)}\right)^q.\end{aligned}$$

The last expression is a geometric series which converges for $m > 0$,

$$\sum_{q=0}^{\infty}\left(\frac{m+2\lambda}{2(m+\lambda)}\right)^q = \frac{1}{1-\frac{m+2\lambda}{2(m+\lambda)}} = \frac{2(m+\lambda)}{m},$$

so we have

$$\begin{aligned}&\max_{-1\leq x\leq 1}\left|f(x) - \sum_{l=0}^{m} \hat{f}_l^\lambda C_l^{(\lambda)}(x)\right| \\ &\quad\leq B_{m+1}\frac{2(m+\lambda)}{m} = K\frac{2(m+\lambda)}{m}\frac{C(\rho)\Gamma\left(\lambda+\frac{1}{2}\right)\Gamma(m+1+2\lambda)}{\sqrt{\lambda}m(2\rho)^{m+1}\Gamma(2\lambda)\Gamma(m+\lambda)}.\end{aligned} \tag{9.25}$$

If we let $\lambda = \gamma m$ for a constant $\gamma > 0$, this bound becomes

$$RE(\gamma m, m) \leq A q^m \tag{9.26}$$

where q is given by

$$q = \frac{(1+2\gamma)^{1+2\gamma}}{2^{1+2\gamma}\rho\gamma^\gamma(1+\gamma)^{1+\gamma}} \tag{9.27}$$

which always satisfies $q < 1$ since $\rho \geq 1$.

In fact, the assumption (9.23) is too strong, and a similar result can be obtained by assuming only that there exists a constant $0 \leq \rho < 1$ and an analytic extension of $f(x)$ onto the elliptic domain

$$D_\rho = \left\{ z : z = \frac{1}{2}\left(re^{i\theta} + \frac{1}{r}e^{-i\theta} \right), \quad 0 \le \theta \le 2\pi, \quad \rho \le r \le 1 \right\} \tag{9.28}$$

This assumption is satisfied by all analytic functions defined on $[-1, 1]$. In fact, the smallest ρ for which Equation (9.23) is satisfied is characterized by the following property

$$\overline{\lim_{m\to\infty}} \left(E_m(f) \right)^{\frac{1}{m}} = \rho \tag{9.29}$$

with $E_m(f)$ defined by

$$E_m(f) = \min_{P\in P_m} \max_{-1\le x\le 1} |f(x) - P(x)| \tag{9.30}$$

where P_m is the set of all polynomials of degree at most m.

If $f(x)$ is analytic not in the whole interval $[-1, 1]$ but in a sub-interval $[a, b] \subset [-1, 1]$, we use the change of variable $x = \epsilon\xi + \delta$ with $\epsilon = (b - a)/2$ and $\delta = (a + b)/2$, such that $g(\xi) = f(x(\xi))$ is defined and is analytic on $-1 \le \xi \le 1$. The previous estimates will then apply to $g(\xi)$, which can be translated back to the function $f(x)$ with a suitable factor involving ϵ.

Now we turn to the verification of the **Gibbs condition** (9.15), which becomes

$$\left| \int_{-1}^{1} (1 - x^2)^{\lambda - \frac{1}{2}} e^{ik\pi x(\xi)} C_l^{(\lambda)}(\xi)\, d\xi \right| \le \left(\frac{\alpha N}{k} \right)^{\lambda}, \tag{9.31}$$

for $k > N, l \le \lambda = \beta N, 0 < \alpha < 1$. Fortunately, there is an explicit formula for the integral in (9.31):

$$\int_{-1}^{1} (1 - x^2)^{\lambda - \frac{1}{2}} e^{ik\pi x(\xi)} C_l^{(\lambda)}(\xi)\, d\xi = \gamma_l^{(\lambda)} \Gamma(\lambda) \left(\frac{2}{\pi k\epsilon} \right)^{\lambda} i^l (l + \lambda) J_{l+\lambda}(\pi k\epsilon), \tag{9.32}$$

where $\epsilon = b - a$ and $J_\nu(x)$ is the Bessel function. Using the fact that $|J_\nu(x)| \le 1$ for all x and $\nu > 0$, and Stirling's formula, the Gibbs condition is satisfied when $\beta \ge 2\pi\epsilon/27$.

Example 9.20 The Legendre polynomial basis is

$$\Psi_k(x) = \frac{1}{\sqrt{\gamma_k^{(1/2)}}} C_k^{(1/2)}(x). \tag{9.33}$$

Once again,

$$\Phi_k^{\lambda}(\xi) = \frac{1}{\sqrt{\gamma_k^{(\lambda)}}} C_k^{(\lambda)}(\xi) \tag{9.34}$$

is a Gibbs complementary basis. To verify this, we note that the orthonormality and the spectral convergence conditions depend only on the basis Φ_k^λ, and were proved above. Unlike the first two conditions, the Gibbs condition depends on both Φ_k^λ and Ψ_k. To show that the Gibbs condition holds we need to prove that

$$\frac{1}{\sqrt{\gamma_k^{(1/2)}}}\left|\int_{-1}^{1}(1-x^2)^{\lambda-\frac{1}{2}}C_k^{(1/2)}(x(\xi))C_l^{(\lambda)}(\xi)\,d\xi\right| \leq \left(\frac{\alpha N}{k}\right)^{\lambda}, \tag{9.35}$$

for $k > N, l \leq \lambda = \beta N, 0 < \alpha < 1$. We use the explicit formula for the Fourier coefficients of $P_k(x)$:

$$P_k(x) = \sum_{l=-\infty}^{\infty} \hat{a}_l^k e^{il\pi x}, \quad \hat{a}_l^k = \frac{i^k}{\sqrt{2l}} J_{k+1/2}(-\pi l).$$

Thus, Equation (9.32) can be used to estimate the Gibbs condition (9.35).

Example 9.21 In general, for the Gegenbauer basis

$$\Psi_k(x) = \frac{1}{\sqrt{\gamma_k^{(\mu)}}} C_k^{(\mu)}(x) \tag{9.36}$$

the same

$$\Phi_k^\lambda(\xi) = \frac{1}{\sqrt{\gamma_k^{(\lambda)}}} C_k^{(\lambda)}(\xi) \tag{9.37}$$

is a Gibbs complementary basis. To verify this, we must show that

$$\frac{1}{\sqrt{\gamma_k^{(\mu)}}}\left|\int_{-1}^{1}(1-x^2)^{\lambda-\frac{1}{2}}C_k^{(\mu)}(x(\xi))C_l^{(\lambda)}(\xi)\,d\xi\right| \leq \left(\frac{\alpha N}{k}\right)^{\lambda},$$

for $k > N, l \leq \lambda = \beta N, 0 < \alpha < 1$. This time, the explicit formula in Equation (9.32) does not apply.

9.4 Linear equations with discontinuous solutions

We have demonstrated that the Gibbs phenomenon can be overcome, and spectral accuracy recovered in the case of a Fourier or polynomial expansion. We consider now a linear hyperbolic equation with discontinuous initial conditions, such as the one used in Example 9.1, and show that a re-expansion of the Fourier–Galerkin method solution in a Gibbs complementary basis is a spectrally accurate solution. This means that we can convect this profile forward in time with a spectral method, and when we have reached our final time, the numerical solution will still contain the necessary information to extract a high-order approximation to the exact solution.

Consider the hyperbolic problem

$$U_t = \mathcal{L}U$$

(where $\mathcal{L}$ is a semi-bounded operator) with initial condition

$$U(x, 0) = U_0(x).$$

The Galerkin approximation is

$$\left(\frac{\partial u}{\partial t} - \mathcal{L}u, \phi_k \right)_{L^2[0,2\pi]} = 0,$$

where ϕ is a trigonometric polynomial in all directions x_j. When $U_0(x)$ is continuous we have the usual error estimate

$$\|u - U\|_{L^2[0,2\pi]} \leq cN^{-s}\|U_0\|_{H_p^s[0,2\pi]}.$$

However, when $U_0(x)$ is discontinuous this error estimate is no longer valid, but we have a corresponding result in the weak sense:

Theorem 9.22 *For every smooth function* $\psi(x)$,

$$|(u - U, \psi)|_{L^2[0,2\pi]} \leq cN^{-s}\|\psi\|_{H_p^s[0,2\pi]}.$$

Proof: Given a smooth function $\psi(x, t)$, we can always find a function $V(x, t)$ which is the solution of

$$\frac{\partial V}{\partial t} = -\mathcal{L}^* V \quad \text{with} \quad V(x, 0) = V_0(x),$$

such that at time $t = \tau$, the two are equal $\psi(x, \tau) = V(x, \tau)$. This is always possible because we can solve the hyperbolic equation backwards and forwards, and so can pick initial conditions which give us the solution we want at time τ. Let $v(x, t)$ be the Galerkin approximation

$$\left(\frac{\partial v}{\partial t} + \mathcal{L}^* v, \phi_k \right)_{L^2[0,2\pi]} = 0,$$

where $\phi_k(x, t)$ is a trigonometric polynomial in x, and $v(x, 0) = \mathcal{P}_N V_0$. The solutions U and V satisfy Green's identity

$$(U(\tau), V(\tau))_{L^2[0,2\pi]} = (U_0, V_0)_{L^2[0,2\pi]},$$

as can be seen from

$$\begin{aligned}\frac{\partial}{\partial t}(U, V)_{L^2[0,2\pi]} &= (U_t, V)_{L^2[0,2\pi]} + (U, V_t)_{L^2[0,2\pi]} \\ &= (\mathcal{L}U, V)_{L^2[0,2\pi]} + (U, -\mathcal{L}^* V)_{L^2[0,2\pi]} \\ &= (\mathcal{L}U, V)_{L^2[0,2\pi]} - (U, \mathcal{L}^* V)_{L^2[0,2\pi]} = 0\end{aligned}$$

Similarly, by observing that both u and v are trigonometric polynomials in x, we have

$$\left(\frac{\partial u}{\partial t}, v\right)_{L^2[0,2\pi]} = (\mathcal{L}u, v)_{L^2[0,2\pi]}$$

and

$$\left(\frac{\partial v}{\partial t}, u\right)_{L^2[0,2\pi]} = -(\mathcal{L}^* v, u)_{L^2[0,2\pi]}$$

and so u and v also satisfy Green's identity

$$(u(\tau), v(\tau))_{L^2[0,2\pi]} = (u_0, v_0)_{L^2[0,2\pi]}.$$

Recall that the terms $(U_0 - u_0, v_0)_{L^2[0,2\pi]}$ and $(u_0, V_0 - v_0)_{L^2[0,2\pi]}$, are zero because $(U_0 - u_0)$ and $(V_0 - v_0)$ are orthogonal to the space of trigonometric polynomials in which u_0 and v_0 live. By using Green's identity and adding and subtracting these zero terms, we obtain

$$\begin{aligned}(U(\tau) - u(\tau), V(\tau)) &= (U(\tau), V(\tau)) - (u(\tau), V(\tau)) \\ &= (U_0, V_0) - (u(\tau), V(\tau) - v(\tau)) - (u_0, v_0) \\ &= (U_0, V_0) - (u_0, v_0) - (U_0 - u_0, v_0) + (u_0, V_0 - v_0) \\ &\quad - (u(\tau), V(\tau) - v(\tau)) \\ &= (U_0, V_0) - (U_0, v_0) + (u_0, V_0 - v_0) \\ &\quad - (u(\tau), V(\tau) - v(\tau)) \\ &= (U_0 + u_0, V_0 - v_0) - (u(\tau), V(\tau) - v(\tau)).\end{aligned}$$

Since $\mathcal{L}$ is semi-bounded, the Galerkin approximation is stable, and spectral accuracy follows:

$$\begin{aligned}(U(\tau) - u(\tau), V(\tau)) &\leq k_1 N^{-s} \|V_0\|_{H_p^s[0,2\pi]} + k_2 N^{-s} \|V(\cdot, \tau)\|_{H_p^s[0,2\pi]} \\ &\leq c N^{-s} \|V_0\|_{H_p^s[0,2\pi]}.\end{aligned}$$

QED

Although spectral methods for scalar nonlinear equations with discontinuous solutions are stable with appropriate filtering, the issue of recovering high-order accuracy is still open. Lax (1978) argued that high order information is retained

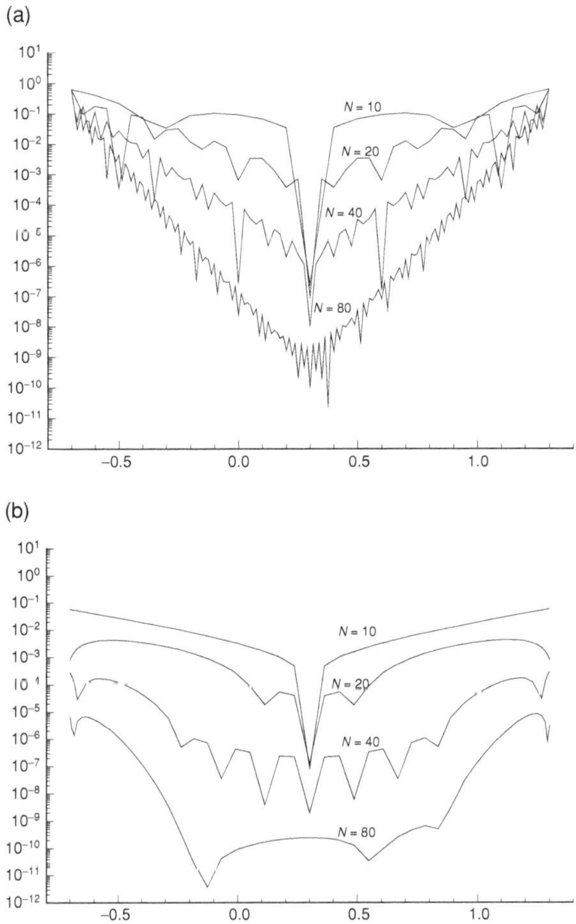

Figure 9.9 Pointwise errors in log scale, Burgers' equation. Fourier–Galerkin using $2N + 1$ modes with exponential solution filters of order r. $r = 4$ for $N = 10$; $r = 6$ for $N = 20$; $r = 8$ for $N = 40$ and $r = 12$ for $N = 80$. (a) Before postprocessing. (b) After postprocessing with parameters $\lambda = 2, m = 1$ for $N = 10$; $\lambda = 3, m = 3$ for $N = 20$; $\lambda = 12, m = 7$ for $N = 40$ and $\lambda = 62, m = 15$ for $N = 80$

in a convergent high resolution scheme. Numerical evidence has confirmed this for the following example.

Example 9.23 Consider the scalar Burgers' equation

$$\frac{\partial u}{\partial t} + \frac{\partial}{\partial x}\left(\frac{u^2}{2}\right) = 0, \quad -1 \leq x < 1$$
$$u(x, 0) = 0.3 + 0.7\sin(\pi x).$$

The solution develops a shock at $t = 1/(0.7\pi)$ and we compute the solution up to $t = 1$. The initial condition is chosen such that the shock is moving with time. We use a Fourier spectral method with filtering to stabilize the method, to approximate the solution to this equation. In Figure 9.9 (a) we plot the pointwise error $u(x,t) - v_N(x,t)$ before postprocessing, which is only first order accurate. Next, we apply the Gegenbauer postprocessor to $v_N(x,t)$. The pointwise errors of the postprocessed solution are plotted in Figure 9.9 (b). We can observe good accuracy everywhere, including at the discontinuity $x = \pm 1 + 0.3$.

These arguments demonstrate that it may be possible to recover high-order accuracy even in the case of nonlinear equations.

9.5 Further reading

The detailed analysis of filters was first pursued by Vandeven (1991) with further studies by Boyd (1998). Similar techniques was proposed by Majda et al (1984). Many applications of such techniques can be found in papers by Don and coworkers (1994, 2003). The use of one-sided filters was first discussed in Gottlieb and Tadmor (1985) and further extensive developed by Tanner and Tadmor (2002) . The theory for the polynomial case is incomplete with partial results by Ould Kaber (1994) and Hesthaven and Kirby (2006). The resolution of the Gibbs phenomenon by reprojection was developed in a series of papers by Gottlieb and Shu (1992, 1994, 1995, 1996, 1998) with a review paper by Gottlieb and Shu (1997). Further developments have been pursued by Gelb and co-workers (1997, 2000, 2003, 2004, 2006) and by Lurati (2006). The result on spectral convergence by reconstruction after propagation was obtained by Abarbanel et al (1986) and demonstrated for the nonlinear case by Shu and Wong (1995) and Gelb and Tadmor (2000).

10

Discrete stability and time integration

All spectral discretizations of the spatial operator in a PDE, result in a system of ODEs of the form

$$\frac{du_N}{dt} = \mathcal{L}_N(u_N(x,t), x, t).$$

This system is then solved numerically. In this chapter, we explore numerical methods which can be used to solve this system of ODEs.

Although numerical ODE solvers are successfully used for nonlinear problems, stability analysis will be performed, as usual, for the linear problem

$$\frac{du_N}{dt} = \mathcal{L}_N u_N(x,t), \tag{10.1}$$

where $u_N(x,t)$ is a vector of length $M(N+1)$ which contains the $(N+1)$ expansion coefficients (in the case of a Galerkin or a tau method) or the grid point values (in the case of a collocation method) for each of the M equations, and $\mathcal{L}_N$ is a $M(N+1) \times M(N+1)$ matrix. It is convenient to write the method in vector form

$$\frac{d\boldsymbol{u}}{dt} = \mathrm{L}\boldsymbol{u}(x,t),$$

which has the exact solution

$$\boldsymbol{u}(x,t) = e^{\mathrm{L}t}\boldsymbol{u}(x,0),$$

where $e^{\mathrm{L}t}$ is the matrix exponential. Methods based on approximating the matrix exponential by ultraspherical polynomials have been successfully used to approximate the solution $\boldsymbol{u}$; however, usually explicit or implicit finite difference schemes are used for this purpose. In either case, the exponential e^z is essentially approximated either as a finite Taylor series

$$e^{\mathrm{L}\Delta t} \simeq \mathrm{K}(\mathrm{L}, \Delta t) = \sum_{i=0}^{m} \frac{(\mathrm{L}\Delta t)^i}{i!},$$

or through a Padé approximation

$$e^{\mathrm{L}\Delta t} \simeq \mathrm{K}(\mathrm{L}, \Delta t) = \frac{\sum_{i=0}^{m} a_i (\mathrm{L}\Delta t)^i}{\sum_{i=0}^{n} b_i (\mathrm{L}\Delta t)^i},$$

with the expansion coefficients a_i and b_i determined by the requirement that the approximation agrees with the Taylor expansion. For a sufficiently small Δt, such an approximation is accurate enough.

The combination of spectral methods in space and finite difference methods in time is generally justified, because the time step is typically sufficiently restricted in size so that the temporal error is comparable to the spatial error.

10.1 Stability of linear operators

The finite difference approximation is

$$\boldsymbol{u}(x, t + \Delta t) = \boldsymbol{u}^{n+1} = \mathrm{K}(\mathrm{L}, \Delta t)\boldsymbol{u}^n,$$

where $t = n\Delta t$, with step size Δt, and the matrix $\mathrm{K}(\mathrm{L}, \Delta t)$ is the approximation to the matrix exponential $e^{\mathrm{L}\Delta t}$. The fully discrete scheme is strongly stable provided that

$$\| [\mathrm{K}(\mathrm{L}, \Delta t)]^n \|_{L^2_w} \leq K(\Delta t),$$

where $\| \cdot \|_{L^2_w}$ is the matrix norm. A sufficient condition for strong stability is that

$$\|\mathrm{K}(\mathrm{L}, \Delta t)\|_{L^2_w} \leq 1 + \kappa\Delta t,$$

for some bounded κ and all sufficiently small values of Δt. In practice, a better, though more restrictive, condition is the above with $\kappa = 0$. In the following, we show how the discrete stability of a scheme is analyzed.

10.1.1 Eigenvalue analysis

The eigenvalues of $\mathrm{K}(\mathrm{L}, \Delta t)$ provide a necessary, but generally not sufficient, condition for stability. In the special case where K is a normal matrix, i.e., $\mathrm{K}^T\mathrm{K} = \mathrm{K}\mathrm{K}^T$, strong stability is ensured in L^2 through the von Neumann stability condition

$$\max |\lambda_K| \leq 1 + \kappa\Delta t,$$

where λ_K are the eigenvalues of $\mathrm{K}(\mathrm{L}, \Delta t)$. In the following we shall discuss the eigenvalue spectrum of various discrete operators obtained using Fourier or polynomial methods.

Fourier methods The eigenvalue spectra of the Fourier–Galerkin approximation of the first and second derivatives are

$$\{-iN/2, \ldots, -i, 0, i, \ldots, N/2\},$$

and

$$\{-N^2/4, \ldots, -1, 0, -1, \ldots, -N^2/4\},$$

respectively. The maximum eigenvalue for the m-order differentiation is

$$\max \left|\lambda^{(m)}\right| = \left(\frac{N}{2}\right)^m.$$

In the Fourier–collocation method the last mode is aliased to zero, and therefore the eigenvalues are $\{(in)^m e^{inx} : \quad n = -N/2 + 1, \ldots, N/2 - 1\}$. In practice the Galerkin and the collocation methods behave in the same way and it is safe to assume that the eigenvalue spectra for the different methods are identical for all practical purposes. Since the Fourier method involves normal differentiation matrices, the von Neumann stability condition is sufficient to ensure stability. For linear problems all we need to do is bound $K(\lambda_k, \Delta t)$ for all λ_k to ensure full discrete stability.

Polynomial methods When using polynomial methods for the approximation of the spatial operators, we cannot obtain the eigenvalues analytically. Furthermore, since the matrices are not normal, the eigenvalues provide only a necessary, but not sufficient, condition for stability.

Consider the eigenvalue spectrum of the diffusive operator

$$\mathcal{L}u = \frac{d^2u}{dx^2}.$$

There is little quantitative difference between the Legendre and Chebyshev methods, or between the tau and collocation approaches. The eigenvalue spectrum is in all cases strictly negative, real and distinct, and bounded by

$$-c_1 N^4 \leq \lambda \leq c_2 < 0,$$

for some constants c_1 and c_2. In all these cases, the maximum eigenvalue scales as N^4 asymptotically.

However, the eigenvalue spectrum of the advective operator

$$\mathcal{L}u = \frac{du}{dx}$$

varies depending on the choice of polynomial and the use of the Galerkin, collocation, or tau approaches. The Chebyshev Gauss–Lobatto collocation

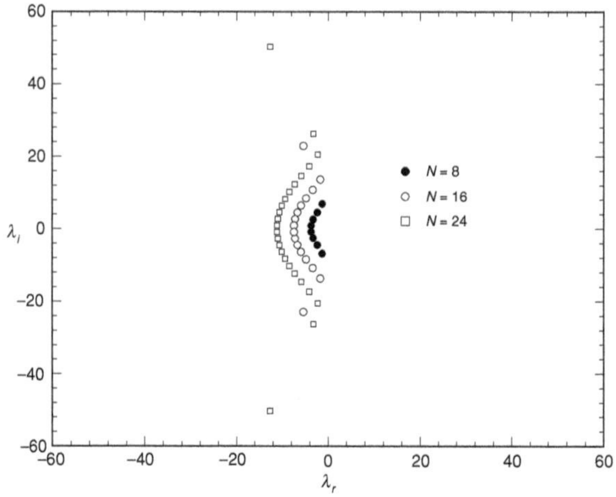

Figure 10.1 Eigenvalue spectra of a Chebyshev Gauss–Lobatto advection operator for increasing resolution, N.

approximation to the advective operator with homogeneous Dirichlet boundary conditions is illustrated in Figure 10.1. Observe that the real parts are strictly negative while the maximum eigenvalue scales as

$$\max\left|\lambda^{(1)}\right| = \mathcal{O}(N^2).$$

In contrast to Fourier methods, here the eigenvalues grow at a rate proportional to N^2 rather than linearly in N. Numerical studies of the spectrum of the Chebyshev tau approximation of the advection operator show that the maximum eigenvalue is slightly smaller than that obtained in the collocation approximation, but overall the behavior is qualitatively the same.

The eigenvalue spectrum for the Legendre Gauss–Lobatto differentiation matrix is shown in Figure 10.2. Note that all real parts of the spectrum are strictly negative (as a result of the stability of the method), albeit with a very small real part for the extreme eigenvalues. Although the spectrum is qualitatively different from that of the Chebyshev Gauss–Lobatto, the maximum eigenvalue also scales as N^2.

The spectrum for the Legendre tau approximation of the advective operator is quite different. Indeed, it is possibly to show analytically that asymptotically the maximum eigenvalue scales like

$$\max\left|\lambda^{(1)}\right| = \mathcal{O}(N).$$

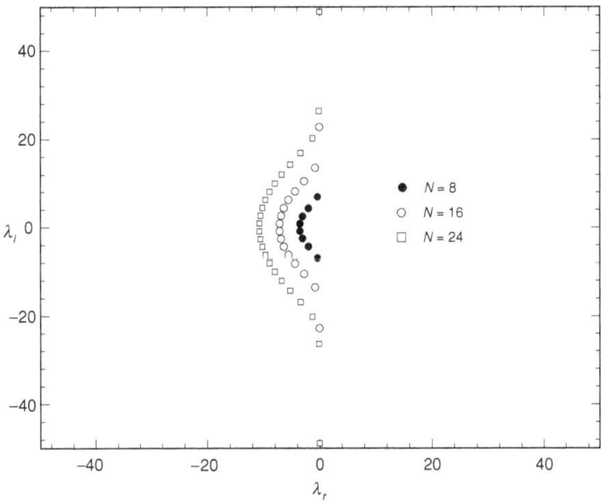

Figure 10.2 Eigenvalue spectra of a Legendre Gauss–Lobatto advection operator for increasing resolution, N.

However, since the matrix is not normal, the eigenvalue analysis does not give us sufficient conditions for stability. Indeed, in the next section we will show that in this case, too, the stability condition is that Δt must scale like $1/N^2$. This is a good example of the shortcomings of eigenvalue analysis for non-normal operators.

10.1.2 Fully discrete analysis

A more relevant analysis for the stability condition of the advective operator is performed on the fully discrete method. Consider the forward Euler scheme for the pseudospectral Legendre method based on the inner collocation points. Let $u = \boldsymbol{u}^{n+1}$ and $v = \boldsymbol{u}^n$, so that Euler's method is

$$u = v + \Delta t v_x - \Delta t v_x(1)\frac{P'_N(x)}{P'_N(1)},$$

where we collocate at the zeroes of $P'_N(x)$ and satisfy the boundary conditions $u(1) = v(1) = 0$. Thus,

$$\begin{aligned} u^2 &= v^2 + 2\Delta t v v_x - 2(\Delta t) v v_x(1)\frac{P'_N(x)}{P'_N(1)} \\ &\quad + (\Delta t)^2 v_x^2 - 2(\Delta t)^2 v_x v_x(1)\frac{P'_N(x)}{P'_N(1)} + (\Delta t)^2 v_x^2(1)\left(\frac{P'_N(x)}{P'_N(1)}\right)^2. \end{aligned}$$

We multiply by $(1+x_j)w_j$ and sum over all j, and use the fact that $P'_N(x_j) = 0$ for the inner points, to obtain

$$\begin{aligned}\sum_{j=0}^{N}(1+x_j)u^2(x_j)w_j &= \sum_{j=0}^{N}(1+x_j)v^2(x_j)w_j \\ &\quad + 2\Delta t\sum_{j=0}^{N}(1+x_j)v(x_j)v_x(x_j)w_j \\ &\quad + (\Delta t)^2\sum_{j=0}^{N}(1+x_j)v_x^2(x_j)w_j - 2(\Delta t)^2 v_x^2(1)w_0,\end{aligned}$$

where w_j are the Gauss quadrature weights. For stability, we need

$$\sum_{j=0}^{N}(1+x_j)u^2(x_j)w_j \leq \sum_{j=0}^{N}(1+x_j)v^2(x_j)w_j,$$

which means that

$$2\Delta t\sum_{j=0}^{N}(1+x_j)v(x_j)v_x(x_j)w_j + (\Delta t)^2\sum_{j=0}^{N}(1+x_j)v_x^2 w_j - 2(\Delta t)^2 v_x^2(1)w_0 \leq 0.$$

Note that using the exactness of the quadrature and integration by parts,

$$2\Delta t\sum_{j=0}^{N}(1+x_j)vv_x w_j = 2\int_{-1}^{1}(1+x)vv_x dx = -\int_{-1}^{1}v^2 dx$$

so that for stability we need

$$\Delta t\left(\int_{-1}^{1}(1+x)v_x^2 dx - 2v_x^2(1)w_0\right) \leq \int_{-1}^{1}v^2 dx.$$

This results in the condition

$$\Delta t \leq \frac{\int_{-1}^{1}v^2 dx}{\int_{-1}^{1}(1+x)v_x^2 dx} \approx \mathcal{O}\left(\frac{1}{N^2}\right).$$

This energy method analysis allows us to obtain an accurate stability condition for the fully discrete method.

10.2 Standard time integration schemes

The choice of time integration method is influenced by several factors such as desired accuracy, available memory and computer speed. In the following we shall briefly discuss the most commonly used numerical methods for integrating time-dependent ODEs resulting from a method of lines formulation where the spatial operators are approximated using spectral methods.

10.2.1 Multi-step schemes

The general multi-step scheme is

$$\sum_{i=0}^{p} \alpha_i \boldsymbol{u}^{n-i} = \Delta t \sum_{i=0}^{p} \beta_i \mathrm{L}^{n-i},$$

where p refers to the order of the scheme. If $\beta_0 = 0$ we can obtain the solution at $t = n\Delta t$ from knowledge of the solution at previous time-steps only, i.e., the scheme is explicit. Multi-step schemes require that solutions at one or more previous time-steps are retained, thus making such schemes memory intensive, in particular when solving large multi-dimensional problems. However, only one evaluation of L^n is required to advance one time-step, thereby reducing the computational workload. The choice of memory storage over computational speed is problem dependent and it is hard to give general guidelines.

Initially, we may not have the solution at the required number of time-steps backward in time. Indeed, we typically have the solution at only one time-step. Thus, it is necessary to start out with a few iterations of a one-step method of appropriate order and use these approximations as starting values. Since one is only taking very few steps with this initial method, the question of stability may be neglected and even unconditionally unstable schemes can be used for initializing multi-step schemes. However, accuracy is still important and we must begin with a start-up method of the same order p as the multi-step method we wish to use. If the old time-steps only enter in L and so are multiplied with a Δt, a start-up method of order $(p - 1)$ is sufficient.

10.2.2 Runge–Kutta schemes

A popular and highly efficient alternative to the multi-step schemes is the family of Runge–Kutta methods. An s-stage explicit Runge–Kutta scheme is

$$\begin{aligned}
\boldsymbol{k}_1 &= \mathrm{L}(\boldsymbol{u}^n, n\Delta t) = \mathrm{L}^n \\
\boldsymbol{k}_i &= \mathrm{L}\left(\boldsymbol{u}^n + \Delta t \sum_{j=1}^{i-1} a_{ij}\boldsymbol{k}_i, (n + c_i)\Delta t\right) \\
\boldsymbol{u}^{n+1} &= \boldsymbol{u}^n + \Delta t \sum_{i=1}^{s} b_i \boldsymbol{k}_i,
\end{aligned}$$

where the choice of the constants a_{ij}, c_i and b_i determines the accuracy and efficiency of the overall scheme.

For linear operators L, the s-stage Runge–Kutta schemes are simply the Taylor expansion of the matrix exponential to order s. The Runge–Kutta schemes require more evaluations of L than the multi-step schemes require to advance a time-step. However, unlike the multi-step schemes, the Runge–Kutta methods require no information from previous time-steps.

Standard schemes A popular second-order scheme is given by $c_2 = a_{21} = 1/2$ and $b_2 = 1$ and the remaining coefficients equal to zero,

$$\begin{aligned} \boldsymbol{k}_1 &= \mathrm{L}(\boldsymbol{u}^n, n\Delta t) = \mathrm{L}^n \\ \boldsymbol{k}_2 &= \mathrm{L}\left(\boldsymbol{u}^n + \tfrac{1}{2}\Delta t \boldsymbol{k}_1, \left(n + \tfrac{1}{2}\right)\Delta t\right) \\ \boldsymbol{u}^{n+1} &= \boldsymbol{u}^n + \Delta t \boldsymbol{k}_2. \end{aligned}$$

This scheme is known as the midpoint method. Alternatively, we can use

$$\begin{aligned} \boldsymbol{k}_1 &= \mathrm{L}(\boldsymbol{u}^n, n\Delta t) = \mathrm{L}^n \\ \boldsymbol{k}_2 &= \mathrm{L}\left(\boldsymbol{u}^n + \Delta t \boldsymbol{k}_1, (n + 1)\Delta t\right) \\ \boldsymbol{u}^{n+1} &= \boldsymbol{u}^n + \frac{\Delta t}{2}(\boldsymbol{k}_1 + \boldsymbol{k}_2). \end{aligned}$$

A popular third-order three-stage scheme is

$$\begin{aligned} \boldsymbol{k}_1 &= \mathrm{L}(\boldsymbol{u}^n, n\Delta t) = \mathrm{L}^n \\ \boldsymbol{k}_2 &= \mathrm{L}\left(\boldsymbol{u}^n + \tfrac{1}{3}\Delta t \boldsymbol{k}_1, \left(n + \tfrac{1}{3}\right)\Delta t\right) \\ \boldsymbol{k}_3 &= \mathrm{L}\left(\boldsymbol{u}^n + \tfrac{2}{3}\Delta t \boldsymbol{k}_2, \left(n + \tfrac{2}{3}\right)\Delta t\right) \\ \boldsymbol{u}^{n+1} &= \boldsymbol{u}^n + \frac{\Delta t}{4}(\boldsymbol{k}_1 + 3\boldsymbol{k}_3), \end{aligned}$$

which requires only two levels of storage.

The classical fourth-order accurate, four-stage scheme is

$$\begin{aligned} \boldsymbol{k}_1 &= \mathrm{L}(\boldsymbol{u}^n, n\Delta t) = \mathrm{L}^n \\ \boldsymbol{k}_2 &= \mathrm{L}\left(\boldsymbol{u}^n + \tfrac{1}{2}\Delta t \boldsymbol{k}_1, \left(n + \tfrac{1}{2}\right)\Delta t\right) \\ \boldsymbol{k}_3 &= \mathrm{L}\left(\boldsymbol{u}^n + \tfrac{1}{2}\Delta t \boldsymbol{k}_2, \left(n + \tfrac{1}{2}\right)\Delta t\right) \\ \boldsymbol{k}_4 &= \mathrm{L}\left(\boldsymbol{u}^n + \Delta t \boldsymbol{k}_3, (n + 1)\Delta t\right) \\ \boldsymbol{u}^{n+1} &= \boldsymbol{u}^n + \frac{\Delta t}{6}(\boldsymbol{k}_1 + 2\boldsymbol{k}_2 + 2\boldsymbol{k}_3 + \boldsymbol{k}_4), \end{aligned}$$

requiring four storage levels.

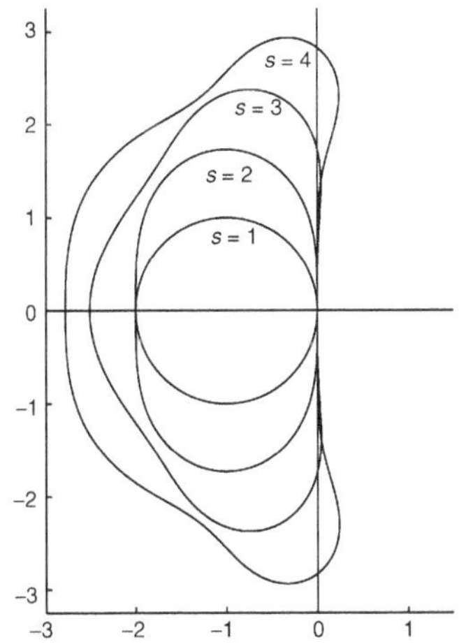

Figure 10.3 Stability regions for Runge–Kutta methods.

To study the linear stability of these schemes we consider the linear scalar equation

$$\frac{du}{dt} = \lambda u,$$

for which the general s-stage scheme can be expressed as a truncated Taylor expansion of the exponential functions

$$u^{n+1} = \sum_{i=1}^{s} \frac{(\lambda \Delta t)^i}{i!} u^n.$$

Thus, the scheme is stable for a region of $\lambda \Delta t$ for which

$$\left| \sum_{i=1}^{s} \frac{(\lambda \Delta t)^i}{i!} \right| \leq 1.$$

In Figure 10.3 we display the linear stability regions for the three Runge–Kutta methods. Note that the stability regions expand with increasing number of stages. However, increasing the order of the scheme also increases the amount of memory and computation required for each step.

Observe that the second-order Runge–Kutta scheme is only marginally stable at the imaginary axis, and is thus ill suited for Fourier and Legendre based schemes for approximating advective operators.

Low-storage methods Runge–Kutta methods require storing the solution at several levels for higher orders (e.g., four for the classical fourth-order Runge–Kutta method). This may lead to excessive memory requirements when dealing with multi-dimensional problems. However, by defining the constants of the Runge–Kutta method in a clever way it is possible to arrive at methods that require only two storage levels. This saving in memory requires the additional computational expense of performing one extra evaluation of L. The introduction of the additional step introduces extra degrees of freedom in the design of the scheme such that the resulting schemes also have a larger stability region, making the work per time unit about the same at the classical methods, but with lower memory requirements.

The **s-stage low-storage method** is

$$
\begin{aligned}
\boldsymbol{u}_0 &= \boldsymbol{u}^n \\
\forall j \in [1, \ldots, s] : \quad & \begin{cases} \boldsymbol{k}_j = a_j \boldsymbol{k}_{j-1} + \Delta t \mathrm{L}\big(\boldsymbol{u}_j, (n + c_j)\Delta t\big) \\ \boldsymbol{u}_j = \boldsymbol{u}_{j-1} + b_j \boldsymbol{k}_j \end{cases} \\
\boldsymbol{u}^{n+1} &= \boldsymbol{u}_s,
\end{aligned}
$$

where the constants a_j, b_j and c_j are chosen to yield the desired order, $s - 1$, of the scheme. For the scheme to be self-starting we require that $a_1 = 0$. We need only two storage levels containing $\boldsymbol{k}_j$ and $\boldsymbol{u}_j$ to advance the solution.

A four-stage third-order Runge–Kutta scheme is obtained using the constants

$$
\begin{array}{llllll}
a_1 = 0 & b_1 = \frac{1}{3} & c_1 = 0 & a_2 = -\frac{11}{15} & b_2 = \frac{5}{6} & c_2 = \frac{1}{3} \\
a_3 = -\frac{5}{3} & b_3 = \frac{3}{5} & c_3 = \frac{5}{9} & a_4 = -1 & b_4 = \frac{1}{4} & c_4 = \frac{8}{9}.
\end{array}
$$

The constants for a five-stage fourth-order Runge–Kutta scheme are

$$
\begin{array}{lll}
a_1 = 0 & b_1 = \frac{1432997174477}{9575080441755} & c_1 = 0 \\
a_2 = -\frac{567301805773}{1357537059087} & b_2 = \frac{5161836677717}{13612068292357} & c_2 = \frac{1432997174477}{9575080441755} \\
a_3 = -\frac{2404267990393}{2016746695238} & b_3 = \frac{1720146321549}{2090206949498} & c_3 = \frac{2526269341429}{6820363962896} \\
a_4 = -\frac{3550918686646}{2091501179385} & b_4 = \frac{3134564353537}{4481467310338} & c_4 = \frac{2006345519317}{3224310063776} \\
a_5 = -\frac{1275806237668}{842570457699} & b_5 = \frac{2277821191437}{14882151754819} & c_5 = \frac{2802321613138}{2924317926251}.
\end{array}
$$

These constants are accurate up to 26 digits, which is sufficient for most implementations.

The asymptotic stability regions for these low storage methods are only slightly larger than those of the classical methods. In particular, both low storage schemes contain the imaginary axis as part of the stability region.

Remark When using multi-stage methods such as the Runge–Kutta methods, we need to decide where to impose the boundary conditions. For spectral methods, boundary conditions can be imposed at each stage or only at the final stage. For hyperbolic equations, it has been our experience that the latter approach results in a slightly larger CFL condition. Even using the penalty method approach, one can add the penalty term at each intermediate stage or only at the final stage.

10.3 Strong stability preserving methods

As discussed above, eigenvalue analysis is not generally sufficient for stability. However, the energy method analysis we performed for the Legendre–collocation method with Euler's method has the drawback that it must be performed for each time discretization we wish to use. It would be convenient to have a way of extending this analysis from Euler's method to higher order methods. Furthermore, if we know of any nonlinear stability properties of the spatial discretization coupled with forward Euler, we would like to be able to extend these to higher order time discretizations as well. Strong stability preserving (SSP) Runge–Kutta and multi-step methods preserve any nonlinear stability property satisfied by forward Euler, allowing us to extend these stability properties to higher order methods.

10.3.1 SSP theory

In this section we review the SSP theory for explicit Runge–Kutta methods which approximate the solution of the ODE

$$\boldsymbol{u}_t = \mathcal{L}(\boldsymbol{u}) \tag{10.2}$$

which arises from the discretization of the spatial derivative in a PDE. The spatial discretization $\mathcal{L}(\boldsymbol{u})$ is chosen so that

$$\boldsymbol{u}^{n+1} = \boldsymbol{u}^n + \Delta t \mathcal{L}(u^n), \tag{10.3}$$

satisfies the strong stability requirement $||\boldsymbol{u}^{n+1}|| \leq ||\boldsymbol{u}^n||$ in some norm $||\cdot||$, under the CFL condition

$$\Delta t \leq \Delta t_{FE}. \tag{10.4}$$

If we write the m-stage Runge–Kutta method in the form

$$\begin{aligned} \boldsymbol{u}^{(0)} &= \boldsymbol{u}^n, \\ \boldsymbol{u}^{(i)} &= \sum_{k=0}^{i-1} \left(\alpha_{i,k}\boldsymbol{u}^{(k)} + \Delta t \beta_{i,k}\mathcal{L}(\boldsymbol{u}^{(k)})\right), \quad \alpha_{i,k} \geq 0, \quad i = 1, \ldots, s \\ \boldsymbol{u}^{n+1} &= \boldsymbol{u}^{(s)}, \end{aligned} \tag{10.5}$$

we observe that if $\alpha_{i,k} \geq 0$ and $\beta_{i,k} \geq 0$, all the intermediate stages $\boldsymbol{u}^{(i)}$ are simply convex combinations of forward Euler operators. If the forward Euler method combined with the spatial discretization $\mathcal{L}$ in (10.3) is strongly stable

$$\|\boldsymbol{u}^n + \Delta t \mathcal{L}(\boldsymbol{u}^n)\| \leq \|\boldsymbol{u}^n\|, \quad \text{for } \Delta t \leq \Delta t_{FE},$$

then the intermediate stages can be bounded

$$\begin{aligned} \left\|\boldsymbol{u}^{(i)}\right\| &= \left\| \sum_{k=0}^{i-1} \alpha_{i,k}\boldsymbol{u}^{(k)} + \Delta t \frac{\beta_{i,k}}{\alpha_{i,k}} \mathcal{L}(\boldsymbol{u}^{(k)}) \right\| \\ &\leq \sum_{k=0}^{i-1} \alpha_{i,k} \left\| \boldsymbol{u}^{(k)} + \Delta t \frac{\beta_{i,k}}{\alpha_{i,k}} \mathcal{L}(\boldsymbol{u}^{(k)}) \right\| \\ &\leq \sum_{k=0}^{i-1} \alpha_{i,k} \left\|\boldsymbol{u}^{(0)}\right\| \quad \text{for } \Delta t \tfrac{\beta_{i,k}}{\alpha_{i,k}} \leq \Delta t_{FE}, \end{aligned}$$

and, since consistency requires that $\sum_{k=0}^{i-1} \alpha_{i,k} = 1$, the method is strongly stable

$$\|\boldsymbol{u}^{n+1}\| \leq \|\boldsymbol{u}^n\|$$

under the CFL restriction

$$\Delta t \leq c \Delta t_{FE}, \quad c = \min_{i,k} \frac{\alpha_{i,k}}{\beta_{i,k}}. \tag{10.6}$$

In fact, any norm, semi-norm or convex function property satisfied by forward Euler will be preserved by the SSP Runge–Kutta method, under a suitable time-step restriction. The research in the field of SSP methods centers around the search for high order SSP methods where the CFL coefficient c in the time-step restriction (10.6) is as large as possible.

10.3.2 SSP methods for linear operators

In this section, we review the optimal SSP Runge–Kutta methods for the case where L is a linear constant coefficient operator. We present the major classes of optimal methods. In the following, (s,p) denotes an s-stage pth-order method:

SSPRK linear (s,s) The class of m stage schemes given by:

$$\begin{aligned} u^{(i)} &= u^{(i-1)} + \Delta t L u^{(i-1)}, \quad i = 1, \ldots, s-1 \\ u^{(s)} &= \sum_{k=0}^{m-2} \alpha_{m,k} u^{(k)} + \alpha_{s,s-1}\left(u^{(s-1)} + \Delta t L u^{(s-1)}\right), \end{aligned}$$

where $\alpha_{1,0} = 1$ and

$$\begin{aligned} \alpha_{s,k} &= \frac{1}{k}\alpha_{s-1,k-1}, \qquad k - 1, \ldots, s-2 \\ \alpha_{s,s-1} &= \frac{1}{s!}, \qquad \alpha_{s,0} = 1 - \sum_{s=1}^{s-1} \alpha_{s,k} \end{aligned}$$

is an m-order linear Runge–Kutta method which is SSP with $c = 1$, which is optimal among all s-stage, $p = s$-order SSPRK methods.

Although the stability limit is the same as forward Euler, the computational cost is m times that. Thus, we find it useful to define the effective CFL as $c_{eff} = c/l$, where l is the number of computations of L required per time step. In the case of SSPRK linear (s,s), the effective CFL is $c_{eff} = 1/s$.

If we increase the number of stages, s, without increasing the order, p, we obtain SSP Runge–Kutta methods with higher CFL coefficients. Although the additional stages increase the computational cost, this is usually more than offset by the larger stepsize that may be taken.

SSPRK linear (m,1) The m-stage, first-order SSP Runge–Kutta method given by

$$\begin{aligned} u^{(0)} &= u^n \\ u^{(i)} &= \left(1 + \frac{\Delta t}{s} L\right) u^{(i-1)} \quad i - 1, \ldots, s \\ u^{n+1} &= u^{(s)}, \end{aligned}$$

has CFL coefficient $c = s$, which is optimal among the class of m-stage, order $p = 1$ methods with non-negative coefficients. The effective CFL is $c_{eff} = 1$.

SSPRK linear (m,2) The m-stage, second-order SSP methods:

$$\begin{aligned} u^{(0)} &= u^n \\ u^{(i)} &= \left(1 + \frac{\Delta t}{m-1} L\right) u^{(i-1)} \quad i = 1, \ldots, s-1 \\ u^{s} &= \frac{1}{m} u^{(0)} + \frac{s-1}{s}\left(1 + \frac{\Delta t}{s-1} L\right) u^{(s-1)} \\ u^{n+1} &= u^{(s)}, \end{aligned}$$

have an optimal CFL coefficient $c = m - 1$ among all methods with non-negative coefficients. Although these methods were designed for linear problems, they are also second-order for nonlinear operators. Each such method uses m stages to attain the order usually obtained by a two-stage method, but has CFL coefficient $c = s - 1$, thus the effective CFL coefficient here is $c_{eff} = (s-1)/s$.

SSPRK linear (s,s-1) The m-stage, order $p = s - 1$ method:

$$
\begin{aligned}
\boldsymbol{u}^{(0)} &= \boldsymbol{u}^n \\
\boldsymbol{u}^{(i)} &= \boldsymbol{u}^{(i-1)} + \frac{1}{2}\Delta t L \boldsymbol{u}^{(i-1)} \quad i = 1, \ldots, s-1 \\
\boldsymbol{u}^{(s)} &= \sum_{k=0}^{m-2} \alpha_{s,k} \boldsymbol{u}^{(k)} + \alpha_{m,m-1}\left(\boldsymbol{u}^{(s-1)} + \frac{1}{2}\Delta t L \boldsymbol{u}^{(s-1)}\right) \\
\boldsymbol{u}^{n+1} &= \boldsymbol{u}^{(s)}.
\end{aligned}
$$

Where the coefficients are defined recursively,

$$
\begin{aligned}
\alpha_{2,0} &= 0 \quad \alpha_{2,1} = 1 \\
\alpha_{s,k} &= \frac{2}{k}\alpha_{s-1,k-1} \quad k = 1, \ldots, s-2 \\
\alpha_{s,s-1} &= \frac{2}{s}\alpha_{s-1,s-2} \quad \alpha_{s,0} = 1 - \sum_{k=1}^{m-1} \alpha_{s,k}
\end{aligned}
$$

is SSP with optimal CFL coefficient $c = 2$. The effective CFL for these methods is $c_{eff} = 2/s$.

These methods can also be extended to linear operators with time-dependent boundary conditions or forcing functions.

10.3.3 Optimal SSP Runge–Kutta methods for nonlinear problems

In this section we present Runge–Kutta methods which are SSP and accurate for nonlinear operators. Unlike in the case of linear operators, we have no stability results for fully discrete methods where $\mathcal{L}(u)$ is nonlinear. However, the SSP methods guarantee that any stability properties expected of the forward Euler method can still be expected of the higher order SSP methods.

SSPRK (2,2) An optimal second-order SSP Runge–Kutta method is

$$
\begin{aligned}
\boldsymbol{u}^{(1)} &= \boldsymbol{u}^n + \Delta t \mathcal{L}(\boldsymbol{u}^n) \\
\boldsymbol{u}^{n+1} &= \frac{1}{2}\boldsymbol{u}^n + \frac{1}{2}\boldsymbol{u}^{(1)} + \frac{1}{2}\Delta t \mathcal{L}\left(\boldsymbol{u}^{(1)}\right)
\end{aligned}
$$

with a CFL coefficient $c = 1$.

SSPRK (3,3) An optimal third-order SSP Runge–Kutta method is

$$\begin{aligned}
u^{(1)} &= u^n + \Delta t \mathcal{L}(u^n) \\
u^{(2)} &= \frac{3}{4}u^n + \frac{1}{4}u^{(1)} + \frac{1}{4}\Delta t \mathcal{L}\big(u^{(1)}\big) \\
u^{n+1} &= \frac{1}{3}u^n + \frac{2}{3}u^{(2)} + \frac{2}{3}\Delta t \mathcal{L}\big(u^{(2)}\big)
\end{aligned}$$

with a CFL coefficient $c = 1$.

Although the computational costs are double and triple (respectively) that of the forward Euler, the increase in the order of the method makes this additional computational cost acceptable. SSPRK(3,3) is widely known as the Shu–Osher method, and is probably the most commonly used SSP Runge–Kutta method. Although this method is only third-order accurate, it is most popular because of its simplicity, its classical linear stability properties, and because finding a fourth-order four-stage SSP Runge–Kutta method proved difficult. In fact, all four-stage, fourth-order Runge–Kutta methods with positive CFL coefficient, c must have at least one negative $\beta_{i,k}$. However, a popular fourth-order method is the $s = 5$-stage method with nonnegative $\beta_{i,k}$, given below. Unfortunately, any SSPRK of order $p > 4$ with nonzero CFL will have negative $\beta_{i,k}$.

SSPRK(5,4) The five-stage fourth-order SSPRK

$$\begin{aligned}
u^{(1)} &= u^n + 0.391752226571890 \Delta t \mathcal{L}(u^n) \\
u^{(2)} &= 0.444370493651235 u^n + 0.555629506348765 u^{(1)} \\
&\quad + 0.368410593050371 \Delta t \mathcal{L}\big(u^{(1)}\big) \\
u^{(3)} &= 0.620101851488403 u^n + 0.379898148511597 u^{(2)} \\
&\quad + 0.251891774271694 \Delta t \mathcal{L}\big(u^{(2)}\big) \\
u^{(4)} &= 0.178079954393132 u^n + 0.821920045606868 u^{(3)} \\
&\quad 0.544974750228521 \Delta t \mathcal{L}\big(u^{(3)}\big) \\
u^{n+1} &= 0.517231671970585 u^{(2)} \\
&\quad + 0.096059710526147 u^{(3)} + 0.063692468666290 \Delta t \mathcal{L}\big(u^{(3)}\big) \\
&\quad + 0.386708617503269 u^{(4)} + 0.226007483236906 \Delta t \mathcal{L}\big(u^{(4)}\big)
\end{aligned}$$

is SSP with CFL coefficient $c = 1.508$, and effective CFL $c_{eff} = 0.377$, which means that this method is more efficient, as well as higher order, than the popular SSPRK(3,3).

Finally, here is a useful low storage SSPRK method.

LS(5,3) The $m = 5$, $p = 3$ Williamson low storage method

$$
\begin{aligned}
U^{(0)} &= \boldsymbol{u}^n \\
dU^{(1)} &= \Delta t \mathcal{L}(U^{(0)}) \\
U^{(1)} &= U^{(0)} + 0.713497331193829 dU^{(0)} \\
dU^{(2)} &= -4.344339134485095 dU^{(1)} + \Delta t \mathcal{L}(U^{(1)}) \\
U^{(2)} &= U^{(1)} + 0.133505249805329 dU^{(1)} \\
dU^{(3)} &= \Delta t \mathcal{L}(U^{(2)}) \\
U^{(3)} &= U^{(2)} + 0.713497331193929 dU^{(2)} \\
dU^{(4)} &= -3.770024161386381 dU^{(3)} + \Delta t \mathcal{L}(U^{(3)}) \\
U^{(4)} &= U^{(3)} + 0.149579395628565 dU^{(3)} \\
dU^{(5)} &= -3.046347284573284 dU^{(4)} + \Delta t \mathcal{L}(U^{(4)}) \\
U^{(5)} &= U^{(4)} + 0.384471116121269 dU^{(4)} \\
u^{n+1} &= U^{(5)}
\end{aligned}
$$

is numerically optimal, with CFL coefficient $c = 1.4$. This method may be used instead of SSPRK(3,3) with almost the same computational cost: SSPRK(3,3) has $c_{\mathit{eff}} = 1/3$ and the low storage method LS(5,3) has $c_{eff} = 0.28$. This increase in cost is reasonable when storage is a critical need.

10.3.4 SSP multi-step methods

Explicit SSP multi-step methods

$$
\boldsymbol{u}^{n+1} = \sum_{i=1}^{s} \left(\alpha_i \boldsymbol{u}^{n+1-i} + \Delta t \beta_i \mathcal{L}(\boldsymbol{u}^{n+1-i}) \right), \quad \alpha_i \geq 0. \tag{10.7}
$$

are also easy to manipulate into convex combinations of forward Euler steps. If the forward Euler method combined with the spatial discretization $\mathcal{L}$ in (10.3) is strongly stable under the CFL restriction (10.4), $\|\boldsymbol{u}^n + \Delta t \mathcal{L}(\boldsymbol{u}^n)\| \leq \|\boldsymbol{u}^n\|$, then the multi-step method is SSP $\|\boldsymbol{u}^{n+1}\| \leq \|\boldsymbol{u}^n\|$, under the CFL restriction

$$
\Delta t \leq c \Delta t_{FE}, \quad c = \min_i \frac{\alpha_i}{|\beta_i|}.
$$

Table 10.1 features the coefficients of some of the most useful SSP multi-step methods. Those denoted with an * were proven optimal.

10.4 Further reading

The results on asymptotic linear growth of the eigenvalues of the Tau method is given by Dubiner (1987) while the general properties of the eigenvalue spectra

Table 10.1 *Coefficients of SSP multi-step methods with order p and s steps. The methods marked with an * were proven optimal.*

steps s	order p	CFL c	α_i	β_i
3*	2*	$\frac{1}{2}$	$\frac{3}{4}, 0, \frac{1}{4}$	$\frac{3}{2}, 0, 0$
4	2	$\frac{2}{3}$	$\frac{8}{9}, 0, 0, \frac{1}{9}$	$\frac{4}{3}, 0, 0, 0$
4	3	$\frac{1}{3}$	$\frac{16}{27}, 0, 0, \frac{11}{27}$	$\frac{16}{9}, 0, 0, \frac{4}{9}$
5	3	$\frac{1}{2}$	$\frac{25}{32}, 0, 0, 0, \frac{7}{32}$	$\frac{25}{16}, 0, 0, 0, \frac{5}{16}$
6*	3*	0.567	$\frac{108}{125}, 0, 0, 0, 0, \frac{17}{125}$	$\frac{36}{25}, 0, 0, 0, 0, \frac{6}{25}$
5	4	0.021	$\frac{1557}{32000}, \frac{1}{32000}, \frac{1}{120}, \frac{2063}{48000}, \frac{9}{10}$	$\frac{5323561}{2304000}, \frac{2659}{2304000}, \frac{904987}{2304000}, \frac{1567579}{768000}, 0$

for the spectral operators have been studied by Gottlieb and Lustman (1983), Welfert (1994), and Trefethen and co-workers (1988, 1990, 1994). A thorough discussion of nonnormality and pseudospectra can be found in the text by Canuto et al (2006).

For the stability analysis of the fully discrete methods, see Gottlieb and Tadmor's work (1991), and Levy and Tadmor's paper (1998). For more about the SSP methods, see Shu and Osher's 1988 paper, and Shu's papers in 1988 and 2002, as well as the 2001 review by Gottlieb, Shu, and Tadmor.

Much recent work in the subject has been done by Ruuth and by Spiteri, and important and exciting connections between SSP theory and contractivity theory have been recently examined by Ferracina and Spijker (2005) and by Higueras (2005).

11
Computational aspects

Up to this point, we have mainly discussed the theoretical aspects of spectral methods. We now turn to some of the computational issues surrounding these methods, and discuss the tools needed for efficient implementation of trigonometric and polynomial spectral methods. We shall also discuss the practical problems of round-off errors and aliasing errors. Finally, we address problems requiring the use of mappings.

11.1 Fast computation of interpolation and differentiation

The appearance of the fast Fourier transform (FFT) in 1965 revolutionized entire fields within science and engineering. By establishing a fast way of evaluating discrete Fourier series and their inverses, this single algorithm allowed the use of methods that were previously impractical due to excessive computational cost.

Fourier spectral methods emerged as efficient methods of computation due to the introduction of the FFT, and a significant part of the fundamental theory of Fourier spectral methods was developed in the decade immediately following the introduction of the FFT. In the following we briefly discuss the key idea behind the FFT. However, the idea behind the FFT is valid only when dealing with trigonometric polynomials, and by extension, Chebyshev polynomials; it is not, in general, applicable to polynomial spectral methods. Therefore, we shall continue by discussing alternative fast methods for the computation of interpolation and differentiation for polynomial based methods. We conclude with a brief section on how to compute the general Gaussian quadrature points and weights needed to compute the discrete polynomial expansion coefficients.

11.1.1 Fast Fourier transforms

The discrete Fourier series expansion of a function $u \in L^2[0, 2\pi]$ using an even number of points,

$$x_j = \frac{2\pi}{N} j, \quad j \in [0, N-1],$$

is

$$\mathcal{I}_N u(x) = \sum_{|n| \le N/2} \tilde{u}_n e^{-inx}, \quad \tilde{u}_n = \frac{1}{c_n N} \sum_{j=0}^{N-1} u(x_j) e^{inx_j}.$$

If we look at the interpolation at $x = x_j$ and recall that $\tilde{u}_{-N/2} = \tilde{u}_{N/2}$ is assumed for uniqueness, we have

$$\mathcal{I}_N u(x_j) = \sum_{n=-N/2}^{N/2-1} \tilde{u}_n e^{-i \frac{2\pi}{N} jn}, \quad \tilde{u}_n = \frac{1}{N} \sum_{j=0}^{N-1} u(x_j) e^{i \frac{2\pi}{N} jn}. \tag{11.1}$$

The cost of computing the discrete Fourier expansion is determined by the fact that there are N terms in each. Direct evaluation of the expansion coefficients or interpolation to the grid, x_j, requires $\mathcal{O}(8N^2)$ real operations since $\tilde{u}_n$ is, in general, a complex number.

Let us now assume that N is a power of two, i.e., $N = 2^M$, and introduce the new index, j_1, as

$$j = \begin{cases} 2j_1 & j \text{ even}, \\ 2j_1 + 1 & j \text{ odd}, \end{cases} \quad j_1 \in [0, \ldots, N/2 - 1].$$

Using this new index, the discrete expansion coefficients are

$$\tilde{u}_n = \frac{1}{N} \left(\sum_{j_1=0}^{N/2-1} u(x_{2j_1}) e^{i \frac{2\pi}{N}(2j_1)n} + \sum_{j_1=0}^{N/2-1} u(x_{2j_1+1}) e^{i \frac{2\pi}{N}(2j_1+1)n} \right)$$

If we let $N_1 = N/2$, this becomes

$$\tilde{u}_n = \frac{1}{N} \left(\sum_{j_1=0}^{N_1-1} u(x_{2j_1}) e^{i \frac{2\pi}{N_1} j_1 n} + e^{i \frac{2\pi}{N} n} \sum_{j_1=0}^{N_1-1} u(x_{2j_1+1}) e^{i \frac{2\pi}{N_1} j_1 n} \right).$$

At this point we realize that the two sums, each of half the length of the original sum, have exactly the same form as the original sum, Equation (11.1). Thus, we may repeat the process and break the computation down to four sums each of length $N/4$. Repeating this process results in an algorithm that computes the discrete expansion coefficients in $\mathcal{O}(5N \log_2 N)$ real operations provided the twiddle factors, $e^{2\pi n/N}$, $e^{2\pi n/N_1}$ and so on, are precomputed. This is the essence of what is known as the fast Fourier transform and it is indeed much faster than the straightforward computation of the Fourier transform.

Along the same lines, the FFT can be used for computing the discrete expansion coefficients as well as the interpolation at the collocation points. Furthermore, a fast transform can be designed also if N is a power of three, by splitting the original sum into three sums each of length $1/3N$; such an algorithm is known as a radix-3 transform. In fact, the algorithm can be formulated for any prime decomposition of a given number.

If $u(x)$ is a real function, the complex FFT is unnecessarily expensive. A real function of length N can be expressed as a complex function, v, of half the length;

$$v_j = u(x_{2j}) + iu(x_{2j+1}), \quad j \in \left[0, \ldots, \frac{N}{2} - 1\right].$$

We can use the FFT to obtain the complex expansion coefficients of v, and then obtain the expansion coefficients of u by

$$\tilde{u}_n = \frac{1}{2}\left(\tilde{v}_n + \overline{\tilde{v}}_{N/2-n}\right) - \frac{i}{2} e^{i\frac{2\pi}{N}n}\left(\tilde{v}_n - \overline{\tilde{v}}_{N/2-n}\right).$$

The fast transformation using the complex FFT yields an $\mathcal{O}(\frac{5}{2}N\log_2 N)$ algorithm for the radix-2 scheme, i.e., twice as fast as just performing the complex FFT with all imaginary components equal to zero.

The FFT may also be applied for the fast computation of sine and cosine transforms. To compute the discrete cosine expansion coefficients

$$\tilde{u}_n = \sum_{j=0}^{N} u(x_j) \cos\left(\frac{\pi}{N} nj\right),$$

we form a new even function $\forall k \in [0, \ldots, 2N-1]$

$$v(x_k) = \begin{cases} u(x_k) & k \leq N \\ u(x_{2N-k}) & k > N, \end{cases}$$

such that the cosine expansion coefficients of $u(x)$ are given by the Fourier coefficients of v,

$$\tilde{v}_n = \sum_{k=0}^{2N-1} v(x_k) e^{i\frac{2\pi}{2N}kn} = 2\sum_{j=0}^{N} u(x_j)\cos\left(\frac{\pi}{N}nj\right).$$

We use the complex transform of a real function to minimize the computational workload, and now have a fast transform for the cosine series which is very close to the Chebyshev expansion. Indeed, if we recall the Chebyshev Gauss–Lobatto quadrature rule and expansion

$$\mathcal{I}_N u(x_j) = \sum_{n=0}^{N} \tilde{u}_n \cos\left(\frac{\pi}{N}nj\right), \quad \tilde{u}_n = \frac{1}{\overline{c}_n N}\sum_{j=0}^{N} \frac{1}{\overline{c}_j} u(x_j)\cos\left(\frac{\pi}{N}nj\right),$$

it is clear that the FFT can be used to compute both sums with a computational cost of $\mathcal{O}(\frac{5}{2}N\log_2 N + 4N)$, where the latter contribution appears from the packing and unpacking required to utilize the FFT. The option of using the FFT for computing the Chebyshev Gauss–Lobatto expansion is yet another reason for the widespread early use of Chebyshev spectral methods.

At this time, implementations of the cosine transform are of very varying quality and it is not possible to estimate, in general, when a fast cosine transform should be used at a specific machine rather than a matrix multiplication directly. However, a good rule of thumb is that if $N > 64$, the fast cosine transform is worth using.

11.1.2 The even-odd decomposition

Unfortunately, the approach that leads to the fast Fourier transform does not extend directly to orthogonal polynomials beyond the Chebyshev polynomials. Hence, if we want to use the expansion coefficients, $\tilde{u}_n$, to compute the derivative at the collocation points, there is usually no way around summing the series directly, unless N is very large.

However, when using the interpolating Lagrange polynomials and the associated differentiation matrices, there is still room for improvement. Consider the vector, $\boldsymbol{u} = (u(x_0), \ldots, u(x_N))$, and the differentiation matrix, D, associated with the chosen set of collocation points, x_j. We recall that the derivative of $\boldsymbol{u}$ at the grid points, $\boldsymbol{u}'$, is

$$\boldsymbol{u}' = \mathrm{D}\boldsymbol{u}.$$

Using ultraspherical polynomials as the basis for the approximation we observed that D is centro-antisymmetric,

$$\mathrm{D}_{ij} = -\mathrm{D}_{N-i,N-j}.$$

This property allows us to develop a fast algorithm for the computation of $\boldsymbol{u}'$. We decompose $\boldsymbol{u}$ into its even ($\boldsymbol{e}$) and odd ($\boldsymbol{o}$) parts,

$$\boldsymbol{u} = \boldsymbol{e} + \boldsymbol{o}.$$

The two new vectors have the entries

$$e_j = \frac{1}{2}(u_j + u_{N-j}), \quad o_j = \frac{1}{2}(u_j - u_{N-j}),$$

where u_j is the jth entry in $\boldsymbol{u}$, and similarly for e_j and o_j. Observe that

$$e_j = e_{N-j}, \quad o_j = -o_{N-j}.$$

The linearity of the differentiation operation implies

$$\boldsymbol{u}' = \mathrm{D}\boldsymbol{u} = \mathrm{D}\boldsymbol{e} + \mathrm{D}\boldsymbol{o} = \boldsymbol{e}' + \boldsymbol{o}'.$$

Let us now first consider the case where N is odd such that we have an even total number of collocation points.

If we compute the derivative of $\boldsymbol{e}$ we obtain

$$\begin{aligned} e'_j &= \sum_{i=0}^{N} \mathrm{D}_{ji} e_i = \sum_{i=0}^{(N-1)/2} \mathrm{D}_{ji} e_i + \mathrm{D}_{j,N-i} e_{N-i} \\ &= \sum_{i=0}^{(N-1)/2} (\mathrm{D}_{ji} + \mathrm{D}_{j,N-i}) e_i. \end{aligned}$$

Moreover,

$$\begin{aligned} e'_{N-j} &= \sum_{i=0}^{N} \mathrm{D}_{N-j,i} e_i = \sum_{i=0}^{(N-1)/2} (\mathrm{D}_{N-j,i} + \mathrm{D}_{N-j,N-i}) e_i \\ &= \sum_{i=0}^{(N-1)/2} -(\mathrm{D}_{j,N-i} + \mathrm{D}_{ji}) e_i = -e'_j, \end{aligned}$$

so it is only necessary to compute the first half of $\boldsymbol{e}'$. If we introduce the matrix, D^e, with the entries

$$\mathrm{D}^e_{ij} = \mathrm{D}_{ij} + \mathrm{D}_{i,N-j} \quad \forall i, j \in [0, \ldots, (N-1)/2],$$

the computation is simply a matrix multiplication

$$\tilde{\boldsymbol{e}}' = \mathrm{D}^e \tilde{\boldsymbol{e}},$$

where $\tilde{\boldsymbol{e}} = (e_0, \ldots, e_{(N-1)/2})^T$.

The computation of $\boldsymbol{o}'$ yields

$$\begin{aligned} o'_j &= \sum_{i=0}^{N} \mathrm{D}_{ji} o_i = \sum_{i=0}^{(N-1)/2} \mathrm{D}_{ji} o_i + \mathrm{D}_{j,N-i} o_{N-i} \\ &= \sum_{i=0}^{(N-1)/2} (\mathrm{D}_{ji} - \mathrm{D}_{j,N-i}) o_i, \end{aligned}$$

and

$$\begin{aligned} o'_{N-j} &= \sum_{i=0}^{N} \mathrm{D}_{N-j,i} o_i = \sum_{i=0}^{(N-1)/2} (\mathrm{D}_{N-j,i} - \mathrm{D}_{N-j,N-i}) o_i \\ &= \sum_{i=0}^{(N-1)/2} (-\mathrm{D}_{j,N-i} + \mathrm{D}_{ji}) o_i = o'_j, \end{aligned}$$

so in this case it also suffices to compute half of the elements in the vector.

Introducing the matrix D^o with the entries

$$\mathrm{D}^o_{ij} = \mathrm{D}_{ij} - \mathrm{D}_{i,N-j}, \quad \forall i,\ j \in [0, \dots, (N-1)/2],$$

differentiation can be written as a matrix multiplication

$$\tilde{\boldsymbol{o}}' = \mathrm{D}^o \tilde{\boldsymbol{o}},$$

where $\tilde{\boldsymbol{o}} = (o_0, \dots, o_{(N-1)/2})^T$. Finally, we can write the derivative

$$\boldsymbol{u}' = \boldsymbol{e}' + \boldsymbol{o}',$$

using

$$u'_j = \tilde{e}'_j + \tilde{o}'_j, \quad u'_{N-j} = e'_{N-j} + o'_{N-j} = -\tilde{e}'_j + \tilde{o}'_j,$$

for $j \in [0, \dots, (N-1)/2]$.

Consequently, to compute the derivative of $\boldsymbol{u}$ at the collocation points we need to construct $\tilde{\boldsymbol{e}}$ and $\tilde{\boldsymbol{o}}$, perform two matrix-vector products of length $N/2$ and reconstruct $\boldsymbol{u}'$. The total operation count for this process becomes

$$2\frac{N}{2} + 2\left(2\left(\frac{N}{2}\right)^2 - \frac{N}{2}\right) + 2\frac{N}{2} = N^2 + N,$$

provided the differentiation matrices are precomputed. This is in contrast to the direct computation of $\boldsymbol{u}'$ which requires $2N^2 - N$ operations. Hence, utilizing the centro-antisymmetry of the differentiation matrix allows for decreasing the computational work by close to a factor of two.

Finally, we consider the case where N is even. If we follow the same approach as above we obtain

$$e'_j = \sum_{i=0}^{N} \mathrm{D}_{ji} e_i = \sum_{i=0}^{N/2} \mathrm{D}_{ji} e_i + \mathrm{D}_{j,N-i} e_{N-i},$$

where the term from $i = N/2$ is computed twice. This, however, is easily fixed by slightly redefining D^e

$$\mathrm{D}^e_{ij} = \begin{cases} \mathrm{D}_{ij} + \mathrm{D}_{i,N-j} & j \neq N/2 \\ \mathrm{D}_{i,N/2} & j = N/2 \end{cases} \quad \forall i \in [0, \dots, N/2-1],\ j \in [0, \dots, N/2].$$

In the same way we define a modified D^o

$$\mathrm{D}^o_{ij} = \mathrm{D}_{ij} - \mathrm{D}_{i,N-j} \quad \forall i \in [0, \dots, N/2],\ j \in [0, \dots, N/2-1],$$

since $\boldsymbol{o}$ is odd, so that the problem of counting the last entry twice does not appear. This also implies that D^o is rectangular, since $\boldsymbol{o}'$ is even and therefore $o'_{N/2}$ needs to be computed.

In the situation where D is centro-symmetric

$$\mathrm{D}_{ij} = \mathrm{D}_{N-i,N-j},$$

the exact same even-odd splitting can be applied with the only difference being that $\boldsymbol{e}'$ is even and $\boldsymbol{o}'$ is odd such that the final reconstruction becomes

$$u'_j = \tilde{e}'_j + \tilde{o}'_j, \quad u'_{N-j} = e'_{N-j} + o'_{N-j} = \tilde{e}'_j - \tilde{o}'_j.$$

Indeed, all even-order differentiation matrices appearing from the ultraspherical polynomials share the property of centro-symmetry, which allows us to apply this splitting technique directly.

11.2 Computation of Gaussian quadrature points and weights

When using polynomial methods we must first establish collocation points, which are the quadrature points of some Gauss quadrature rule. However, with the exception of a few special cases, like the Chebyshev polynomials, no closed form expressions for the quadrature nodes are known. Nevertheless, there is a simple and elegant way of computing these nodes as well as the corresponding weights, although the latter can be computed using explicit forms.

We restrict our presentation to the case of ultraspherical polynomials, $P_n^{(\alpha)}(x)$, although everything generalizes to the general case of Jacobi polynomials.

First, let's recall the three-term recurrence relation for ultraspherical polynomials

$$x P_n^{(\alpha)}(x) = a_{n-1,n} P_{n-1}^{(\alpha)}(x) + a_{n+1,n} P_{n+1}^{(\alpha)}(x), \tag{11.2}$$

where the recurrence coefficients are

$$a_{n-1,n} = \frac{n+2\alpha}{2n+2\alpha+1}, \quad a_{n+1,n} = \frac{n+1}{2n+2\alpha+1}.$$

If we normalize the polynomials slightly differently and introduce the modified polynomials

$$\tilde{P}_n^{(\alpha)}(x) = \frac{1}{\sqrt{\gamma_n}} P_n^{(\alpha)}(x),$$

such that $(\tilde{P}_n^{(\alpha)}, \tilde{P}_k^{(\alpha)})_w = \delta_{nk}$, the recurrence coefficients become

$$\tilde{a}_{n-1,n} = \sqrt{\frac{\gamma_{n-1}}{\gamma_n}} a_{n-1,n} = \sqrt{\frac{n(n+2\alpha)}{(2n+2\alpha+1)(2n+2\alpha-1)}},$$

and

$$\tilde{a}_{n+1,n} = \sqrt{\frac{\gamma_{n+1}}{\gamma_n}} a_{n+1,n} = \sqrt{\frac{(n+1)(n+2\alpha+1)}{(2n+2\alpha+3)(2n+2\alpha+1)}}.$$

A key observation is that

$$\tilde{a}_{n+1,n} = \tilde{a}_{n,n+1}.$$

We define

$$\beta_n = \tilde{a}_{n+1,n}$$

so that Equation (11.2) reduces to

$$x\tilde{P}_n^{(\alpha)}(x) = \beta_{n-1}\tilde{P}_{n-1}^{(\alpha)}(x) + \beta_n\tilde{P}_{n+1}^{(\alpha)}(x). \tag{11.3}$$

We introduce the vector $\tilde{\boldsymbol{P}}^{(\alpha)}(x) = (\tilde{P}_0^{(\alpha)}(x), \ldots, \tilde{P}_N^{(\alpha)}(x))^T$, and the symmetric bi-diagonal matrix

$$\mathrm{J}_N = \begin{bmatrix} 0 & \beta_1 & 0 & 0 & 0 & \cdots & 0 \\ \beta_1 & 0 & \beta_2 & 0 & 0 & \cdots & 0 \\ 0 & \beta_2 & 0 & \beta_3 & 0 & \cdots & 0 \\ 0 & 0 & \beta_3 & 0 & \beta_4 & \cdots & 0 \\ 0 & 0 & 0 & \beta_4 & 0 & \ddots & 0 \\ \vdots & \vdots & \vdots & \vdots & \ddots & \ddots & \beta_{N-1} \\ 0 & 0 & 0 & 0 & 0 & \beta_{N-1} & 0 \end{bmatrix},$$

so that Equation (11.3) becomes

$$x\tilde{\boldsymbol{P}}^{(\alpha)}(x) = \mathrm{J}_N\tilde{\boldsymbol{P}}^{(\alpha)}(x) + \beta_N\tilde{P}_{N+1}^{(\alpha)}(x).$$

Gauss Since the Gauss quadrature points, z_j, are defined as the roots of $P_{N+1}^{(\alpha)}(x)$, and therefore also of $\tilde{P}_{N+1}^{(\alpha)}$ we realize that the real grid points, z_j, are the eigenvalues of the symmetric bi-diagonal matrix, J_N. The eigenvalue problem may be solved using the QR algorithm and the corresponding weights may be computed using their exact formulas. However, we may alternatively recover the weights from the eigenvectors of J_N. To understand this, recall that the formula for the interpolating Lagrange polynomial associated with the Gauss quadrature points is

$$\tilde{l}_j(z) = u_j \sum_{n=0}^{N} \frac{P_n^\alpha(z)P_n^\alpha(z_j)}{\gamma_n} = u_j\left(\tilde{\boldsymbol{P}}^{(\alpha)}(z)\right)^T\tilde{\boldsymbol{P}}^{(\alpha)}(z_j).$$

Using the Christoffel–Darboux identity we established that

$$\tilde{l}_j(z_j) = u_j\left(\tilde{\boldsymbol{P}}^{(\alpha)}(z_j)\right)^T\tilde{\boldsymbol{P}}^{(\alpha)}(z_j) = 1.$$

In other words, the normalized eigenvector $\boldsymbol{Q}(z_j)$ corresponding to the eigenvalue z_j of J_N is

$$\boldsymbol{Q}(z_j) = \sqrt{u_j}\tilde{\boldsymbol{P}}^{(\alpha)}(z_j),$$

from which, by equating the first components of the two vectors, we obtain the expression of the weight

$$u_j = \left(\frac{Q_0(z_j)}{\tilde{P}_0^{(\alpha)}(z_j)}\right)^2 = \gamma_0(Q_0(z_j))^2 = \sqrt{\pi}\frac{\Gamma(\alpha+1)}{\Gamma(\alpha+3/2)}(Q_0(z_j))^2,$$

since $\tilde{P}_0^{(\alpha)}(z_j) = 1/\sqrt{\gamma_0}$. Here $Q_0(z_j)$ signifies the first component of the eigenvector, $\boldsymbol{Q}(z_j)$. Hence, the quadrature points as well as the weights may be obtained directly by computing the $N+1$ eigenvalues and the first component of the corresponding eigenvectors.

Gauss–Radau The algorithm for computing the Gauss–Radau quadrature points and weights is very similar, the main difference is due to the definition of the Gauss–Radau quadrature points, which are the roots of the polynomial

$$q(y) = P_{N+1}^{(\alpha)}(y) + \alpha_N P_N^{(\alpha)}(y) = 0,$$

where α_N is chosen such that $q(y)$ vanish at one of the two boundaries,

$$\alpha_N = -\frac{P_{N+1}^{(\alpha)}(\pm 1)}{P_N^{(\alpha)}(\pm 1)} = (\mp 1)\frac{N+1+2\alpha}{N+1},$$

where the upper sign corresponds to $q(1) = 0$ while the lower sign yields α_N for $q(-1) = 0$.

The three-term recurrence relation, Equation (11.3), yields

$$y\tilde{\boldsymbol{P}}^{(\alpha)}(y) = \mathrm{J}_N\tilde{\boldsymbol{P}}^{(\alpha)}(y) + \beta_N\tilde{P}_{N+1}^{(\alpha)}(y).$$

Using the definition of the Gauss–Radau quadrature points,

$$\tilde{P}_{N+1}^{(\alpha)}(\pm 1) = \tilde{\alpha}_N\tilde{P}_N^{(\alpha)}(\pm 1),$$

where

$$\tilde{\alpha}_N = -\sqrt{\frac{\gamma_N}{\gamma_{N+1}}}\alpha_N = (\pm 1)\sqrt{\frac{(N+1+2\alpha)(2N+2\alpha+3)}{(N+1)(2N+2\alpha+1)}}.$$

Thus, the last row of J_N is modified so that

$$\begin{aligned}(\pm 1)\tilde{P}_N^{(\alpha)}(\pm 1) &= \beta_{N-1}\tilde{P}_{N-1}^{(\alpha)}(\pm 1) + \beta_N\tilde{P}_{N+1}^{(\alpha)}(\pm 1)\\ &= \beta_{N-1}\tilde{P}_{N-1}^{(\alpha)}(\pm 1) + \beta_N\tilde{\alpha}_N\tilde{P}_N^{(\alpha)}(\pm 1),\end{aligned}$$

i.e., only the element $(i, j) = (N, N)$ of J_N needs to be modified while the remaining part of the algorithm follows the description above. The Gauss–Radau quadrature points are thus found by solving the modified eigenvalue problem while the corresponding weights, v_j, are found from the first components of the corresponding eigenvectors as discussed in connection with the Gauss quadratures.

Gauss–Lobatto Finally, we consider the modifications necessary to compute the Gauss–Lobatto quadrature points and weights. In this case, the quadrature points are the roots of the polynomial

$$q(x) = P_{N+1}^{(\alpha)}(x) + \alpha_N P_N^{(\alpha)}(x) + \alpha_{N-1} P_{N-1}^{(\alpha)}(x),$$

where the coefficients, α_{N-1} and α_N, are found by requiring $q(\pm 1) = 0$, i.e., by solving the system

$$\alpha_N P_N^{(\alpha)}(-1) + \alpha_{N-1} P_{N-1}^{(\alpha)}(-1) = -P_{N+1}^{(\alpha)}(-1)$$
$$\alpha_N P_N^{(\alpha)}(1) + \alpha_{N-1} P_{N-1}^{(\alpha)}(1) = -P_{N+1}^{(\alpha)}(1).$$

If we then normalize the polynomials as above, these constants become

$$\tilde{\alpha}_N = \sqrt{\frac{\gamma_N}{\gamma_{N+1}}}\alpha_N, \quad \tilde{\alpha}_{N-1} = \sqrt{\frac{\gamma_{N-1}}{\gamma_{N+1}}}\alpha_{N-1},$$

and the equation for the quadrature points is

$$\tilde{P}_{N+1}^{(\alpha)}(x) + \tilde{\alpha}_N \tilde{P}_N^{(\alpha)}(x) + \alpha_{\tilde{N}-1} P_{N-1}^{(\alpha)}(x) = 0.$$

This requirement is enforced on the eigenvalue problem by changing two elements of J_N

$$(\mathrm{J}_N)_{N,N-1} = \beta_{N-1} - \beta_N \tilde{\alpha}_{N-1}, \quad (\mathrm{J}_N)_{N,N} = -\beta_N \tilde{\alpha}_N,$$

while the remaining part of the algorithm remains unchanged. Unfortunately, using this approach for computing the Gauss–Lobatto quadrature points, J_N looses its symmetry, thereby making the solution of the resulting eigensystem slightly harder.

A different approach can be taken by recalling that

$$\frac{dP_N^{(\alpha)}(x)}{dx} = (2\alpha + 1) P_{N-1}^{(\alpha+1)}(x),$$

i.e., the interior Gauss–Lobatto quadrature points for $P_N^{(\alpha)}(x)$ are the roots of the polynomial $P_{N-1}^{(\alpha+1)}(x)$, which are also the Gauss quadrature points of the polynomial $P_{N-2}^{(\alpha+1)}(x)$. The latter may be computed using the symmetric approach discussed above, and the weights computed using the exact formula.

11.3 Finite precision effects

In the idealized world of approximation and stability theory, the accuracy of spectral methods depends, as we have seen, only on the regularity of the function being approximated and the operators involved. However, in the non-ideal world of computation, the finite precision of the computer has a very significant effect on any algorithm being implemented.

For spectral methods the effect of round-off errors is most pronounced when derivatives are computed, with a very significant difference between the behavior of derivatives computed using continuous expansion coefficients and discrete expansion coefficients or interpolating Lagrange polynomials.

For polynomial spectral methods in particular, the effects of the finite precision can have a significant impact on the overall accuracy of the scheme. Indeed, for certain problems the results are essentially useless due to the overwhelming amplification of round-off errors.

11.3.1 Finite precision effects in Fourier methods

Consider a given function $u(x) \in L^2[0, 2\pi]$, approximated using a continuous Fourier series

$$\mathcal{P}_N u(x) = \sum_{|n| \leq N/2} \hat{u}_n e^{inx}, \quad \hat{u}_n = \frac{1}{2\pi} \int_0^{2\pi} u(x) e^{-inx} \, dx.$$

The approximation to the m-derivative of $u(x)$ is

$$\mathcal{P}_N u^{(m)}(x) = \sum_{|n| \leq N/2} (in)^m \hat{u}_n e^{inx},$$

which means that the highest modes are amplified. Let us study the effect of this phenomenon through an example.

Example 11.1 Let us consider the $C^\infty[0, 2\pi]$ and periodic function

$$u(x) = \frac{3}{5 - 4\cos(x)},$$

with the continuous expansion coefficients

$$\hat{u}_n = 2^{-|n|},$$

from which we directly obtain the approximation to the m-derivative

$$\mathcal{P}_N u^{(m)}(x) = \sum_{|n| \leq N/2} (in)^m 2^{-|n|} e^{inx}.$$

Table 11.1 *Maximum pointwise error of a spatial Fourier m-derivative for increasing resolution, N, as obtained using the continuous expansion coefficients of Example 11.1. The number, N_0, represents the number of modes required to approximate the function, $u(x)$, to $\mathcal{O}(\varepsilon_M)$, i.e., such that $\hat{u}_{N_0/2} \simeq \varepsilon_M$. For this case $N_0 \approx 100$ with a machine accuracy of $\varepsilon_M \sim 1.0E{-}16$.*

N	$m = 1$	$m = 2$	$m = 3$	$m = 4$
8	0.438E+00	0.575E+01	0.119E+02	0.255E+03
16	0.477E−01	0.105E+01	0.522E+01	0.106E+03
32	0.360E−03	0.139E−01	0.114E+00	0.434E+01
64	0.104E−07	0.778E−06	0.119E−04	0.873E−03
128	0.222E−14	0.160E−13	0.524E−13	0.335E−11
256	0.311E−14	0.160E−13	0.782E−13	0.398E−12
512	0.355E−14	0.160E−13	0.782E−13	0.568E−12
1024	0.355E−14	0.195E−13	0.853E−13	0.568E−12
Accuracy	$\mathcal{O}(\varepsilon_M(N_0/2))$	$\mathcal{O}(\varepsilon_M(N_0/2)^2)$	$\mathcal{O}(\varepsilon_M(N_0/2)^3)$	$\mathcal{O}(\varepsilon_M(N_0/2)^4)$

In Table 11.1 we show the maximum pointwise error of this approximation to $u^{(m)}(x)$ with increasing number of modes N.

We observe that once the function is well resolved the error approaches machine zero, ε_M, faster than any algebraic order of N. However, a close inspection reveals that the accuracy of the approximation decays with increasing order of the derivative

$$\max_{x\in[0,2\pi]} \left|u^{(m)}(x) - \mathcal{P}_N u^{(m)}(x)\right| \sim \mathcal{O}(\varepsilon_M(N_0/2)^m) \quad \text{as} \quad N \to \infty.$$

Here N_0 corresponds to the number of modes required to approximate the function, $u(x)$, to $\mathcal{O}(\varepsilon_M)$. For $N \gg N_0$ we have $\hat{u}_n \ll \varepsilon_M$ and, since $\hat{u}_n$ decays exponentially in this limit, $\hat{u}_k^{(m)} \ll \varepsilon_M$, i.e., the last term that contributes to the accuracy is $\hat{u}_{N_0/2} \sim \mathcal{O}(\varepsilon_M)$. This is the limiting factor on accuracy.

The key observation to make is that, in the continuous case, the effect of the finite precision is independent of N once the function is well resolved and only a slight dependency of the order of the derivative on the accuracy is observed. Unfortunately, such behavior does not carry over to the discrete case.

We now consider the case where the same function $u(x) \in L^2[0, 2\pi]$ is approximated using the discrete expansion coefficients

$$\mathcal{I}_N u(x) = \sum_{n=-N/2}^{N/2} \tilde{u}_n e^{inx}, \quad \tilde{u}_n = \frac{1}{N c_n} \sum_{j=0}^{N-1} u(x_j) e^{-inx_j},$$

Table 11.2 *Maximum pointwise error of a spatial Fourier m-derivative for increasing resolution, N, as obtained using the discrete expansion coefficients of Example 11.2. Machine accuracy is* $\varepsilon_M \sim 1.0E\text{–}16$.

N	$m = 1$	$m = 2$	$m = 3$	$m = 4$
8	0.654E+00	0.402E+01	0.134E+02	0.233E+03
16	0.814E−01	0.500E+00	0.618E+01	0.770E+02
32	0.648E−03	0.391E−02	0.173E+00	0.210E+01
64	0.198E−07	0.119E−06	0.205E−04	0.247E−03
128	0.380E−13	0.216E−11	0.116E−09	0.715E−08
256	0.657E−13	0.705E−11	0.680E−09	0.954E−07
512	0.272E−12	0.605E−10	0.132E−07	0.110E−05
1024	0.447E−12	0.253E−09	0.844E−07	0.157E−04
Accuracy	$\mathcal{O}(\varepsilon_M(N/2))$	$\mathcal{O}(\varepsilon_M(N/2)^2)$	$\mathcal{O}(\varepsilon_M(N/2)^3)$	$\mathcal{O}(\varepsilon_M(N/2)^4)$

where we use the even grid

$$x_j = \frac{2\pi}{N} j, \quad j \in [0, \ldots, N-1].$$

The actual computation of the expansion coefficients may be performed using the FFT or by simply summing the series. Once the expansion coefficients are obtained, computation of the approximation to the m-derivative of $u(x)$ is obtained, just as for the continuous expansion,

$$\mathcal{I}_N u^{(m)}(x) = \sum_{|n| \le N/2} (in)^m \tilde{u}_n e^{inx}.$$

Let's see the effect of finite precision on the discrete expansion in the following example.

Example 11.2 Consider, again, the $C^\infty[0, 2\pi]$, and periodic, function

$$u(x) = \frac{3}{5 - 4\cos(x)}.$$

The expansion coefficients are now found using an FFT, from which we immediately obtain the expansion coefficients for the mth derivative

$$\tilde{u}_n^{(m)} = (in)^m \tilde{u}_n.$$

In Table 11.2 we show the maximum pointwise error of this approximation to $u^{(m)}(x)$ with increasing number of modes, N.

Note the pronounced difference in the accuracy of the derivatives as compared to the results in Table 11.1, where the error is constant for a given m once the function is well resolved. We find that when using the FFT to compute the coefficients, the accuracy deteriorates with increasing order of the derivative, as in Table 11.1, but also for increasing N. Indeed, we observe that the error scales like

$$\max_{x\in[0,2\pi]} \left|u^{(m)}(x) - \mathcal{I}_N u^{(m)}(x)\right| \sim \mathcal{O}(\varepsilon_M (N/2)^m) \quad \text{as} \quad N \to \infty,$$

which is a significant reduction in the accuracy, in particular for high order derivatives. This is a consequence of the uniform error of $\mathcal{O}(\varepsilon_M)$ introduced by the numerical computation of the discrete expansion coefficients. As a result, $\tilde{u}_n \sim \mathcal{O}(\varepsilon_M)$ even for $N \gg N_0$, where we would expect the expansion coefficients to decay exponentially. However, due to the finite accuracy, it is impossible to obtain these very small numbers. Thus, the maximum error is obtained from the maximum mode number $N/2$ and the effect of this term, introduced by the uniform noise level, is clearly seen in Table 11.2.

There is no way of avoiding the round-off error in the calculation of the expansion coefficients, but its effect can be reduced by computing the FFT with the highest possible accuracy. It is also important to attempt to formulate the differential equations using the lowest order derivatives possible.

In the discussion above, we illustrated the problem of round-off errors using the FFT and the discrete expansion coefficients. The general picture remains the same when differentiation matrices are used instead. Indeed, it is usually found that using the FFT and the expansion coefficients results in an algorithm which suffers least from effects of the finite precision. However, if the entries of the differentiation matrices are carefully computed, i.e., the exact entries are computed for high-order derivatives rather than obtained by multiplying several first-order differentiation matrices, the two different algorithms yield a comparable accuracy.

11.3.2 Finite precision in polynomial methods

Polynomial methods are considerably more sensitive to round-off error than the discrete Fourier expansion. For simplicity, we restrict our discussion to the case of Chebyshev expansions and derivatives. However, unless otherwise stated, the results carry over to the case of general ultraspherical polynomials.

Let's begin by considering the continuous Chebyshev expansion of $u(x) \in L^2_w[-1, 1]$

$$\mathcal{P}_N u(x) = \sum_{n=0}^{N} \hat{u}_n T_n(x), \quad \hat{u}_n = \frac{2}{c_n \pi} \int_{-1}^{1} u(x) T_n(x) \frac{1}{\sqrt{1-x^2}}\, dx.$$

The approximation to the m-derivative of $u(x)$ is

$$\mathcal{P}_N u^{(m)}(x) = \sum_{n=0}^{N} \hat{u}_n^{(m)} T_n(x),$$

where the continuous expansion coefficients for $u^{(m)}(x)$ are found by repeated use of the backward recursion formula

$$c_{n-1}\hat{u}_{n-1}^{(m)} = \hat{u}_{n+1}^{(m)} + 2n\hat{u}_n^{(m-1)} \quad \forall n \in [N, \ldots, 1],$$

with the assumption that $\hat{u}_N^{(m)} = \hat{u}_{N+1}^{(m)} = 0$. Let's examine the accuracy of this transform-recurrence-transform method through an example.

Example 11.3 Consider the function $u(x) \in C^\infty[-1, 1]$

$$u(x) = \frac{1}{x+a}, \quad a > 1,$$

for which the continuous expansion coefficients are

$$\hat{u}_n = \frac{2}{c_n} \frac{1}{\sqrt{a^2-1}} \left(\sqrt{a^2-1} - a\right)^n.$$

Since the function is smooth we see that the expansion coefficients decay exponentially in n. Note that when a approaches 1 the function develops a strong gradient at $x = -1$ and becomes singular in the limit. In this example we used $a = 1.1$.

Using the backward recursion formula for Chebyshev polynomials, we calculated the expansion coefficients for higher derivatives. In Table 11.3 we list the maximum pointwise error of the expansion as a function of N, the order of the approximating polynomial.

Again we find that once the function is well approximated the error is close to machine zero, ε_M. However, a closer look shows the rate of decay of the maximum pointwise error

$$\max_{x\in[-1,1]} \left|u^{(m)}(x) - \mathcal{P}_N u^{(m)}(x)\right| \sim \mathcal{O}\left(\varepsilon_M N_0^m\right) \quad \text{as} \quad N \to \infty,$$

where N_0 corresponds to the minimum mode number required to approximate $u(x)$ to $\mathcal{O}(\varepsilon_M)$. Due to the rapid decay of the expansion coefficients, we know that for $N \gg N_0$, $\hat{u}_n \ll \varepsilon_M$, i.e., the last term that contributes to the accuracy is $2N_0\hat{u}_{N_0}$ which is $\mathcal{O}(\varepsilon_M)$. Unlike the Fourier series, the expansion coefficient $\hat{u}_n^{(m)}$ depends on all coefficients with higher n. However, due to the rapid decay of the coefficients, the backward recursion is stable, i.e., the last term of order $\mathcal{O}(\varepsilon_M)$ is carried backwards in the recursion without being amplified, thus leading to the observed scaling.

Table 11.3 *Maximum pointwise error of a spatial Chebyshev m-derivative for increasing resolution, N, as obtained using the analytic Chebyshev expansion coefficients in Example 11.3. The number N_0 represents the number of modes required to approximate the function, $u(x)$, such that $2N_0\hat{u}_{N_0} \simeq \mathcal{O}(\varepsilon_M)$. For this case $N_0 \approx 100$ and the machine accuracy is $\varepsilon_M \sim 1.0E{-}16$.*

N	$m = 1$	$m = 2$	$m = 3$	$m = 4$
8	0.273E+02	0.136E+04	0.552E+05	0.237E+07
16	0.232E+01	0.295E+03	0.247E+05	0.168E+07
32	0.651E−02	0.268E+01	0.678E+03	0.126E+06
64	0.164E−07	0.245E−04	0.221E−01	0.143E+02
128	0.568E−13	0.125E−11	0.582E−10	0.931E−09
256	0.568E−13	0.296E−11	0.873E−10	0.349E−08
512	0.568E−13	0.341E−11	0.873E−10	0.442E−08
1024	0.853E−13	0.341E−11	0.873E−10	0.442E−08
Accuracy	$\mathcal{O}(\varepsilon_M N_0)$	$\mathcal{O}(\varepsilon_M N_0^2)$	$\mathcal{O}(\varepsilon_M N_0^3)$	$\mathcal{O}(\varepsilon_M N_0^4)$

The situation for the discrete expansion is rather different. Consider the Chebyshev Gauss–Lobatto expansion

$$\mathcal{I}_N u(x) = \sum_{n=0}^{N} \tilde{u}_n T_n(x), \quad \tilde{u}_n = \frac{2}{\bar{c}_n N} \sum_{j=0}^{N} \frac{1}{\bar{c}_j} u(x_j) T_n(x_j),$$

where the Gauss–Lobatto quadrature points are

$$x_j = -\cos\left(\frac{\pi}{N} j\right), \quad j \in [0, N].$$

The discrete approximation to the mth derivative of $u(x)$ is obtained, just as for the continuous expansion, using the backward recursion repeatedly. In the following example we consider the accuracy of this approach.

Example 11.4 Let us again consider the function, $u(x) \in C^\infty[-1, 1]$,

$$u(x) = \frac{1}{x + a}, \quad a > 1,$$

where we compute the discrete Chebyshev expansion coefficients using a standard fast cosine transform algorithm and we use the analytic backward recursion formulas to calculate the expansion coefficients for the higher derivatives. In Table 11.4 we list the maximum pointwise errors, for increasing resolution and order of derivative, for $a = 1.1$. Compare these results with those in Table 11.3.

Table 11.4 *Maximum pointwise error of a discrete spatial Chebyshev m-derivative for increasing resolution, N. The test function is given in Example 11.4. Machine accuracy is* $\varepsilon_M \sim 1.0E{-}16$.

N	$m = 1$	$m = 2$	$m = 3$	$m = 4$
8	0.571E+01	0.403E+02	0.276E+04	0.119E+06
16	0.485E+00	0.936E+01	0.436E+04	0.443E+06
32	0.912E−03	0.771E−01	0.129E+03	0.404E+05
64	0.130E−08	0.514E−06	0.289E−02	0.333E+01
128	0.154E−09	0.474E−06	0.104E−02	0.179E+01
256	0.527E−08	0.636E−04	0.560E+00	0.390E+04
512	0.237E−08	0.374E−03	0.203E+02	0.723E+06
1024	0.227E−07	0.362E−01	0.458E+04	0.457E+09
Accuracy	$\mathcal{O}(\varepsilon_M N^3)$	$\mathcal{O}(\varepsilon_M N^5)$	$\mathcal{O}(\varepsilon_M N^7)$	$\mathcal{O}(\varepsilon_M N^9)$

The errors resulting from the use of the discrete Chebyshev expansion coefficients are significantly larger than those resulting from the use of the continuous coefficients. The error increases rapidly with the number of modes and with the order of the derivative and we see that the decay of the error

$$\max_{x\in[-1,1]} \left|u^{(m)}(x) - \mathcal{I}_N u^{(m)}(x)\right| \sim \mathcal{O}(\varepsilon_M N^{2m+1}) \quad \text{as} \quad N \to \infty.$$

The strong dependence on N observed in Table 11.4 implies that for high values of N and/or m it is impossible to approximate the derivative of the function with any reasonable accuracy.

The problem lies in the combination of the cosine transform and the backward recursion used to determine the expansion coefficients for the higher derivatives. The backward recursion leads to an $\mathcal{O}(N^2)$ amplification of the initial round-off error of $\mathcal{O}(\varepsilon_M N)$ resulting from the transform. This last term could be avoided by using a direct summation, but this may be prohibitively expensive for large N. The ill-conditioning of the backward recursion has the unfortunate consequence that the approximation of high-order derivatives remains a non-trivial task when using polynomial methods and must be done carefully. Comparing the results illustrated in Example 11.3 and Example 11.4 it becomes clear that the most important issue here is the accuracy of the computation of the discrete expansion coefficients. Once again we conclude that the discrete expansion coefficients should always be computed with the highest possible accuracy.

Using the expansion coefficients and backward recursion is mathematically equivalent to using the differentiation matrices, but the two methods are numerically very different. Let us consider the discrete Chebyshev approximation using the interpolating Lagrange polynomials

$$\mathcal{I}_N u(x) = \sum_{j=0}^{N} u(x_j) l_j(x) = \sum_{j=0}^{N} u(x_j) \frac{(-1)^{N+j+1}(1-x^2)T_N'(x)}{\bar{c}_j N^2 (x - x_j)},$$

based on the Chebyshev Gauss–Lobatto quadrature points

$$x_j = -\cos\left(\frac{\pi}{N} j\right), \quad j \in [0, \ldots, N].$$

Differentiation is then accomplished through a matrix vector product

$$\left.\frac{d\mathcal{I}_N u}{dx}\right|_{x_j} = \sum_{i=0}^{N} \mathrm{D}_{ji} u(x_i),$$

where the entries of the differentiation matrix are

$$\mathrm{D}_{ij} = \begin{cases} -\frac{2N^2+1}{6} & i = j = 0 \\ \frac{\bar{c}_i}{\bar{c}_j} \frac{(-1)^{i+j}}{x_i - x_j} & i \neq j \\ -\frac{x_i}{2\left(1-x_i^2\right)} & i = j \in [1, \ldots, N-1] \\ \frac{2N^2+1}{6} & i = j = N. \end{cases} \tag{11.4}$$

We consider the accuracy of this approach in the following example.

Example 11.5 Consider the function,

$$u(x) = \frac{1}{x + a}, \quad a > 1.$$

We compute its derivatives using the differentiation matrix D,

$$\boldsymbol{u}^{(1)} = \mathrm{D}\boldsymbol{u}, \quad \boldsymbol{u}^{(m)} = \mathrm{D}^m \boldsymbol{u}.$$

In Table 11.5 we list the maximum pointwise error, for increasing resolution and order of derivative, for $a = 1.1$.

The decay of the errors is clearly very disappointing,

$$\max_{x \in [-1,1]} \left| u^{(m)}(x) - \mathcal{I}_N u^{(m)}(x) \right| \sim \mathcal{O}\left(\varepsilon_M N^{2m+2}\right) \quad \text{as} \quad N \to \infty.$$

Fortunately, there are several things that can be done to improve on this result. But first we must understand what causes the sensitivity to round-off errors.

All off-diagonal entries of the matrix D are

$$\mathrm{D}_{ij} \sim C \frac{1}{x_i - x_j}.$$

Table 11.5 *Maximum pointwise error of a discrete spatial Chebyshev m-derivative for increasing resolution, N. The test function is given in Example 11.5. Machine accuracy is $\varepsilon_M \sim 1.0E{-}16$.*

N	$m = 1$	$m = 2$	$m = 3$	$m = 4$
8	0.201E+02	0.127E+04	0.545E+05	0.237E+07
16	0.116E+01	0.221E+03	0.218E+05	0.160E+07
32	0.191E−02	0.134E+01	0.441E+03	0.956E+05
64	0.276E−08	0.741E−05	0.917E−02	0.731E+01
128	0.633E−08	0.458E−04	0.161E+00	0.386E+03
256	0.139E−06	0.406E−02	0.589E+02	0.578E+06
512	0.178E−05	0.178E+00	0.983E+04	0.379E+09
1024	0.202E−04	0.837E+01	0.200E+07	0.325E+12
Accuracy	$\mathcal{O}(\varepsilon_M N^4)$	$\mathcal{O}(\varepsilon_M N^6)$	$\mathcal{O}(\varepsilon_M N^8)$	$\mathcal{O}(\varepsilon_M N^{10})$

Close to the boundary, $x = \pm 1$, this term scales like

$$\begin{aligned}
\mathrm{D}_{ij} &\sim \frac{1}{x_i - x_j} \sim \frac{1}{1-\cos(\pi/N)} \sim \frac{1}{\mathcal{O}(N^{-2}) + \varepsilon_M} \\
&\sim \frac{\mathcal{O}(N^2)}{1 + \mathcal{O}\left(\varepsilon_M N^2\right)} \sim \mathcal{O}(N^2)(1 - \mathcal{O}\left(\varepsilon_M N^2\right)) \\
&\sim \mathcal{O}(N^2) + \mathcal{O}\left(\varepsilon_M N^4\right),
\end{aligned}$$

i.e., the differentiation matrix scales as $\mathcal{O}(\varepsilon_M N^4)$ reflecting the observed scaling in Table 11.5. What can be done about this? The subtraction of almost equal numbers can be avoided by using trigonometric identities so that the entries of D are initialized

$$\mathrm{D}_{ij} = \begin{cases} -\frac{2N^2+1}{6} & i = j = 0 \\ \frac{\bar{c}_i}{2\bar{c}_j} \frac{(-1)^{i+j}}{\sin\left(\frac{i+j}{2N}\pi\right)\sin\left(\frac{i-j}{2N}\pi\right)} & i \neq j \\ -\frac{x_i}{2\sin^2\left(\frac{\pi}{N} i\right)} & i = j \in [1, \ldots, N-1] \\ \frac{2N^2+1}{6} & i = j = N. \end{cases}$$

A direct implementation of this matrix reduces the error, making the accuracy scale like

$$\max_{x \in [-1,1]} \left| u^{(m)}(x) - \mathcal{I}_N u^{(m)}(x) \right| \sim \mathcal{O}\left(\varepsilon_M N^{2m+1}\right) \quad \text{as} \quad N \to \infty,$$

which is similar to the scaling using the fast cosine transform and the backward recursion. However, it is in fact possible to make the matrix-vector method perform even better, as indicated by the scaling which is $\mathcal{O}(N^2)$. The solution lies hidden in the computation of the elements for $i \sim j \sim N$. For these elements, the computation we perform is similar to an evaluation of the function $\sin(\pi - \delta)$, where $\delta \ll \pi$, i.e., this operation is sensitive to round-off error effects. Indeed, by making an estimate of the condition number, as above, we obtain

$$
\begin{aligned}
\mathrm{D}_{ij} &\sim \frac{1}{\sin(\delta)\sin(\pi-\delta)} \sim \frac{1}{\mathcal{O}(N^{-1})(\mathcal{O}(N^{-1})+\varepsilon_M)} \\
&\sim \frac{\mathcal{O}(N^2)}{1+\mathcal{O}(\varepsilon_M N)} \sim \mathcal{O}(N^2)+\mathcal{O}(\varepsilon_M N^3),
\end{aligned}
$$

and recover a condition number of $\mathcal{O}(\varepsilon_M N^3)$ as expected. This analysis also suggests a way to avoid this problem since it happens only for $i \sim j \sim N$. The remedy is to use the centro-antisymmetry of D so that the entries of the differentiation matrix are

$$
\mathrm{D}_{ij} = \begin{cases}
-\frac{2N^2+1}{6} & i=j=0 \\
\frac{\bar{c}_i}{2\bar{c}_j}\frac{(-1)^{i+j}}{\sin\left(\frac{i+j}{2N}\pi\right)\sin\left(\frac{i-j}{2N}\pi\right)} & i \in [0,\ldots,N/2] \neq j \in [0,\ldots,N] \\
-\frac{x_i}{2\sin^2\left(\frac{\pi}{N}i\right)} & i=j \in [1,N/2] \\
-\mathrm{D}_{N-i,N-j} & i \in [N/2+1,\ldots,N],\ j \in [0,\ldots,N]
\end{cases} \qquad (11.5)
$$

i.e., only the upper half of the matrix is computed while the lower half is obtained by using the centro-antisymmetry. We illustrate the accuracy of this approach in the next example.

Example 11.6 Consider the function

$$
u(x) = \frac{1}{x+a}, \qquad a > 1,
$$

where we compute derivatives using the differentiation matrix, D, implemented using the trigonometric identities and the centro-antisymmetry, Equation (11.5). Higher derivatives are computed by repeatedly multiplying by the differentiation matrix

$$
\boldsymbol{u}^{(1)} = \mathrm{D}\boldsymbol{u}, \quad \boldsymbol{u}^{(m)} = \mathrm{D}^m\boldsymbol{u}.
$$

In Table 11.6 we list the maximum pointwise error as obtained for increasing resolution and order of derivative for $a = 1.1$.

From Table 11.6 we recover an estimate of error decay

$$
\max_{x\in[-1,1]} \left|u^{(m)}(x) - \mathcal{I}_N u^{(m)}(x)\right| \sim \mathcal{O}\left(\varepsilon_M N^{2m}\right) \quad \text{as} \quad N \to \infty,
$$

which is even better than using a standard cosine transform and the backward recursion. It also illustrates the care that has to be exercised when initializing the entries of differentiation matrices for polynomial methods.

Table 11.6 *Maximum pointwise error of a discrete spatial Chebyshev m-derivative for increasing resolution, N. The test function is given in Example 11.6. Machine accuracy is $\varepsilon_M \sim 1.0E\text{–}16$.*

N	$m = 1$	$m = 2$	$m = 3$	$m = 4$
8	0.201E+02	0.127E+04	0.545E+05	0.237E+07
16	0.116E+01	0.221E+03	0.218E+05	0.160E+07
32	0.191E−02	0.134E+01	0.441E+03	0.956E+05
64	0.262E−08	0.721E−05	0.901E−02	0.721E+01
128	0.203E−10	0.467E−07	0.437E−04	0.196E+00
256	0.554E−10	0.113E−05	0.128E−01	0.109E+03
512	0.354E−09	0.201E−04	0.871E+00	0.304E+05
1024	0.182E−08	0.744E−03	0.153E+03	0.214E+08
Accuracy	$\mathcal{O}(\varepsilon_M N^2)$	$\mathcal{O}(\varepsilon_M N^4)$	$\mathcal{O}(\varepsilon_M N^6)$	$\mathcal{O}(\varepsilon_M N^8)$

For methods other than the Chebyshev, the situation is worse. Trigonometric identities can only be used for the Chebyshev case. The centro-antisymmetry of D, on the other hand, is shared among all the differentiation matrices. However, the effect of using this for more general polynomials remains unknown. Nevertheless, using the even-odd splitting for computing derivatives results in an error that scales somewhat like that seen in Example 11.6 also for Legendre differentiation matrices, provided care is exercised in computing the entries of D. This suggests that the use of the centro-antisymmetry, which is implicit in the even-odd splitting, does indeed result in a smaller condition number of the differentiation matrices. We also emphasize that, whenever available, the exact entries of $\mathrm{D}^{(m)}$ should be used rather than computed by multiplying matrices.

For the general polynomial case, one could also use the assumption that the differentiation of a constant has to vanish, i.e.,

$$\sum_{j=0}^{N} \mathrm{D}_{ij} = 0.$$

One may then compute the diagonal elements of the differentiation matrix

$$\mathrm{D}_{ii} = -\sum_{\substack{j=0 \\ j \neq i}}^{N} \mathrm{D}_{ij} \quad \forall i \in [0, \ldots, N],$$

so that the round-off errors incurred in the computation of the entries are somehow accounted for in the diagonal elements. However, for Chebyshev polynomials the techniques listed above are superior, and in general this last technique should only be used when nothing more specific is available.

11.4 On the use of mappings

Mappings are used most frequently for the treatment of geometries different from the standard $(0, 2\pi)$ or $(-1, 1)$ intervals. Mappings can also be used for improving accuracy in the computation of high-order derivatives.

Consider the function, $u(x) \in L^2[a, b]$, where $a < b$ while a and/or b may be infinite. We make a change of variables through the mapping function $\psi(\xi)$

$$\psi(\xi) : \mathsf{I} \to [a, b] \quad \text{as} \quad x = \psi(\xi),$$

where $\mathsf{I} = [\xi_{\min}, \xi_{\max}]$ represents the Fourier methods interval $[0, 2\pi]$, or the polynomial expansions interval $[-1, 1]$; a differentiation gives

$$dx = \psi'(\xi) d\xi,$$

so that the magnitude of ψ' supplies a measure of how the nodes are distorted relative to the standard grid given in I. For $\psi' < 1$ the grid is compressed, whereas it is dilated when $\psi' > 1$.

When mapping in spectral methods we need to compute derivatives in x, but we only know how to compute derivatives with respect to ξ, so we use the chain rule for differentiation

$$\frac{d}{dx} = \frac{1}{\psi'} \frac{d}{d\xi}, \quad \frac{d^2}{dx^2} = \frac{1}{(\psi')^3} \left(\psi' \frac{d^2}{d\xi^2} - \psi'' \frac{d}{d\xi} \right),$$

and similarly for higher derivatives.

The simplest example of a mapping function is for mapping I onto the finite interval $[a, b]$,

$$x = \psi(\xi) = a + \frac{\xi_{\max} + \xi}{\xi_{\max} - \xi_{\min}} (b - a), \qquad \psi'(\xi) = \frac{b - a}{2}.$$

Before discussing more general mapping functions, we need to address the issue of implementation. We will restrict our attention to polynomial expansions, but note that everything in this discussion carries over to Fourier methods.

In the case of Galerkin or tau methods, we consider approximations of the form

$$\mathcal{P}_N u(\psi(\xi)) = \sum_{n=0}^{N} \hat{u}_n P_n^{(\alpha)}(\xi),$$

where

$$\hat{u}_n = \frac{1}{\gamma_n} \int_{-1}^{1} u(\psi(\xi)) P_n^{(\alpha)}(\xi) w(\xi) \, d\xi.$$

Within this formulation, we can get an approximation of the derivative of the mapped function $u(x)$

$$\mathcal{P}_N u^{(1)}(\psi(\xi)) = \frac{1}{\psi'(\xi)} \sum_{n=0}^{N} \hat{u}_n^{(1)} P_n^{(\alpha)}(\xi),$$

where $\hat{u}_n^{(1)}$ are the expansion coefficients for the first derivative, which are obtained through the backward recursion. In the case of Galerkin and tau methods the unknowns are the expansion coefficients, $\hat{u}_n$, for which we need to obtain equations. So we also need to expand the mapping function, $\psi'(\xi)$

$$\mathcal{P}_N \left(\frac{1}{\psi'(\xi)} \right) = \sum_{n=0}^{N} \hat{\psi}_n P_n^{(\alpha)}(\xi),$$

such that

$$\mathcal{P}_N u^{(1)}(\psi(\xi)) = \sum_{n,l=0}^{N} \hat{\psi}_l \hat{u}_n P_l^{(\alpha)}(\xi) P_n^{(\alpha)}(\xi),$$

which is a convolution sum. Although there is an expression for the convolution of the polynomials, it is usually very complicated, with the exception of a few special cases such as the Chebyshev polynomials. Thus, the expression for the expansion coefficients becomes complicated and general mappings are, for this reason, rarely used in Galerkin and tau methods. The exception is in cases where the mapping is particularly simple, e.g., the linear mapping, where the mapping derivative becomes a constant and the convolution reduces to a multiplication. Another example where the mapping is reasonably simple is the case where the mapping function, $1/\psi'(\xi)$, is a non-singular rational function in ξ, in which case the convolution operator becomes a banded and sparse matrix operator as a consequence of the three-term recurrence relations valid for orthogonal polynomials.

In collocation methods, the use of mapping is much simpler. In this case we seek approximations of the form

$$\mathcal{I}_N u(\psi(\xi)) = \sum_{n=0}^{N} \tilde{u}_n P_n^{(\alpha)}(\xi) = \sum_{j=0}^{N} u(\psi(\xi_j)) l_j(\xi),$$

where ξ_j are gridpoints in the standard interval I. The discrete expansion coefficients, $\tilde{u}_n$, are found using a quadrature formula and $l_j(\xi)$ is the interpolating Lagrange polynomial associated with the grid points. In this case we wish to compute derivatives of $u(x)$ at the grid points

$$\mathcal{I}_N u^{(1)}(\psi(\xi_i)) = \frac{1}{\psi'(\xi_i)} \sum_{n=0}^{N} \tilde{u}_n^{(1)} P_n^{(\alpha)}(\xi_i) = \frac{1}{\psi'(\xi_i)} \sum_{j=0}^{N} u(\psi(\xi_j)) \mathrm{D}_{ij},$$

where D represents the differentiation matrix associated with $l_j(\xi)$ and the grid points, ξ_i. However, since everything is done in point-space the mapping just corresponds to multiplying with a constant at each grid point. If we introduce the diagonal matrix

$$\mathrm{M}^{(m)}_{ii} = \psi^{(m)}(\xi_i),$$

the mapping is accomplished by multiplying the inverse of this matrix with the solution vector following the computation of the derivative, at the grid points. When using the m-order differentiation matrix, $\mathrm{D}^{(m)}$, for the computation of the derivative, we obtain

$$\boldsymbol{u}^{(1)} = \left(\mathrm{M}^{(1)}\right)^{-1} \mathrm{D}\boldsymbol{u}, \quad \boldsymbol{u}^{(2)} = \left(\mathrm{M}^{(1)}\right)^{-3} \left(\mathrm{M}^{(1)}\mathrm{D}^{(2)} - \mathrm{M}^{(2)}\mathrm{D}^{(1)}\right) \boldsymbol{u},$$

where $\boldsymbol{u}$ is the solution vector at the grid points and similarly $\boldsymbol{u}^{(1)}$ and $\boldsymbol{u}^{(2)}$ are the first and second derivative, respectively, at the nodal points. Mapped higher derivatives may be obtained similarly, in this easy way. Note that since $\mathrm{M}^{(m)}$ is diagonal very little computational overhead is introduced.

11.4.1 Local refinement using Fourier methods

What are the general properties that the mapping functions must satisfy so that the mapped functions retain spectral accuracy?

Consider the general function, $u(x) \in L^2[a, b]$ and the mapping, $\psi(\xi) : \mathsf{I} \to [a, b]$; the continuous expansion coefficients are

$$\begin{aligned} 2\pi \hat{u}_n &= \int_0^{2\pi} u(\psi(\xi))e^{-in\xi}\, d\xi \\ &= \frac{-1}{in}\left[u(\psi(2\pi)) - u(\psi(0))\right] + \frac{1}{in}\int_0^{2\pi} \psi'(\xi)u'(\psi(\xi))e^{-in\xi}\, d\xi. \end{aligned}$$

To maintain spectral accuracy $\psi(\xi)$ must satisfy the same requirements as $u(x)$, i.e., $\psi(\xi)$ and its derivatives have to be periodic and smooth.

Now let's consider two mappings which are useful in connection with Fourier methods. Many problems have periodic solutions which are rapidly changing in some regions and slowly changing in others. For such problems it seems natural to cluster the grid points around the steep gradients of the solution. This can be done by mapping the equidistant grid to increase the local resolution.

One choice of a mapping function having this effect on a periodic function defined on the interval $\xi \in [-\pi, \pi]$, is the arctan mapping

$$x = \psi(\xi) = 2 \arctan\left(L \tan \frac{\xi - \xi_0}{2}\right), \quad \psi'(\xi) = \frac{L\left(1 + \tan^2 \frac{\xi - \xi_0}{2}\right)}{1 + L^2 \tan^2 \frac{\xi - \xi_0}{2}}, \qquad (11.6)$$

where $L \leq 1$ is a control parameter for the amount of clustering that is required

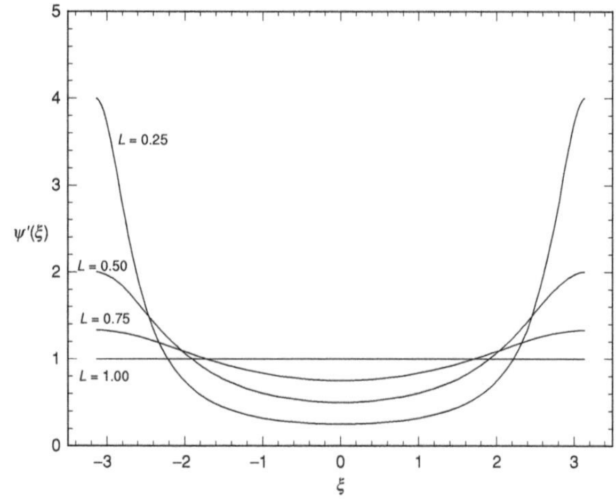

Figure 11.1 Illustration of clustering of grid points around $\xi_0 = 0$ using the arctan mapping, Equation (11.6), for different values of the mapping parameter, L.

around ξ_0. Clearly, for $L = 1$ the mapping reduces to a unity mapping. The best way to understand the effect of this mapping is to recall that

$$dx = \psi'(\xi)d\xi,$$

i.e., since $d\xi$ is a constant on the equidistant grid we obtain clustering where $\psi'(\xi) < 1$ and stretching of the grid where $\psi'(\xi) > 1$. This is illustrated in Figure 11.1 where we plot the value of $\psi'(\xi)$ for different values of the mapping parameter, L.

We observe a clear clustering of the grid points around $\xi_0 = 0$ and find that the size of L controls the amount of clustering, with increasing clustering around ξ_0 for decreasing L.

Since $\psi'(\xi)$ consists of trigonometric functions, periodicity of $u(x)$ is preserved through the mapping, for any order of differentiation. Moreover, the mapping function is smooth and introduces no singularities in the domain. Consequently, we expect the mapping to preserve spectral accuracy of the approximation of a smooth function.

An alternative mapping with similar properties is

$$x = \psi(\xi) = \arctan\left(\frac{(1-\beta^2)\sin(\xi-\xi_0)}{(1+\beta^2)\cos(\xi-\xi_0)+2\beta}\right), \tag{11.7}$$

$$\psi'(\xi) = \frac{1-\beta^2}{1+\beta^2+2\beta\cos(\xi-\xi_0)},$$

where $|\beta| < 1$ controls the clustering around $\xi_0 \in [-\pi, \pi]$. For reasons of comparison we plot in Figure 11.2 the mapping derivative for several values of β. Note that the mapping becomes singular for $\beta = 1$, while $\beta = 0$ corresponds to

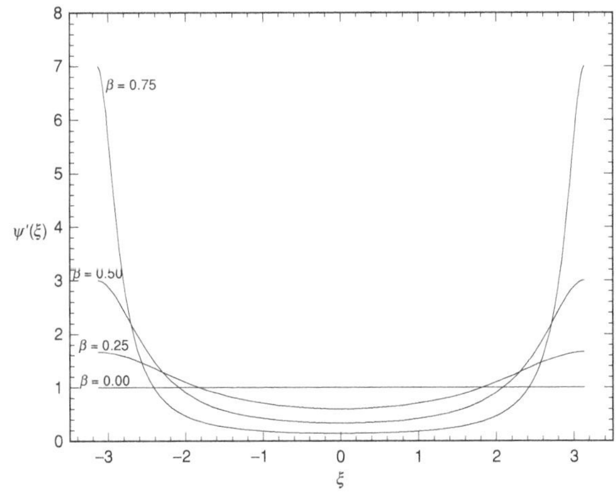

Figure 11.2 Illustration of clustering of grid points around $\xi_0 = 0$ using the mapping (11.7) for different values of the mapping parameter, β.

a unity mapping. This mapping also preserves periodicity and spectral accuracy of the approximation. Comparing the two schemes as illustrated in Figure 11.1 and Figure 11.2, we observe that the latter mapping leads to a less localized clustering around ξ_0 which is, in many cases, a desirable property.

11.4.2 Mapping functions for polynomial methods

In the following we examine mappings for methods based on the ultraspherical polynomials.

Consider a function $u(x) \in L^2_w[a, b]$ expanded in ultraspherical polynomials

$$\mathcal{I}_N u(\psi(\xi)) = \sum_{n=0}^{N} \tilde{u}_n P_n^{(\alpha)}(\xi) = \sum_{j=0}^{N} u(\psi(\xi_j)) l_j(\xi).$$

Performing the usual integration by parts procedure to study the rate of decay of the expansion coefficients, it is easily established that $\psi'(\xi)$ must be a smooth function to maintain spectral convergence and $\psi'(\pm 1)$ must be bounded to avoid boundary effects.

In the following we shall discuss mappings that allow for the use of ultraspherical polynomials, and in particular Chebyshev polynomials, for the approximation of problems in the semi-infinite and infinite interval. However, we shall also discuss a different use of mappings that results in an increased accuracy with which derivatives may be computed.

Treatment of semi-infinite intervals The most straightforward way of approximating problems in the semi-infinite interval is to use expansions of Laguerre polynomials. However, the lack of fast Laguerre transforms and the poor convergence properties of these polynomials suggests that alternative methods should be considered.

While methods based on ultraspherical polynomials are attractive, they are limited to finite domains. Attention has to be paid to the approximation of the mapped function since we have to impose a singular mapping in order to map infinity into a finite value. An appropriate guideline is that uniform spectral convergence may be maintained if the function, $u(x)$, and the mapping function, $x = \psi(\xi)$, both are sufficiently smooth, and the function, $u(x)$, decays fast enough without severe oscillations towards infinity.

A widely used mapping is the exponential mapping, $\psi(\xi) : \mathsf{I} \to [0, \infty)$, as

$$x = \psi(\xi) = -L \ln\left(\frac{1-\xi}{2}\right), \quad \psi'(\xi) = \frac{L}{1-\xi},$$

where L is a scale length of the mapping. We note that the grid is distorted with a linear rate towards infinity and that no singular behavior at $x = 0$ is introduced. This mapping has been given significant attention due to its rather complicated behavior. It has been shown that we can expect to maintain the spectral convergence only for functions that decay at least exponentially towards infinity. This result, however, is based on asymptotic arguments and good results have been reported. The reason is the logarithmic behavior which results in a strong stretching of the grid. This may behave well in many cases, while in other situations it may result in a slowly convergent approximation.

An alternative to the exponential map is the algebraic mapping

$$x = \psi(\xi) = L\frac{1+\xi}{1-\xi}, \quad \psi'(\xi) = \frac{2L}{(1-\xi)^2},$$

where L again plays the role of a scale length. This mapping has been studied in great detail and, used in Chebyshev approximations, the mapped basis functions have been dubbed **rational Chebyshev polynomials**, defined as

$$TL_n(x) = T_n\left(\frac{x-L}{x+L}\right).$$

This family is defined for $x \in [0, \infty)$ and orthogonality can be shown. Consequently, we may expect spectral accuracy for approximation of smooth functions. The advantage of using the algebraic mapping is that the function, $u(x)$, only needs to decay algebraically towards infinity or asymptotically towards a constant value in order for the approximation to maintain spectral accuracy. This is contrary to the exponential mapping which requires at least exponential decay. Several authors have found that algebraic mappings are more accurate and robust than exponential mappings, which is the reason for their widespread use.

One should observe that when applying a singular mapping it is not always convenient to choose the Gauss–Lobatto points as collocation points as they include the singular point. The proper choice may be the Gauss–Radau points for the polynomial family.

As an alternative to using a singular mapping, one may truncate the domain and apply a mapping. At first it may seem natural to just apply a linear mapping after the truncation. However, this has the effect of wasting a significant amount of resolution towards infinity where only little is needed. If this is not the case, truncation becomes obsolete.

The idea behind domain truncation is that if the function decays exponentially fast towards infinity, then we will only make an exponentially small error by truncating the interval. This approach yields spectral convergence of the approximation for increasing resolution.

An often-used mapping is the logarithmic mapping function, $\psi(\xi) : \mathsf{I} \to [0, L_{\max}]$, defined as

$$x = \psi(\xi) - L_{\max} \frac{e^{a(1-\xi)} - e^{2a}}{1 - e^{2a}}, \quad \psi'(\xi) = -a\psi(\xi),$$

where a is a tuning parameter.

However, the problem with domain truncation in that for increasing resolution we need to increase the domain size so that the error introduced by truncating the domain will not dominate over the error of the approximation.

Treatment of infinite intervals When approximating functions defined on the infinite interval, we can develop singular mappings which may be used to map the infinite interval into the standard interval such that ultraspherical polynomials can be applied for approximating the function. Similar to the guidelines used for choosing the mapping function on the semi-infinite interval, we can expect that spectral convergence is conserved under the mapping provided the function, $u(x)$, is exponentially decaying and non-oscillatory when approaching infinity. Clearly, the mapping function needs to be singular at both endpoints to allow for mapping of the infinite interval onto the finite standard interval.

As in the semi-infinite case, we can construct an exponential mapping function, $\psi(\xi) : \mathsf{I} \to (-\infty, \infty)$,

$$x = \psi(\xi) = L \tanh^{-1} \xi, \quad \psi'(\xi) = \frac{L}{1 - \xi^2},$$

where L plays the role of a scale length. This mapping requires exponential decay of the function towards infinity to yield spectral accuracy.

Alternatively, we use an algebraic mapping

$$x = \psi(\xi) = L \frac{\xi}{\sqrt{1 - \xi^2}}, \quad \psi'(\xi) = \frac{L}{\sqrt{(1 - \xi^2)^3}},$$

where L again plays the role of a scale length. This mapping has been given significant attention and, used in Chebyshev approximations, a special symbol has been introduced for the **rational Chebyshev polynomials**

$$TB_n(x) = T_n\left(\frac{x}{\sqrt{L^2+x^2}}\right),$$

for which one may prove orthogonality as well as completeness. The advantage of applying this mapping is that spectral accuracy of the approximation may be obtained even when the function decays only algebraically or asymptotically converges towards a constant value at infinity.

We note that the proper choice of collocation points on the infinite interval may, in certain cases, not be the usual Gauss–Lobatto points but rather the Gauss quadrature points.

Mappings for accuracy improvement As a final example of the use of mappings we return to the problem of round-off errors in pseudospectral methods. In many problems in physics the partial differential equation includes derivatives of high order, e.g., third-order derivatives in the Korteweg–de Vries equation. Additionally, such equations often introduce very complex behavior, thus requiring a large number of modes in the polynomial expansion. For problems of this type, the effect of round-off error becomes a significant issue as the polynomial differential operators are ill conditioned. Even for moderate values of m and N, this problem can ruin the numerical scheme.

To alleviate this problem, at least partially, one may apply the mapping, $\psi(\xi) : \mathsf{I} \to \mathsf{I}$, as

$$x = \psi(\xi) = \frac{\arcsin(\alpha\xi)}{\arcsin\alpha}, \quad \psi'(\xi) = \frac{\alpha}{\arcsin\alpha}\frac{1}{\sqrt{1-(\alpha\xi)^2}}, \tag{11.8}$$

where α controls the mapping. This mapping is singular for $\xi = \pm\alpha^{-1}$. It may be shown that the error, ε, introduced by applying the mapping is related to α by

$$\alpha = \cosh^{-1}\left(\frac{|\ln\varepsilon|}{N}\right),$$

i.e., by choosing $\varepsilon \sim \varepsilon_M$, the error introduced by the mapping is guaranteed to be harmless.

The effect of the mapping is to stretch the grid close to the boundary points. This is easily realized by considering the two limiting values of α;

$$\begin{aligned} \alpha \to 0 \quad & \Delta_{\min}x \to 1 - \cos\tfrac{\pi}{N}, \\ \alpha \to 1 \quad & \Delta_{\min}x \to \tfrac{2}{N}, \end{aligned}$$

where $\Delta_{\min}x$ represents the minimum grid spacing. We observe that for α approaching one, the grid is mapped to an equidistant grid. In the opposite

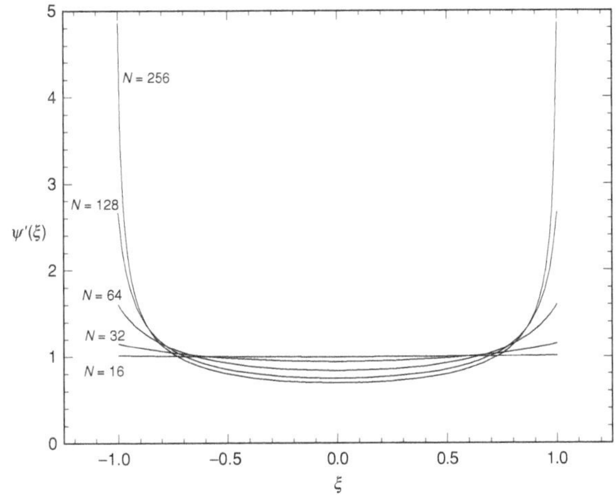

Figure 11.3 Illustration of the effect of the mapping used for accuracy improvement (Equation (11.8)) when evaluating spatial derivatives at increasing resolution.

limit, the grid is equivalent to the well known Chebyshev Gauss–Lobatto grid. One should note that the limit of one is approached when increasing N, i.e., it is advantageous to evaluate high-order derivatives with high resolution at an almost equidistant grid. In Figure 11.3 we plot the mapping derivative for different resolution with the optimal value of α. This clearly illustrates that the mapping gets stronger for increasing resolution. Another strategy is to choose ε to be of the order of the approximation, hence balancing the error.

Example 11.7 Consider the function

$$u(x) = \sin(2x), \quad x \in [-1, 1].$$

We wish to evaluate the first four derivatives of this function using a standard Chebyshev collocation method with the entries given in Equation (11.5). In Table 11.7 we list the maximum pointwise error that is obtained for increasing resolution.

We clearly observe the effect of the round-off error and it is obvious that only very moderate resolution can be used in connection with the evaluation of high-order derivatives.

We apply the singular mapping in the hope that the accuracy of the derivatives improve. In Table 11.8 we list the maximum pointwise error for derivatives with increasing resolution. For information we also list the optimal value for α as found for a machine accuracy of $\varepsilon_M \simeq 1.0E\text{–}16$.

The effect of applying the mapping is to gain at least an order of magnitude in accuracy and significantly more for high derivatives and large N.

Table 11.7 *Maximum pointwise error of a spatial derivative of order m, for increasing resolution, N, for the function in Example 11.7, as obtained using a standard Chebyshev collocation method.*

N	$m = 1$	$m = 2$	$m = 3$	$m = 4$
8	0.155E−03	0.665E−02	0.126E+00	0.142E+01
16	0.316E−12	0.553E−10	0.428E−08	0.207E−06
32	0.563E−13	0.171E−10	0.331E−08	0.484E−06
64	0.574E−13	0.159E−09	0.174E−06	0.111E−03
128	0.512E−12	0.331E−08	0.124E−04	0.321E−01
256	0.758E−12	0.708E−08	0.303E−03	0.432E+01
512	0.186E−10	0.233E−05	0.143E+00	0.587E+04
1024	0.913E−10	0.361E−04	0.756E+01	0.109E+07

Table 11.8 *Maximum pointwise error of a spatial derivative of order m, for increasing resolution, N, as obtained using a mapped Chebyshev collocation method. The mapping is given in Equation 11.8.*

N	α	$m = 1$	$m = 2$	$m = 3$	$m = 4$
8	0.0202	0.154E−03	0.659E−02	0.124E+00	0.141E+01
16	0.1989	0.290E−13	0.383E−11	0.236E−09	0.953E−08
32	0.5760	0.211E−13	0.847E−11	0.231E−08	0.360E−06
64	0.8550	0.180E−12	0.225E−09	0.118E−06	0.436E−04
128	0.9601	0.138E−12	0.227E−09	0.334E−06	0.282E−03
256	0.9898	0.549E−12	0.201E−08	0.262E−05	0.521E−02
512	0.9974	0.949E−11	0.857E−07	0.467E−03	0.180E+01
1024	0.9994	0.198E−10	0.379E−06	0.433E−02	0.344E+02

11.5 Further reading

The even-odd splitting of the differentiation matrices was introduced by Solomonoff (1992) while the classic approach to the computation of Gaussian weights and nodes is due to Golub and Welsch (1969). The study of round-off effects has been initiated by Breuer and Everson (1992), Bayliss et al (1994) and Don and Solomonoff (1995) where the use of the mapping by Kosloff and Tal-Ezer (1993) is also introduced. Choices of the mapping parameter was discussed in Hesthaven et al (1999). The general use of mappings, their analysis and properties is discussed in detail in the text by Boyd (2000) with some early analysis by Bayliss and Turkel (1992).

12

Spectral methods on general grids

So far, we have generally sought to obtain an approximate solution, u_N, by requiring that the residual R_N vanishes in a certain way. Imposing boundary conditions is then done by special choice of the basis, as in the Galerkin method, or by imposing the boundary conditions strongly, i.e., exactly, as in the collocation method.

For the Galerkin method, this causes problems for more complex boundary conditions as one is required to indentify a suitable basis. This is partially overcome in the collocation method, in particular if we have collocation points at the boundary points, although imposing more general boundary operators is also somewhat complex in this approach. A downside of the collocation method is, however, the complexity often associated with establishing stability of the resulting schemes.

These difficulties are often caused by the requirement of having to impose the boundary conditions exactly. However, as we have already seen, this can be circumvented by the use of the penalty method in which the boundary condition is added later. Thus, the construction of u_N and R_N are done independently, e.g., we do not need to use the same points to construct u_N and to require R_N to vanish at.

This expansion of the basic formulation highlights the individual importance of how to approximate the solution, enabling accuracy, and how to satisfy the equations, which accounts for stability, and enables new families of schemes, e.g., stable spectral methods on general grids.

12.1 Representing solutions and operators on general grids

In the previous chapters, we have considered two different ways to represent the approximation $u_N(x)$ the modal Nth order polynomial expansion

$$u_N(x) = \sum_{n=0}^{N} a_n \phi_n(x),$$

where $\phi_n(x)$ is some suitably chosen basis, most often a Legendre or Chebyshev polynomial, and a_n are the expansion coefficients; and alternatively, the nodal representation

$$u_N(x) = \sum_{i=0}^{N} u_N(x_i) l_i(x),$$

where $l_i(x)$ is the Nth order Lagrange polynomial based on any $(N+1)$ independent grid points, x_i.

If we require the modal and nodal expansion to be identical, e.g., by defining a_n to be the discrete expansion coefficents, we have the connection between the expansion coefficients a_n and the grid values $u_N(x_i)$,

$$\mathrm{V}\boldsymbol{a} = \boldsymbol{u}_N,$$

where $\boldsymbol{a} = [a_0, \ldots, a_N]^T$ and $\boldsymbol{u} = [u_N(x_0), \ldots, u_N(x_N)]^T$. The matrix V is defined as

$$\mathrm{V}_{ij} = \phi_i(x_j),$$

and we recover the identity

$$\boldsymbol{a}^T \boldsymbol{\phi}(x) = \boldsymbol{u}_N^T \boldsymbol{l}(x) \Rightarrow \mathrm{V}^T \boldsymbol{l}(x) = \boldsymbol{\phi}(x),$$

where again $\boldsymbol{\phi}(x) = [\phi_0(x), \ldots, \phi_N(x)]^T$ and $\boldsymbol{l}(x) = [l_0(x), \ldots, l_N(x)]^T$. Thus, the matrix, V, transforms between the modal and nodal bases, and can likewise be used to evaluate $\boldsymbol{l}(x)$.

An advantage of this approach is that one can define all operators and operations on general selections of points in higher dimensions, e.g., two or three dimensions, as long as the nodal set allows unique polynomial interpolation. Computing derivatives of the general expansion at the grid points, x_i, amounts to

$$\left.\frac{du_N}{dx}\right|_{x_i} = \sum_{j=0}^{N} u_N(x_j) \left.\frac{dl_j}{dx}\right|_{x_i} = \sum_{j=0}^{N} u_N(x_j) \mathrm{D}_{ij},$$

where D is the differentiation matrix. Instead of explicitly computing $\frac{dl_j}{dx}$ to obtain the differentiation matrix, we can use the relation

$$\mathrm{V}^T \boldsymbol{l}'(x) = \boldsymbol{\phi}'(x) \;\Rightarrow\; \mathrm{V}^T \mathrm{D}^T = (\mathrm{V}')^T \;\Rightarrow\; \mathrm{D} = \mathrm{V}'\mathrm{V}^{-1},$$

where

$$\mathrm{V}'_{ij} = \left.\frac{d\phi_j}{dx}\right|_{x_i}.$$

In the formulation of the Legendre Galerkin method, to be explained later, the symmetric mass matrix, M, is

$$\mathrm{M}_{ij} = \int_{-1}^{1} l_i(x) l_j(x)\, dx.$$

Given an $(N + 1)$ long vector, $\boldsymbol{u}$,

$$\begin{aligned}
\boldsymbol{u}^T \mathrm{M} \boldsymbol{u} &= \sum_{i=0}^{N} \sum_{j=0}^{N} u_i \mathrm{M}_{ij} u_j \\
&= \int_{-1}^{1} \sum_{i=0}^{N} u_i l_i(x) \sum_{j=0}^{N} u_j l_j(x)\, dx \\
&= \int_{-1}^{1} u^2\, dx = \|u\|^2_{L^2[-1,1]}.
\end{aligned}$$

Hence, M is also positive definite and $\boldsymbol{u}^T \mathrm{M} \boldsymbol{u}$ is the L^2 norm of the Nth-order polynomial, u_N, defined by the values u_i at the grid points, x_i, i.e.,

$$u(x) = \sum_{i=0}^{N} u_i l_i(x).$$

In one dimension, one can easily perform this integration in order to obtain the elements M_{ij}. This is less easy in multiple dimensions on complex domains. In such cases, we want to avoid this integration. To efficiently compute the mass matrix we would like to use the Legendre basis $\phi(x)$. We will use V to go between the two bases. Thus we have

$$\begin{aligned}
(\mathrm{V}^T \mathrm{M} \mathrm{V})_{ij} &= \sum_{k=0}^{N} \sum_{l=0}^{N} \mathrm{V}_{ki} \mathrm{M}_{kl} \mathrm{V}_{lj} \\
&= \int_{-1}^{1} \sum_{k=0}^{N} \phi_i(x_k) l_k(x) \sum_{l=0}^{N} \phi_i(x_l) l_l(x)\, dx \\
&= \int_{-1}^{1} \phi_i(x) \phi_j(x)\, dx.
\end{aligned}$$

Since $\phi_i(x) = P_i(x)$ is the Legendre polynomial, which is orthogonal in $L^2[-1, 1]$, the entries in M can be computed using only the transformation matrix, V, and the Legendre normalization, $\gamma_n = 2/(2n + 1)$.

The stiffness matrix is

$$\mathrm{S}_{ij} = \int_{-1}^{1} l_i(x) l_j'(x)\, dx.$$

To compute this, consider

$$\begin{aligned}
(\mathrm{MD})_{ij} &= \sum_{k=0}^{N} \mathrm{M}_{ik}\mathrm{D}_{kj} \\
&= \int_{-1}^{1} l_i(x) \left(\sum_{k=0}^{N} l_k(x) \frac{dl_j(x_k)}{dx} \right) dx \\
&= \int_{-1}^{1} l_i(x) l_j'(x)\, dx = \mathrm{S}_{ij}.
\end{aligned}$$

A property of the stiffness matrix is that

$$\begin{aligned}
\boldsymbol{u}^T \mathrm{S}\boldsymbol{u} = & \sum_{i=0}^{N}\sum_{j=0}^{N} u_i \mathrm{S}_{ij} u_j \\
&= \int_{-1}^{1} \sum_{i=0}^{N} u_i l_i(x) \sum_{j=0}^{N} u_j l_j'(x)\, dx \\
&= \int_{-1}^{1} u u'\, dx = \frac{1}{2}\left(u_N^2 - u_0^2\right).
\end{aligned}$$

Hence, $\boldsymbol{u}^T \mathrm{S}\boldsymbol{u}$ plays the role of an integration by parts of the polynomial, $u_N(x)$.

Using the above formulation, one is free to use any set of points which may be found acceptable, e.g., they may cluster quadratically close to the edges but otherwise be equidistant with the aim of having a more uniform resolution as compared to the Gauss–Lobatto quadrature points.

12.2 Penalty methods

As mentioned before, the penalty method enables us to use spectral methods on general grids in one dimension, and in complex domains in multiple dimensions. We illustrate this with the following example of the simple wave equation.

$$\begin{aligned}
&\frac{\partial u}{\partial t} + a\frac{\partial u}{\partial x} = 0, \quad x \in [-1, 1], \qquad (12.1) \\
&u(-1, t) = g(t), \\
&u(x, 0) = f(x),
\end{aligned}$$

where $a \geq 0$.

The residual

$$R_N(x,t) = \frac{\partial u_N}{\partial t} + a\frac{\partial u_N}{\partial x},$$

must satisfy

$$\int_{-1}^{1} R_N \psi_i^{(1)}(x)\,dx = \oint_{-1}^{1} \boldsymbol{n}[u_N(-1,t) - g(t)]\psi_i^{(2)}(x)\,dx, \tag{12.2}$$

where we have extended the usual approach by now allowing two families of $N+1$ test functions, $\psi_i^{(1)}(x)$ and $\psi_i^{(2)}(x)$. The vector $\boldsymbol{n}$ represents an outward pointing normal vector and takes, in the one-dimensional case, the simple values of $\boldsymbol{n} = \pm 1$ at $x = \pm 1$. Equation (12.2) is the most general form of spectral penalty methods.

Observe that the methods we are familiar with from previous chapters can be found as a subset of the statement in Equation (12.2) by ensuring that a total of $N+1$ test functions are different from zero. Indeed, a classic collocation approach is obtained by defining $\psi_i^{(1)}(x) = \delta(x - x_i)$ and $\psi_i^{(2)}(x) = 0$ for $i = 1, \ldots, N$ and $\psi_0^{(1)}(x) = 0$, $\psi_0^{(2)}(x) = \delta(x+1)$.

The possibilities in the generalization are, however, realized when we allow both the test functions to be nonzero simultaneously. The effect of this is that we do not enforce the equation or boundary conditions separately but rather we enforce both terms at the same time. As we shall see shortly, this leads to schemes with some interesting properties.

These schemes are clearly consistent since the exact solution satisfies the boundary condition, hence making the right hand side vanish.

12.2.1 Galerkin methods

In the Galerkin method we seek a solution of the form

$$u_N(x,t) = \sum_{i=0}^{N} u_N(x_i,t) l_i(x)$$

such that

$$\mathrm{M}\frac{d\boldsymbol{u}}{dt} + a\mathrm{S}\boldsymbol{u} = -\tau \boldsymbol{l}_L[u_N(-1,t) - g(t)], \tag{12.3}$$

where $\boldsymbol{l}_L = [l_0(-1), \ldots, l_N(-1)]^T$, and $\boldsymbol{u} = [u_N(x_0), \ldots, u_N(x_N)]^T$ are the unknowns at the grid points, x_i.

This scheme is stable provided

$$\tau \geq \frac{a}{2}.$$

In practice, taking $\tau = a/2$ gives the best CFL condition. The scheme is

$$\mathrm{M}\frac{d\boldsymbol{u}}{dt} + a\mathrm{S}\boldsymbol{u} = -\frac{a}{2}\boldsymbol{l}_L[u_N(-1,t) - g(t)]. \tag{12.4}$$

The choice of arbitrary gridpoints is possible, since we have separated the spatial operators from the boundary conditions. At the point $x = -1$ we do not enforce the boundary condition exactly but rather weakly in combination with the equation itself.

Upon inverting the matrix M in Equation (12.4) we have

$$\frac{d\boldsymbol{u}}{dt} + a\mathrm{D}\boldsymbol{u} = -\frac{a}{2}\mathrm{M}^{-1}\boldsymbol{l}_L[u_N(-1,t) - g(t)]. \tag{12.5}$$

The connection between this method and the Legendre Galerkin method is explained in the following theorem.

Theorem 12.1 *Let* M *and* $\boldsymbol{l}_L$ *be defined as above. Then*

$$\mathrm{M}^{-1}\boldsymbol{l}_L = \boldsymbol{r} = [r_0, \ldots, r_N]^T,$$

for

$$r_i = (-1)^N \frac{P'_{N+1}(x_i) - P'_N(x_i)}{2},$$

where x_i are the grid points on which the approximation is based.

Proof: We shall prove the theorem by showing that $\boldsymbol{r}$ satisfies

$$\mathrm{M}\boldsymbol{r} = \boldsymbol{l}_L \;\Rightarrow\; (\mathrm{M}\boldsymbol{r})_i = l_i(-1).$$

In fact,

$$\begin{aligned}
(\mathrm{M}\boldsymbol{r})_i &= \int_{-1}^{1} l_i(x) \sum_{k=0}^{N} l_k(x) r_k \, dx \\
&= \int_{-1}^{1} l_i(x)(-1)^N \frac{P'_{N+1}(x) - P'_N(x)}{2} \, dx \\
&= (-1)^N l_i(1) \frac{P_{N+1}(1) - P_N(1)}{2} - (-1)^N l_i(-1) \frac{P_{N+1}(-1) - P_N(-1)}{2} \\
&\quad - \int_{-1}^{1} l'_i(x) \frac{P_{N+1}(x) - P_N(x)}{2} \, dx.
\end{aligned}$$

However, since $P_N(\pm 1) = (\pm 1)^N$ and P_N and P_{N+1} are orthogonal to all polynomials of order less than N, we have

$$(\mathrm{M}\boldsymbol{r})_i = l_i(-1),$$

as stated.

QED

We see from this theorem that the polynomial $u_N(x,t)$ satisfies the equation

$$\frac{\partial u_N}{\partial t} + a\frac{\partial u_N}{\partial x} = -\tau(-1)^N \frac{P'_{N+1}(x) - P'_N(x)}{2}[u_N(-1,t) - g(t)].$$

Since the right hand side in the above is orthogonal to any polynomial which vanishes at the boundary, $x = -1$, the solution, $u_N(x,t)$ is identical to that of the Legendre Galerkin method. This formulation of the Galerkin method enables one to impose complex boundary conditions without having to rederive the method and/or seek a special basis as in the classical Galerkin method.

12.2.2 Collocation methods

The solution of the Legendre collocation penalty method satisfies the following error equation

$$\frac{\partial u_N}{\partial t} + a\frac{\partial u_N}{\partial x} = -\tau \frac{(1-x)P'_N(x)}{2P'_n(-1)}[u_N(-1,t) - g(t)]. \tag{12.6}$$

Note that the residual vanishes at all interior Gauss–Lobatto points and the boundary conditions are satisfied weakly via the penalty formulation. This scheme is stable provided

$$\tau \geq \frac{a}{2\omega_0} = a\frac{N(N+1)}{4},$$

as we showed in Chapter 8.

In our new formulation, the Legendre collocation method can be obtained by approximating the mass and stiffness matrices as follows;

$$\mathrm{M}^c_{ij} = \sum_{k=0}^{N} l_i(y_k) l_j(y_k)\omega_k,$$

and

$$\mathrm{S}^c_{ij} = \sum_{k=0}^{N} l_i(y_k) l'_j(y_k)\omega_k.$$

Note that since $l_i l'_j$ is a polynomial of degree $2N-1$, the quadrature is exact and so $\mathrm{S}^c = \mathrm{S}$.

The resulting scheme then becomes

$$\mathrm{M}^c \frac{d\boldsymbol{u}}{dt} + a\mathrm{S}\boldsymbol{u} = -\tau \boldsymbol{l}_L[u_N(-1,t) - g(t)], \tag{12.7}$$

where, as usual, $\boldsymbol{u}$ is the vector of unknowns at the grid points, x_i.

The following theorem shows that Equation (12.7) is the Legendre collocation method.

Theorem 12.2 *Let* M^c *and* $\boldsymbol{l}_L$ *be defined as in the above. Then*

$$(\mathrm{M}^c)^{-1}\boldsymbol{l}_L = \boldsymbol{r} = [r_0, \ldots, r_N]^T,$$

for

$$r_i = \frac{(1-x_i)P_N'(x_i)}{2\omega_0 P_N'(-1)},$$

where x_i *are the grid points on which the approximation is based.*

Proof: We shall prove this by considering

$$\mathrm{M}^c\boldsymbol{r} = \boldsymbol{l}_L \;\; \Rightarrow \;\; (\mathrm{M}\boldsymbol{r})_i = l_i(-1).$$

Consider

$$\begin{aligned}
2\omega_0 P_N'(-1)(\mathrm{M}^c\boldsymbol{r})_i &= (-1)^N \sum_{k=0}^{N}\sum_{l=0}^{N} l_i(y_l)l_k(y_l)\omega_l(1-x_k)P_N'(x_k)\\
&= (-1)^N \sum_{l=0}^{N} l_i(y_l)\omega_l\left(\sum_{k=0}^{N}(1-x_k)P_N'(x_k)\right)\\
&= (-1)^N \sum_{l=0}^{N} l_i(y_l)(1-y_l)P_N'(y_l)\omega_l\\
&= 2\omega_0 P_N'(-1)l_i(-1).
\end{aligned}$$

QED

The scheme written at the arbitrary points, x_i, is

$$\frac{du_i(t)}{dt} + a(\mathrm{D}\boldsymbol{u})_i = -\tau\frac{(1-x_i)P_N'(x_i)}{2\omega_0 P_N'(-1)}[u_N(-1,t) - g(t)], \tag{12.8}$$

with $\tau \geq a/2$ as the stability condition.

An interesting special case of this latter scheme is obtained by taking x_i to be the Chebyshev Gauss–Lobatto quadrature points while y_i, i.e., the points at which the equation is satisfied, are the Legendre Gauss–Lobatto points. In this case we have

$$\frac{du_i(t)}{dt} + a(\mathrm{D}\boldsymbol{u})_i = -\tau\frac{(1-x_i)P_N'(x_i)}{2P_N'(-1)\omega_0}[u_N(-1,t) - g(t)],$$

known as a Chebyshev–Legendre method. In this case, the penalty term is added at each grid point, not only the boundary. It is in fact a Legendre collocation method although it computes derivatives at Chebyshev Gauss–Lobatto points.

This latter operation can benefit from the fast Fourier transform for large values of N. Fast transform methods do also exist for Legendre transforms but they are less efficient than FFT-based methods.

12.2.3 Generalizations of penalty methods

The penalty formulation treats, with equal ease, complex boundary conditions. Consider the parabolic equation

$$\frac{\partial u}{\partial t} = \frac{\partial^2 u}{\partial x^2}, \quad x \in [-1, 1], \tag{12.9}$$
$$u(-1, t) = g(t), \quad \frac{\partial u}{\partial x}(1, t) = h(t),$$
$$u(x, 0) = f(x).$$

In the penalty formulation, we seek solutions of the form

$$u_N(x, t) = \sum_{i=0}^{N} u_N(x_i, t) l_i(x),$$

and require u_N to satisfy

$$\frac{\partial u_N}{\partial t} - \frac{\partial^2 u_N}{\partial x^2} = -\tau_1 \frac{(1 - x)P_N'(x)}{2P_N'(-1)}[u_N(-1, t) - g(t)]$$
$$+ \tau_2 \frac{(1 + x)P_N'(x)}{2P_N'(1)}[(u_N)_x(1, t) - h(t)].$$

Assume, for simplicity, that the approximation is based on the Legendre Gauss–Lobatto points and we also choose to satisfy the equation at these points, then the scheme is stable for

$$\tau_1 \geq \frac{1}{4\omega^2}, \quad \tau_2 = \frac{1}{\omega},$$

with

$$\omega = \frac{2}{N(N + 1)}.$$

We can easily consider more complex boundary operators, e.g., for the problem in Equation (12.9) one could encounter boundary conditions of the form

$$u(1, t) + \frac{\partial u}{\partial x}(1, t) = g(t).$$

The implementation via penalty methods is likewise straightforward as one only needs to change the scheme at the boundaries.

The ease of the penalty formulation carries over to nonlinear problems. Consider, for example,

$$\frac{\partial u}{\partial t} + \frac{\partial f(u)}{\partial x} = 0, \quad x \in [-1, 1],$$

with suitable initial conditions. Assume for simplicity that

$$\frac{\partial f}{\partial u}(\pm 1, t) \geq 0,$$

such that $x = -1$ is an inflow and $x = 1$ an outflow boundary and assume that $u(-1, t) = g(t)$ is the boundary condition. Following the formulation above, we seek a solution, $u_N(x, t)$, which satisfies the equation

$$\frac{\partial u_N}{\partial t} + \frac{\partial f_N(u_N)}{\partial x} = -\tau a(t) \frac{(1-x)P_N'(x)}{2P_N'(-1)}[u_N(-1, t) - g(t)],$$

where

$$a(t) = \frac{\partial f}{\partial u}(-1, t).$$

If the solution $u(x, t)$ is smooth, a necessary (but not sufficient) condition for stability can be obtained by considering the linearized method, resulting in the condition

$$\tau \geq \frac{1}{2\omega},$$

as discussed previously.

The extension to nonlinear systems follows the exact same path, i.e., one enforces the boundary conditions via the penalty method and establishes necessary conditions for stability by analyzing the linearized constant coefficient case.

Example 12.3 To highlight the flexibilities of this approach, let us consider a slightly more complex case, in which we wish to solve the advection problem

$$\frac{\partial u}{\partial t} + \boldsymbol{v} \cdot u = 0, \quad \boldsymbol{x} \in D = \{x, y \geq -1, x + y \leq 0\},$$

i.e., a two-dimensional triangular domain. As usual, we assume that we have an initial condition, $u(\boldsymbol{x}, 0) = f(\boldsymbol{x})$, as well as a boundary condition, $u(\boldsymbol{x}, t) = g(t)$ at all points of the triangular boundary where $\boldsymbol{n} \cdot \boldsymbol{v} \leq 0$. Let us now seek solutions of the usual kind,

$$u_N(\boldsymbol{x}, t) = \sum_{i=1}^{N_p} u_i(t) L_i(\boldsymbol{x}),$$

where $L_i(\boldsymbol{x})$ are the two-dimensional Lagrange polynomials defined on the triangular domain through the grid points, x_i, of which there are

$$N_p = \frac{(N+1)(N+2)}{2}.$$

This is exactly the number of terms in a two-dimensional polynomial of order N.

Just as in the one-dimensional case, we choose some grids, preferably good for interpolation, as well as a polynomial basis, $\phi_i(\boldsymbol{x})$, which is orthogonal on the triangle. The latter can be found by, e.g., using a Gram–Schmidt orthogonalization on the simple basis, $x^i y^j$. With this, we can follow the one-dimensional approach and define a transformation matrix, V, with the entries

$$\mathrm{V}_{ij} = \phi_j(\boldsymbol{x}_i).$$

This allows us to evaluate the Lagrange polynomial, which in this general case is not known in closed form, as well as initialize the mass matrix, M, and the differentiation matrices, D_x and D_y.

Let us consider the following penalty scheme

$$\int_D \left(\frac{\partial u_N}{\partial t} + \boldsymbol{v} \cdot u_N \right) L_i(\boldsymbol{x})\, d\boldsymbol{x}$$
$$= \tau \int_{\partial D} \left(\frac{|\boldsymbol{n} \cdot \boldsymbol{v}| - \boldsymbol{n} \cdot \boldsymbol{v}}{2} \right) [u_N(\boldsymbol{x}, t) - g(t)] L_i(\boldsymbol{x})\, d\boldsymbol{x}.$$

Note that the right hand side is constructed such that the penalty term is only imposed at the inflow boundaries where $\boldsymbol{n} \cdot \boldsymbol{v} \leq 0$.

In a more compact form this becomes

$$\frac{d\boldsymbol{u}}{dt} + (v_x \mathrm{D}_x + v_y \mathrm{D}_y)\boldsymbol{u}$$
$$= \tau \mathrm{M}^{-1} \int_{\partial D} \left(\frac{|\boldsymbol{n} \cdot \boldsymbol{v}| - \boldsymbol{n} \cdot \boldsymbol{v}}{2} \right) [u_N(\boldsymbol{x}, t) - g(t)] \boldsymbol{L}(\boldsymbol{x})\, d\boldsymbol{x}.$$

Using the exact same approach as for the one-dimensional Galerkin case, e.g., expanding MD_x by integration by parts along x, one easily proves that this scheme is stable for

$$\tau \geq \frac{1}{2}.$$

This last example highlights the strength of the combination of the general grids and the penalty methods in that one can now formulate stable and spectrally accurate schemes on very general domains, requiring only that a robust local interpolation is possible. This extends to general dimensions and, at least in principle, to general domains, although finding interpolations points in general domains is a challenge.

12.3 Discontinuous Galerkin methods

The discontinuous Galerkin method is a different approach to the derivation of the schemes. Consider the nonlinear scalar problem

$$\frac{\partial u}{\partial t} + \frac{\partial f(u)}{\partial x} = 0, \quad x \in [-1, 1],$$

again, we assume that $x = -1$ is an inflow boundary with $u(-1, t) = g(t)$. We seek solutions, $u_N(x, t)$, of the form

$$u_N(x, t) = \sum_{i=0}^{N} u_N(x_i, t) l_i(x),$$

where x_i represents an arbitrary set of grid points.

In the Galerkin approach, we require that the residual is orthogonal to the set of test functions, $l_i(x)$,

$$\int_{-1}^{1} \left(\frac{\partial u_N}{\partial t} + \frac{\partial f_N(u_N)}{\partial x} \right) l_i(x)\, dx = 0.$$

Integration by parts yields

$$\int_{-1}^{1} \left(\frac{\partial u_N}{\partial t} l_i(x) - f_N(u_N) l_i'(x) \right) dx = -[f_N(1) l_i(1) - f_N(-1) l_i(-1)],$$

where we denote $f_N(\pm 1) = f_N(u_N(\pm 1))$. We now need to specify $f_N(u_N(\pm 1))$. A reasonable choice could be

$$f_N(u_N(-1, t)) = f(g(t)),$$

while we leave $f_N(u_N(1, t))$ unchanged. This would lead to the scheme

$$\int_{-1}^{1} \frac{\partial u_N}{\partial t} l_i(x) - f_N(u_N) l_i'(x)\, dx = -[f_N(1) l_i(1) - f(g(t)) l_i(-1)].$$

Using the mass and stiffness matrix formulation, this becomes

$$\mathrm{M}\frac{d\boldsymbol{u}}{dt} - \mathrm{S}^T \boldsymbol{f} = \boldsymbol{l}_L f(g(t)) - \boldsymbol{l}_R f_N(1),$$

where $\boldsymbol{l}_L = [l_0(-1), \ldots, l_N(-1)]^T$ and $\boldsymbol{l}_R = [l_0(1), \ldots, l_N(1)]^T$. This is the standard weak form of the discontinuous Galerkin method with a pure upwind flux. To associate it with the schemes we have discussed in the above, repeat the integration by parts to obtain

$$\mathrm{M}\frac{d\boldsymbol{u}}{dt} + \mathrm{S}\boldsymbol{f} = -\boldsymbol{l}_L [f_N(-1, t) - f(g(t))],$$

as the strong form of the discontinuous Galerkin method. Clearly, this is a special case of the general class of schemes in Equation (12.2) with the minor difference that the penalty term is on the flux function, $f(u)$, rather than on u itself.

Let us denote the choices we made in defining $f_N(\pm 1, t)$, as the general numerical flux, f^*. Clearly, f^* depends on both the local solution and the kind of boundary being considered. With this general form the schemes become

$$\mathrm{M}\frac{d\boldsymbol{u}}{dt} - \mathrm{S}^T \boldsymbol{f} = \boldsymbol{l}_L f^*(-1) - \boldsymbol{l}_R f^*(-1),$$

and

$$\mathrm{M}\frac{d\boldsymbol{u}}{dt} + \mathrm{S}\boldsymbol{f} = \boldsymbol{l}_N(f_N(1) - f^*(1)) - \boldsymbol{l}_0(f_N(-1) - f^*(-1)),$$

as the weak and strong form, respectively.

The choice of the numerical flux, $f^*(u^-, u^+)$, with u^- the local solution and u^+ the exterior solution/boundary condition, is now the key issue to address. Many options are available for the construction of numerical fluxes. A generally suitable and robust choice is the Lax–Friedrich flux

$$f^*(a, b) = \frac{f(a) + f(b)}{2} + \frac{C}{2}(a - b),$$

where $C = \max |f_u|$, and the maximum is taken globally or locally (a local Lax–Friedrich flux).

Discontinuous Galerkin methods have been analyzed extensively, and have been shown to have many attractive properties, in particular for nonlinear conservation laws. The brief discussion in the above can be generalized to systems of nonlinear equations as well as higher order equations by introducing additional equations to maintain the first-order nature of the spatial operator.

The boundary conditions, $g(t)$, which we assume to be specified could also be a solution computed in a neighboring element, i.e., we can construct methods based on multiple spectral elements, coupled by the penalty term and/or the numerical flux in the discontinuous Galerkin formulation. With no restrictions on local polynomial approximation and the position of the grid points, this approach can be extended straightforwardly to multiple dimensions and fully unstructured grids, simply by defining elements of a desired type and sets of grid points which are well suited for interpolation on this element. This leads to the development of stable and robust spectral methods on fully unstructured grids, hence bringing this class of methods to within reach of being applied to solve large scale applications of realistic complexity.

12.4 Further reading

The use of weakly imposed boundary conditions, ultimately leading to the penalty methods, was first proposed for spectral methods in Funaro (1986) for a simple elliptic problem and extended to hyperbolic problems in Funaro and Gottlieb (1988 and 1991). In a sequence of papers by Hesthaven and Gottlieb (1996), and Hesthaven (1997 and 1999), this technique was vastly expanded to include nonlinear systems and complex boundary operators. In Don and Gottlieb (1994) the Chebyshev–Legendre method is proposed and in Trujillo and Karniadakis (1999) it is demonstrated how to impose constraints using a penalty term. A recent review of the development of these techniques can be found in Hesthaven (2000).

The use of these techniques to formulate stable spectral methods on general grids was first proposed in Carpenter and Gottlieb (1996) for simple linear wave equations, and extended to general elements and fully unstructured grids in Hesthaven and Gottlieb (1999), and Hesthaven and Teng (2000), where the distribution of grid points is also discussed. The general form of defining operators on general grids as discussed here was introduced in Hesthaven and Warburton (2002).

The design and analysis of discontinuous Galerkin methods already has a vast literature. A good mathematical introduction can be found in Cockburn, Karniadakis and Shu (2000), which also has a vast list of references and a historical overview. Warburton, Lomtev, Du, Sherwin, and Karniadakis (1999), and Karniadakis and Sherwin (2005) illustrate the general formulation and large scale application of such methods using spectral elements.

Appendix A

Elements of convergence theory

The single most important property of a numerical scheme for PDEs, is that the numerical solution approximates the exact solution and that the level of accuracy improves as we refine the grid in time and space. Such behavior is known as convergence. In this appendix we define the concepts and state the theorems which are necessary for convergence analysis of all numerical methods. These are the terms employed in Chapter 3 and in Chapter 8.

It is simpler to present these concepts in terms of a one-dimensional, scalar, linear, constant coefficient initial boundary value problem.

$$\begin{aligned} \frac{\partial u(x,t)}{\partial t} &= \mathcal{L}u(x,t), & x \in \mathrm{D}, \quad & t \geq 0, \\ \mathcal{B}u(x,t) &= 0, & x \in \delta\mathrm{D}, \quad & t > 0, \\ u(x,0) &= g(x), & x \in \mathrm{D}, \quad & t = 0, \end{aligned} \tag{A.1}$$

where $\mathcal{L}$ is independent of time and space. We can assume that the boundary operator $\mathcal{B}[\mathrm{D}]$ is included in the operator $\mathcal{L}$.

Definition A.1 (Wellposedness) *Equation (A.1) is* ***wellposed*** *if, for every $g \in C_0^r$ and for each time $T_0 > 0$ there exists a unique solution $u(x,t)$, which is a classical solution, and such that*

$$\|u(t)\| \leq Ce^{\alpha t}\|g\|_{H^p[D]}, \quad 0 \leq t \leq T_0$$

for $p \leq r$ and some positive constants C and α. It is ***strongly well posed*** *if this is true for $p = 0$, i.e., when $\|\cdot\|_{H^p[D]}$ is the L^2 norm.*

Let us here assume that the solution, $u(x,t)$, belongs to a Hilbert space, H, with norm $\|\cdot\|_{L^2_w[\mathrm{D}]}$, in which the problem is wellposed. The boundary operator, $\mathcal{B}$, restricts the allowable solution space to the Hilbert space $\mathrm{B} \subset \mathrm{H}$, consisting of all $u(x,t) \in \mathrm{H}$ for which $\mathcal{B}u(x,t) = 0$ on $\delta\mathrm{D}$. Wellposedness implies that the operator, $\mathcal{L}$, is a bounded operator from H into B, i.e., $\mathcal{L}[\mathrm{D}] : \mathrm{H} \to \mathrm{B}$.

The formulation of any numerical scheme for the solution of partial differential equations begins with the choice of finite dimensional space, B_N which approximates the continuous space, B. Next, we define a projection operator, $\mathcal{P}_N[\mathsf{D}] : \mathsf{H} \to \mathsf{B}_N$. This choice specifies the way in which the equation is satisfied, e.g., Galerkin, collocation, tau. Here N is the dimension of the subspace, $\mathsf{B}_N \in \mathsf{B}$, and of the projection operator, $\mathcal{P}_N$.

The projection is often defined through the method of weighted residuals (MWR), by enforcing the requirement that the numerical solution, $u_N(x, t) \in \mathsf{B}_N$, satisfies

$$\begin{aligned} \frac{\partial u_N}{\partial t} - \mathcal{L}_N u_N &= 0, \\ u_N(0) - g_N &= 0, \end{aligned} \tag{A.2}$$

where we have introduced the approximated operator, $\mathcal{L}_N[\mathsf{D}] : \mathsf{B} \to \mathsf{B}_N$, defined as $\mathcal{L}_N = \mathcal{P}_N \mathcal{L} \mathcal{P}_N$, and $g_N = \mathcal{P}_N g$.

The aim of the convergence analysis is to derive conditions under which u_N approaches u as N tends to infinity for any $t \in [0, T]$. However, since u_N and u occupy different spaces, it is unclear how to measure the difference between them. It is more natural to compare u_N and the projection, $\mathcal{P}_N u$, as they both belong to the space B_N. This approach makes sense when we consider the inequality

$$\|u(t) - u_N(t)\|_{L^2_w[\mathsf{D}]} \le \|u(t) - \mathcal{P}_N u(t)\|_{L^2_w[\mathsf{D}]} + \|u_N(t) - \mathcal{P}_N u(t)\|_{L^2_w[\mathsf{D}]}$$

and the fact that, due to the regularity of the solution and the nature of the space chosen,

$$\|u(t) - \mathcal{P}_N u(t)\|_{L^2_w[\mathsf{D}]} \to 0 \quad \text{as} \quad N \to \infty \quad \forall t \in [0, T].$$

However, the convergence rate of the first term may be different from that of the second term.

The projection of Equation (A.1) is

$$\frac{\partial \mathcal{P}_N u}{\partial t} = \mathcal{P}_N \mathcal{L} u. \tag{A.3}$$

If $u_N \in \mathsf{B}_N$, then the projection $\mathcal{P}_N u_N = u_N$. Thus, combining Equation (A.2) and Equation (A.3) yields the error equation

$$\frac{\partial}{\partial t} (\mathcal{P}_N u - u_N) = \mathcal{L}_N (\mathcal{P}_N u - u_N) + \mathcal{P}_N \mathcal{L} (\mathrm{I} - \mathcal{P}_N) u. \tag{A.4}$$

If $\mathcal{P}_N u(0) - u_N(0) = 0$ and the truncation error,

$$\mathcal{P}_N \mathcal{L} (\mathrm{I} - \mathcal{P}_N) u, \tag{A.5}$$

vanishes, we see that the error, $\mathcal{P}_N u - u_N$, is zero for all $t \in [0, T]$.

We now define the concept of convergence.

Definition A.2 (Convergence) *An approximation is convergent if*

$$\|\mathcal{P}_N u(t) - u_N(t)\|_{L^2_w[\mathsf{D}]} \to 0 \quad as \quad N \to \infty,$$

for all $t \in [0, T]$, $u(0) \in \mathsf{B}$, *and* $u_N(0) \in \mathsf{B}_N$.

It is generally very difficult to prove convergence of a specific scheme directly. However, there fortunately is an alternative avenue along which to proceed. For this, we need the concepts of *consistency* and *stability*.

Definition A.3 (Consistency) *An approximation is consistent if*

$$\left.\begin{array}{l} \|\mathcal{P}_N \mathcal{L}\,(\mathrm{I} - \mathcal{P}_N)\,u\|_{L^2_w[\mathsf{D}]} \to 0 \\ \|\mathcal{P}_N u(0) - u_N(0)\|_{L^2_w[\mathsf{D}]} \to 0 \end{array}\right. \quad as \quad N \to \infty,$$

for all $u(0) \in \mathsf{B}$ *and* $u_N(0) \in \mathsf{B}_N$.

Consistency requires that the truncation error introduced by the approximation vanishes as N approaches infinity.

Definition A.4 (Stability) *An approximation is stable if*

$$\|e^{\mathcal{L}_N t}\|_{L^2_w[\mathsf{D}]} \leq C(t), \quad \forall N,$$

with the associated operator norm

$$\|e^{\mathcal{L}_N t}\|_{L^2_w[\mathsf{D}]} = \sup_{u \in \mathsf{B}} \frac{\|e^{\mathcal{L}_N t} u\|_{L^2_w[\mathsf{D}]}}{\|u\|_{L^2_w[\mathsf{D}]}},$$

and $C(t)$ *is independent of* N *and bounded for any* $t \in [0, T]$.

Stability guarantees that the solution remains bounded as N approaches infinity. Stability of the approximation is closely related to the question of wellposedness for the partial differential equation.

These concepts are connected through one of the principal results in the convergence theory of the numerical approximation of linear partial differential equations.

Theorem A.5 (Lax–Richtmyer equivalence theorem) *A consistent approximation to a linear wellposed partial differential equation is convergent if and only if it is stable.*

Note that the consistency, stability and wellposedness of the problem has to be established in equivalent spaces, i.e., using equivalent norms, for the theorem to remain valid.

Appendix B

A zoo of polynomials

B.1 Legendre polynomials

The Legendre polynomials, $P_n(x)$, are defined as the solution to the Sturm–Liouville problem with $p(x) = 1 - x^2$, $q(x) = 0$ and $w(x) = 1$,

$$\frac{d}{dx}(1 - x^2)\frac{dP_n(x)}{dx} + n(n+1)P_n(x) = 0,$$

where $P_n(x)$ is assumed bounded for $x \in [-1, 1]$.

The Legendre polynomials are given as $P_0(x) = 1$, $P_1(x) = x$, $P_2(x) = \frac{1}{2}(3x^2 - 1)$, $P_3(x) = \frac{1}{2}(5x^3 - 3x)$ and are orthogonal in $L^2_w[-1, 1]$ with $w(x) = 1$,

$$\int_{-1}^{1} P_n(x)P_m(x)\,dx = \frac{2}{2n+1}\delta_{mn}.$$

B.1.1 The Legendre expansion

The continuous expansion is given as

$$u(x) = \sum_{n=0}^{N} \hat{u}_n P_n(x), \quad \hat{u}_n = \frac{2n+1}{2}\int_{-1}^{1} u(x)P_n(x)\,dx.$$

The discrete expansion coefficients depend on what family of Gauss points are chosen.

Legendre Gauss quadrature

$$z_j = \{z \,|\, P_{N+1}(z) = 0\}, \quad u_j = \frac{2}{(1 - z_j^2)[P'_{N+1}(z_j)]^2}, \quad j \in [0, \ldots, N].$$

The normalization constant is given as

$$\tilde{\gamma}_n = \frac{2}{2n+1},$$

resulting in the expansion coefficients,

$$\tilde{u}_n = \frac{1}{\tilde{\gamma}_n} \sum_{j=0}^{N} u(z_j) P_n(z_j) u_j.$$

Legendre Gauss–Radau quadrature

$$y_j = \{y \,|P_N(y) + P_{N+1}(y) = 0\},$$
$$v_j = \begin{cases} \frac{2}{(N+1)^2} & j = 0 \\ \frac{1}{(N+1)^2} \frac{1-y_j}{[P_N(y_j)]^2} & j \in [1, \dots, N]. \end{cases}$$

The normalization constant is given as

$$\tilde{\gamma}_n = \frac{2}{2n+1},$$

yielding the discrete expansion coefficients

$$\tilde{u}_n = \frac{1}{\tilde{\gamma}_n} \sum_{j=0}^{N} u(y_j) P_n(y_j) v_j.$$

Legendre Gauss–Lobatto quadrature

$$x_j = \{x \,|(1-x^2)P'_N(x) = 0\}, \quad w_j = \frac{2}{N(N+1)} \frac{1}{[P_N(x_j)]^2}.$$

The normalization constant is given as

$$\tilde{\gamma}_n = \begin{cases} \frac{2}{2n+1} & j \in [0, N-1] \\ \frac{2}{N} & j = N, \end{cases}$$

from which the discrete expansion coefficients become

$$\tilde{u}_n = \frac{1}{\tilde{\gamma}_n} \sum_{j=0}^{N} u(x_j) P_n(x_j) w_j.$$

B.1.2 Recurrence and other relations

Here we give a number of useful recurrence relations.

$$(n+1)P_{n+1}(x) = (2n+1)xP_n(x) - nP_{n-1}(x).$$

$$P_n(x) = \frac{1}{2n+1}P'_{n+1}(x) - \frac{1}{2n+1}P'_{n-1}(x), \quad P_0(x) = P'_1(x).$$

We also have

$$\int P_n(x)\,dx = \begin{cases} P_1(x) & n=0 \\ \frac{1}{6}(2P_2(x)+1) & n=1 \\ \frac{1}{2n+1}P_{n+1}(x) - \frac{1}{2n+1}P_{n-1}(x) & n \geq 2 \end{cases}$$

$$P_n(-x) = (-1)^n P_n(x).$$

B.1.3 Special values

The Legendre polynomials have the following special values.

$$|P_n(x)| \leq 1, \quad |P'_n(x)| \leq \frac{1}{2}n(n+1).$$

$$P_n(\pm 1) = (\pm 1)^n, \quad P'_n(\pm 1) = \frac{(\pm 1)^{n+1}}{2}n(n+1).$$

The values of P_n at the center $x=0$ behave as

$$P_{2n}(0) = (-1)^n \frac{(n-1)!}{\left(\prod_{i=1}^{n/2} 2i\right)^2}, \quad P_{2n+1}(0) = 0.$$

Finally, we obtain the results for integration,

$$\int_{-1}^{1} P_n(x)\,dx = 2\delta_{0n}.$$

B.1.4 Operators

In the following we will consider this question for Legendre expansions: Given a polynomial approximation,

$$f(x) = \sum_{n=0}^{\infty} \hat{a}_n P_n(x), \quad \mathcal{L}f(x) = \sum_{n=0}^{\infty} \hat{b}_n P_n(x),$$

where $\mathcal{L}$ is a given operator, what is the relation between $\hat{a}_n$ and $\hat{b}_n$? We will give the result for the most commonly used operators, $\mathcal{L}$.

$$\mathcal{L} = \frac{d}{dx}\colon\ \hat{b}_n = (2n+1) \sum_{\substack{p=n+1 \\ p+n \text{ odd}}}^{\infty} \hat{a}_p.$$

$$\mathcal{L} = \frac{d^2}{dx^2}\colon\ \hat{b}_n = \frac{2n+1}{2} \sum_{\substack{p=n+2 \\ p+n \text{ even}}}^{\infty} \big(p(p+1) - n(n+1)\big)\hat{a}_p.$$

$$\mathcal{L} = x\colon\ \hat{b}_n = \frac{n}{2n-1}\hat{a}_{n-1} + \frac{n+1}{2n+3}\hat{a}_{n+1}.$$

Finally, if we have

$$\frac{d^q}{dx^q}u(x) = \sum_{n=0}^{\infty} \hat{u}_n^{(q)} P_n(x),$$

then

$$\frac{1}{2n-1}\hat{u}_{n-1}^{(q)} - \frac{1}{2n+3}\hat{u}_{n+1}^{(q)} = \hat{u}_n^{(q-1)}.$$

B.2 Chebyshev polynomials

The Chebyshev polynomials of the first kind, $T_n(x)$, appear as a solution to the singular Sturm–Liouville problem with $p(x) = \sqrt{1-x^2}, q(x) = 0$ and $w(x) = (\sqrt{1-x^2})^{-1}$,

$$\frac{d}{dx}\left(\sqrt{1-x^2}\frac{dT_n(x)}{dx}\right) + \frac{n^2}{\sqrt{1-x^2}}T_n(x) - 0,$$

where $T_n(x)$ is assumed bounded for $x \in [-1, 1]$.

The Chebyshev polynomials may be given on explicit form as

$$T_n(x) = \cos(n \arccos x).$$

Thus, $T_0(x) = 1$, $T_1(x) = x$, $T_2(x) = 2x^2 - 1$, $T_3(x) = 4x^3 - 3x$, etc.

The Chebyshev polynomials are orthogonal in $L^2_w[-1, 1]$,

$$\int_{-1}^{1} T_n(x)T_m(x)\frac{1}{\sqrt{1-x^2}}\,dx = \frac{\pi}{2}c_n\delta_{mn},$$

where

$$c_n = \begin{cases} 2 & n = 0, \\ 1 & \text{otherwise.} \end{cases}$$

B.2.1 The Chebyshev expansion

The continuous expansion is given as

$$u(x) = \sum_{n=0}^{N} \hat{u}_n T_n(x), \quad \hat{u}_n = \frac{2}{\pi c_n} \int_{-1}^{1} u(x) T_n(x) \frac{1}{\sqrt{1-x^2}} dx,$$

where the details of the discrete expansion depend on which family of Gauss points are chosen.

Chebyshev Gauss quadrature

$$z_j = -\cos\left(\frac{(2j+1)\pi}{2N+2}\right), \quad u_j = \frac{\pi}{N+1}, \quad j \in [0, \ldots, N].$$

The normalization constant is given as

$$\tilde{\gamma}_n = \begin{cases} \pi & n = 0 \\ \frac{\pi}{2} & n \in [1, \ldots, N], \end{cases}$$

with the discrete expansion coefficients being

$$\tilde{u}_n = \frac{1}{\tilde{\gamma}_n} \sum_{j=0}^{N} u(z_j) T_n(z_j) u_j.$$

Chebyshev Gauss–Radau quadrature

$$y_j = -\cos\left(\frac{2j\pi}{2N+1}\right), \quad v_j = \begin{cases} \frac{\pi}{2N+1} & j = 0 \\ \frac{2\pi}{2N+2} & j \in [1, \ldots, N]. \end{cases}$$

The normalization constant is given as

$$\tilde{\gamma}_n = \begin{cases} \pi & n = 0 \\ \frac{\pi}{2} & n \in [1, \ldots, N], \end{cases}$$

yielding the discrete expansion coefficients,

$$\tilde{u}_n = \frac{1}{\tilde{\gamma}_n} \sum_{j=0}^{N} u(y_j) T_n(y_j) v_j.$$

Chebyshev Gauss–Lobatto quadrature

$$x_j = -\cos\left(\frac{\pi j}{N}\right), \quad w_j = \begin{cases} \frac{\pi}{2N} & j = 0, N \\ \frac{\pi}{N} & j \in [1, \ldots, N-1]. \end{cases}$$

The normalization constant is given as

$$\tilde{\gamma}_n = \begin{cases} \frac{\pi}{2} & j \in [1, \ldots, N-1] \\ \pi & j = 0, N, \end{cases}$$

resulting in the discrete expansion coefficients being

$$\tilde{u}_n = \frac{1}{\tilde{\gamma}_n} \sum_{j=0}^{N} u(x_j) T_n(x_j) w_j.$$

B.2.2 Recurrence and other relations

The number of recurrence relations is large and we will only give the most useful ones.

$$T_{n+1}(x) = 2x T_n(x) - T_{n-1}(x).$$

$$T_n = \frac{1}{2(n+1)} T'_{n+1}(x) - \frac{1}{2(n-1)} T'_{n-1}(x), \quad T_0(x) = T'_1(x).$$

Other useful relations are:

$$2T_n^2(x) = 1 + T_{2n}(x).$$

$$2T_n(x)T_m(x) = T_{|n+m|}(x) + T_{|n-m|}(x).$$

$$\int T_n(x)\,dx = \begin{cases} T_1(x) & n = 0 \\ \frac{1}{4}(T_2(x) + 1) & n = 1 \\ \frac{1}{2(n+1)} T_{n+1}(x) - \frac{1}{2(n-1)} T_{n-1}(x) & n \geq 2. \end{cases}$$

$$T_n(-x) = (-1)^n T_n(x).$$

B.2.3 Special values

From the definition of the Chebyshev polynomials we may make the following observations.

$$|T_n(x)| \leq 1, \quad |T'_n(x)| \leq n^2.$$

$$\frac{d^q}{dx^q} T_n(\pm 1) = (\pm 1)^{n+q} \prod_{k=0}^{q-1} \frac{n^2 - k^2}{2k + 1},$$

with the special cases

$$T_n(\pm 1) = (\pm 1)^n, \quad T_n'(\pm 1) = (\pm 1)^{n+1} n^2.$$

The values of T_n at the center $x = 0$ behave as

$$T_{2n}(0) = (-1)^n, \quad T_{2n+1}(0) = 0.$$
$$T_{2n}'(0) = 0, \quad T_{2n+1}'(0) = (-1)^n n.$$

Finally, we obtain the results for integration,

$$\int_{-1}^{1} T_n(x)\,dx = \begin{cases} -\frac{2}{n^2-1} & n \text{ even} \\ 0 & n \text{ odd.} \end{cases}$$

B.2.4 Operators

Now we consider the following question for Chebyshev expansions: Given a polynomial approximation,

$$f(x) = \sum_{n=0}^{\infty} \hat{a}_n T_n(x), \quad \mathcal{L} f(x) = \sum_{n=0}^{\infty} \hat{b}_n T_n(x),$$

where $\mathcal{L}$ is a given operator, what is the relation between $\hat{a}_n$ and $\hat{b}_n$? We give the result for the most commonly used operators, $\mathcal{L}$.

$$\mathcal{L} = \frac{d}{dx}\colon\ \hat{b}_n = \frac{2}{c_n} \sum_{\substack{p=n+1 \\ p+n \text{ odd}}}^{\infty} p\hat{a}_p.$$

$$\mathcal{L} = \frac{d^2}{dx^2}\colon\ \hat{b}_n = \frac{1}{c_n} \sum_{\substack{p=n+2 \\ p+n \text{ even}}}^{\infty} p(p^2 - n^2)\hat{a}_p.$$

$$\mathcal{L} = \frac{d^3}{dx^3}\colon\ \hat{b}_n = \frac{1}{4c_n} \sum_{\substack{p=n+3 \\ p+n \text{ odd}}}^{\infty} p\left(p^2(p^2-2) - 2p^2n^2 + (n^2-1)^2\right)\hat{a}_p.$$

$$\mathcal{L} = \frac{d^4}{dx^4}\colon\ \hat{b}_n = \frac{1}{24c_n} \sum_{\substack{p=n+4 \\ p+n \text{ even}}}^{\infty} p\left(p^2(p^2-4)^2 - 3p^4n^2 + 3p^2n^4 - n^2(n^2-4)^2\right)\hat{a}_p.$$

$$\mathcal{L} = x\colon\ \hat{b}_n = \frac{1}{2}(c_{n-1}\hat{a}_{n-1} + \hat{a}_{n+1}).$$

$$\mathcal{L} = x^2\colon\ \hat{b}_n = \frac{1}{4}(c_{n-2}\hat{a}_{n-2} + (c_n + c_{n-1})\hat{a}_n + \hat{a}_{n+2}).$$

Finally, if we have

$$\frac{d^q}{dx^q}u(x) = \sum_{n=0}^{\infty} \hat{u}_n^{(q)} T_n(x),$$

then

$$c_{n-1}\hat{u}_{n-1}^{(q)} - \hat{u}_{n+1}^{(q)} = 2n\hat{u}_n^{(q-1)}.$$

Bibliography

S. Abarbanel and D. Gottlieb, Information content in spectral calculations. In *Progress and Supercomputing in Computational Fluid Dynamics*, eds. E. M. Murman and S. S. Abarbanel, No. 6 in Proceedings of U.S.–Israel Workshop. Boston: Birkhauser (1985), pp. 345–356.

S. Abarbanel, D. Gottlieb, and E. Tadmor, Spectral methods for discontinuous problems. In *Numerical Methods for Fluid Dynamics II*, eds. K. W. Morton and M. J. Baines. Oxford: Clarendon Press (1986), pp. 129–153.

O. Andreassen, I. Lie, and C. E. Wasberg, The spectral viscosity method applied to simulation of waves in a stratified atmosphere. *Journal of Computational Physics*, **110** (1994), 257–273.

O. Andreassen and I. Lie, Simulation of acoustical and elastic waves and interaction. *Journal of the Acoustical Society of America*, **95** (1994), 171–186.

R. Archibald and A. Gelb, A method to reduce the gibbs ringing artifact in MRI scans while keeping tissue boundary integrity. *IEEE Medical Imaging*, **21**: 4 (2002).

R. Archibald and A. Gelb, Reducing the effects of noise in image reconstruction. *Journal of Scientific Computing*, **17** (2002), 167–180.

R. Archibald, K. Chen, A. Gelb, and R. Renaut, Improving tissue segmentation of human brain MRI through pre-processing by the Gegenbauer reconstruction method. *NeuroImage*, **20**:1 (2003), 489–502.

J. M. Augenbaum, An adaptive pseudospectral method for discontinuous problems. *Applied Numerical Mathematics*, **5** (1989), 459–480.

M. D. Baker, Endre Süli, and Antony F. Ware, *Stability and convergence of the spectral Lagrange–Galerkin method for periodic and non-periodic convection-dominated diffusion problems.* Technical Report 92/19, Oxford University Computing Laboratory, Numerical Analysis Group, (1992).

S. Balachandar and D. A. Yuen, Three-dimensional fully spectral numerical method for mantle convection with depth-dependent properties. *Journal of Computational Physics*, **113** (1994), 62–74.

C. Basdevant, M. Deville, P. Haldenvvang, J. M. Lacroix, D. Orlandi, A. Patera, R. Peyret, and J. Quazzani, Spectral and finite difference solutions of Burgers' equation. *Comput & Fluids*, **14** (1986), 23–41.

A. Bayliss, A. Class, and B. J. Matkowsky, Roundoff error in computing derivatives using the Chebyshev differentiation matrix. *Journal of Computational Physics*, **116** (1994), 380–383.

A. Bayliss, D. Gottlieb, B. J. Matkowsky, and M. Minkoff, An adaptive pseudo-spectral method for reaction-diffusion problems. *Journal of Computational Physics*, **81** (1989), 421–443.

A. Bayliss and B. Matkowsky, Fronts, relaxation oscillations, and period-doubling in solid fuel combustion. *Journal of Computational Physics*, **71** (1987), 147–168.

A. Bayliss and E. Turkel, Mappings and accuracy for Chebyshev pseudospectral approximations. *Journal of Computational Physics*, **101** (1992), 349–359.

A. Bayliss, A. Class, and B. J. Matkowsky, Adaptive approximation of solutions to problems with multiple layers by Chebyshev pseudo-spectral methods. *Journal of Computational Physics*, **116** (1995), 160–172.

F. Ben Belgacem and Y. Maday, A spectral element methodology tuned to parallel implementations. In *Analysis, Algorithms and Applications of Spectral and High Order Methods for Partial Differential Equations*, Christine Bernardi and Yvon Maday, eds. Selected Papers from the International Conference on Spectral and High Order Methods (ICOSAHOM '92). Amsterdam, (1994).

C. Bernardi and Y. Maday, A collocation method over staggered grids for the Stokes problem. *International Journal of Numerical Methods in Fluids*, **8** (1988), 537–557.

C. Bernardi and Y. Maday, Properties of some weighted sobolev spaces and application to spectral approximations. *SIAM Journal on Numerical Analysis*, **26** (1989), 769–829.

C. Bernardi and Y. Maday, *Approximations Spectrales de Problemes aux Limites Elliptiques*. Paris: Springer-Verlag, (1992).

C. Bernardi and Y. Maday, Polynomial interpolation results in Sobolev spaces. *Journal of Computational Applied Mathematics*, **43** (1992), 53–80.

C. Bernardi and Y. Maday, Spectral methods. In *Handbook of Numerical Analysis V*, ed. P. G. Ciarlet and J. L. Lions. Amsterdam: Elsevier Sciences, (1999).

M. Berzins and P. M. Dew, A note on the extension of the Chebyshev method to quasi-linear parabolic PDEs with mixed boundary conditions. *International Journal on Computer Mathematics*, **8** (1980), 249–263.

M. Berzins and P. M. Dew, A generalized Chebyshev method for non-linear parabolic equations in one space variable. *IMA Journal of Numerical Analysis*, **1** (1981), 469–487.

R. H. Bisseling and R. Kosloff, Optimal choice of grid points in multidimensional pseudospectral Fourier methods. *Journal of Computational Physics*, **76** (1988), 243–262.

K. Black, Polynomial collocation using a domain decomposition solution to parabolic PDEs via the penalty method and explicit/implicit time-marching. *Journal of Scientific Computing*, **7** (1992), 313–338.

P. Bontoux, B. Bondet De La Bernardie, and B. Roux, Spectral methods for natural convection problems. In *Numerical Methods for Coupled Problems*, eds. E. Hinton, P. Bettess, and R. W. Lewis. Swansea: Pineridge, (1981), pp. 1018–1030.

J. P. Boyd, Two comments on filtering (artificial viscosity) for Chebyshev and Legendre spectral and spectral element methods: preserving boundary conditions and interpretation of the filter as a diffusion. *Journal of Computational Physics*, **143** (1998), 283–288.

J. P. Boyd, *Chebyshev and Fourier Spectral Methods*, 2nd edn. New York: Dover Publishers, (2000).

N. Bressan and Alfio Quarteroni, Analysis of Chebyshev collocation methods for parabolic equations. *SIAM Journal of Numerical Analysis*, **23** (1986), 1138–1154.

K. S. Breuer and R. M. Everson, On the errors incurred calculating derivatives using Chebyshev polynomials. *Journal of Computational Physics*, **99** (1992), 56–67.

D. L. Brown and M. L. Minion, Performance of under-resolved two-dimensional flow simulations. *Journal of Computational Physics*, **122** (1995), 165–183.

W. Cai, D. Gottlieb, and C.-W. Shu, Essentially nonoscillatory spectral Fourier methods for shock wave calculation. *Mathematics of Computation*, **52** (1989), 389–410.

W. Cai, D. Gottlieb, and C. W. Shu, On one-sided filters for spectral Fourier approximation of discontinuous functions. *SIAM Journal of Numerical Analysis*, **29** (1992), 905–916.

W. Cai and C.-W. Shu, Uniform high-order spectral methods for one- and two-dimensional Euler equations. *Journal of Computational Physics*, **104** (1993), 427–443.

C. Canuto, Boundary conditions in Legendre and Chebyshev methods. *SIAM Journal of Numerical Analysis*, **23** (1986), 815–831.

C. Canuto and A. Quarteroni, *Approximation Results for Orthogonal Polynomials in Sobolev Spaces*, Math. Comp. **38**(1982), pp. 67–86.

C. Canuto, and A. Quarteroni, *Error Estimates for Spectral and Pseudospectral Approximations of Hyperbolic Equations*, SIAM J. Numer. Anal. **19**(1982), pp. 629–642.

C. Canuto and A. Quarteroni, Spectral and pseudo-spectral methods for parabolic problems with nonperiodic boundary conditions. *Calcolo*, **18** (1981), 197–218.

C. Canuto and A. Quarteroni, Spectral methods for hyperbolic equations. *Rendiconti del Seminario Matermatico Universitae Politecnico di Torino*, **39** (1981), 21–31.

C. Canuto, M. Y. Hussaini, A. Quarteroni, and T. A. Zang, *Spectral Methods in Fluid Dynamics. Springer Series in Computational Physics*. New York: Springer-Verlag, (1988).

C. Canuto, M. Y. Hussaini, A. Quarteroni, and T. A. Zang, *Spectral Methods – Fundamentals in Single Domains*. Springer Verlag, 2006.

A. Canuto and A. Quarteroni, (eds.), Spectral and high order methods for partial differential equations. *Proceedings of the Icosahom 1989 Conference*. Como, Italy. Elsevier Science (1990).

W.-M. Cao and B.-Y. Guo, A Fourier–Chebyshev spectral method for three-dimensional vorticity equations with unilaterally periodic boundary conditions. *Applied Mathematics Journal of Chinese Universities*, **7** (1992), 350–366.

M. H. Carpenter and D. Gottlieb, Spectral methods on arbitrary grids. *Journal of Computational Physics*, **129** (1996), 74–86.

G. Q. Chen, Q. Du, and E. Tadmor, Spectral viscosity approximations to multidimensional scalar conservation laws. *Mathematics of Computation*, **61** (1993), 629–643.

B. Cockburn, G. Karniadakis and C.-W. Shu (eds.), Discontinuous Galerkin Methods: Theory, Computation and Applications. *Lecture Notes in Computational Science and Engineering*, **11**. Springer, (2000).

L. Dettori and B. Yang, On the Chebyshev method for parabolic and hyperbolic equations. *M^2 AN*, **30** (1996), 907–920.

M. Deville, P. F. Fischer, and E. Mund, *High-Order Methods in Incompressible Fluid Flows.* Cambridge: Cambridge University Press, (2002).

P. G. Dinesen, J. S. Hesthaven, and J. P. Lynov, A pseudospectral collocation time-domain method for diffractive optics. *Applied Numerical Mathematics*, **33** (2000), 199–206.

W. S. Don, Numerical study of pseudospectral methods in shock wave applications. *Journal of Computational Physics*, **110** (1994), 103–111.

W. S. Don and D. Gottlieb, Spectral simulation of unsteady flow past a cylinder. *Computer Methods in Applied Mechanics and Engineering*, **80** (1990), 39–58.

W. S. Don and D. Gottlieb, The Chebyshev–Legendre method: implementing Legendre methods on Chebyshev points. *SIAM Journal on Numerical Analysis*, **31** (1994), 1519–1534.

W. S. Don and D. Gottlieb, *High order methods for complicated flows interacting with shock waves.* AIAA 97-0538, Reno Nevada, (1997).

W. S. Don and D. Gottlieb, Spectral simulation of supersonic reactive flows. *SIAM Journal on Numerical Analysis*, **35** (1998), 2370–2384.

W. S. Don and C. B. Quillen, Numerical Simulation of Reactive Flow. Part I: Resolution. *Journal of Computional Physics*, **122** (1995), 244–265.

W. S. Don and A. Solomonoff, Accuracy enhancement for higher derivatives using Chebyshev collocation and a mapping technique. *SIAM Journal on Scientific Computing*, (1995).

W. S. Don, D. Gottlieb, and J. H. Jung, A multidomain spectral method for supersonic reactive flows. *Journal of Computational Physics*, **192** (2003), 325–354.

M. Dubiner, Asymptotic analysis of spectral methods. *Journal of Scientific Computing*, **2** (1987), 3–32.

M. Dubiner, Spectral methods on triangles and other domains. *Journal of Scientific Computing*, **6** (1991), 345–390.

L. Ferracina and M. N. Spijker, An extension and analysis of the Shu–Osher representation of Runge–Kutta Methods. *Mathematics of Computation*, **74** (2005), 201–219.

P. F. Fischer and A. T. Patera, Parallel spectral element methods for the incompressible Navier–stokes equations. In *Solution of Super Large Problems in Computational Mechanics*, eds. J. H. Kane and A. D. Carlson. New York: Plenum, (1989).

P. F. Fischer and A. T. Patera, Parallel spectral element solutions of eddy-promoter channel flow. In *Proceedings of the European Research Community on Flow Turbulence and Combustion Workshop, Laussane, Switzerland.* Cambridge: Cambridge University Press, (1992).

D. Fishelov, Spectral methods for the small disturbance equation of transonic flows. *SIAM Journal of Scientific and Statistical Computing*, **9** (1988), 232–251.

D. Fishelov, The spectrum and stability of the Chebyshev collocation operator for transonic flow. *Mathematics of Computation*, **51** (1988), 599–579.

B. Fornberg, *A Practical Guide to Pseudospectral Methods.* Cambridge: Cambridge University Press, (1996).

B. Fornberg, *On A Fourier Method for the Integration of Hyperbolic Problems*, SIAM J. Numer. Anal. **12**(1975), pp. 509–528.

D. Funaro, A multidomain spectral approximation of elliptic equations. *Numerical Method for Partial Differential Equations*. **2**:3 (1986), 187–205.

D. Funaro, Computing the inverse of the Chebyshev collocation derivative matrix. *SIAM Journal of Scientific and Statistical Computing*, **9** (1988), 1050–1058.

D. Funaro, Domain decomposition methods for pseudo spectral approximations. Part I. Second order equations in one dimension. *Numerische Mathematik*, **52** (1988), 329–344.

D. Funaro, Pseudospectral approximation of a PDE defined on a triangle. *Applied Mathematics and Computation*, **42** (1991), 121–138.

D. Funaro, Polynomial Approximation of Differential Equations. *Lecture Notes in Physics*, **8**. Berlin: Springer-Verlag, (1992).

D. Funaro, Some remarks about the collocation method on a modified Legendre grid. *Computers and Mathematics with Applications*, **33** (1997), 95–103.

D. Funaro and D. Gottlieb, A new method of imposing boundary conditions in pseudospectral approximations of hyperbolic equations. *Mathematics of Composition*, **51** (1988), 599–613.

D. Funaro and D. Gottlieb, Convergence results for pseudospectral approximations of hyperbolic systems by a penalty-type boundary treatment. *Mathematics of Computation*, **57** (1991), 585–596.

D. Funaro and W. Heinrichs, Some results about the pseudospectral approximation of one dimensional fourth order problems. *Numerische Mathematik*, **58** (1990), 399–418.

A. Gelb, Parameter optimization and reduction of round off error for the Gegenbauer reconstruction method. *Journal of Scientific Computing*, **20** (2004), 433–459.

A. Gelb and D. Gottlieb, The resolution of the Gibbs phenomenon for spliced functions in one and two dimensions, *Computers and Mathematics with Applications*, **33** (1997), 35–58.

A. Gelb and E. Tadmor, Enhanced spectral viscosity approximations for conservation laws. *Applied Numerical Mathematics*, **33** (2000), 1–21.

A. Gelb and J. Tanner, Robust reprojection methods for the resolution of Gibbs phenomenon. *Applied and Computational Harmonic Analysis*, **20** (2006), 3–25.

F. X. Giraldo, J. S. Hesthaven, and T. Warburton, Nodal high order discontinuous Galerkin method for the spherical shallow water equations. *Journal of Computational Physics*, **181** (2002), 499–525.

G. H. Golub and J. H. Welsch, *Calculation of Gauss Quadraturs Rules*, Math. Comp **23**(1969), 221–230.

J. Goodman, T. Hou, and E. Tadmor, On the stability of the unsmoothed Fourier method for hyperbolic equations. *Numerische Mathematik*, **67** (1994), 93–129.

D. Gottlieb, The stability of pseudospectral Chebyshev methods. *Mathematics of Computation*, **36** (1981), 107–118.

D. Gottlieb, Spectral methods for compressible flow problems. In *Proceedings of the 9th International Conference on Numerical Methods in Fluid Dynamics, Saclay, France*, eds. Soubbaramayer and Boujot. *Lecture Notes in Physics*. New York: Springer-Verlag, **218** (1984), pp. 48–61.

D. Gottlieb and P. F. Fischer, On the optimal number of subdomains for hyperbolic problems on parallel computers. *International Journal of Supercomputer Applications and High Performance Computing*, **11** (1997), 65–76.

D. Gottlieb and S. Gottlieb, Spectral methods for discontinuous problems. *Proceedings of the 20th Biennial Conference on Numerical Analysis*, eds. D. F. Griffiths and G. A. Watson, (2003).

S. Gottlieb and L.-J. Gottlieb, Strong stability preserving properties of Runge–Kutta time discretization methods for linear constant coefficient operators. *Journal of Scientific Computing*, **18** (2003), 83–110.

D. Gottlieb and S. Gottlieb, Spectral methods for compressible reactive flows. *Comptes Rendus Mecanique*, **333** (2005), 3–16.

D. Gottlieb and J. S. Hesthaven, Spectral methods for hyperbolic problems. *Journal of Computational and Applied Mathematics*, **128** (2001), 83–181.

D. Gottlieb and L. Lustman, The Dufort–Frankel Chebyshev method for parabolic initial value problems. *Computers and Fluids*, **11** (1983), 107–120.

D. Gottlieb and L. Lustman, The spectrum of the Chebyshev collocation operator for the heat equation. *SIAM Journal of Numerical Analysis*, **20** (1983), 909–921.

D. Gottlieb and S. A. Orszag, *Numerical Analysis of Spectral Methods: Theory and Applications*. CBMS-NSF **26**. Philadelphia: SIAM, (1977).

D. Gottlieb and S. A. Orszag, High resolution spectral calculations of inviscid compressible flows. In *Approximation Methods for Navier–Stokes Problems*. New York: Springer-Verlag, (1980), pp. 381–398.

D. Gottlieb and C.-W. Shu, Resolution properties of the Fourier method for discontinuous waves. *Computer Methods in Applied Mechanics and Engineering*, **116** (1994), 27–37.

D. Gottlieb and C.-W. Shu, On the Gibbs phenomenon IV: Recovering exponential accuracy in a subinterval from a Gegenbauer partial sum of a piecewise analytic function. *Mathematics of Computation*, **64** (1995), 1081–1095.

D. Gottlieb and C.W. Shu, On the Gibbs phenomenon V: Recovering exponential accuracy from collocation point values of a piecewise analytic function. *Numerische Mathematik*, **71** (1995), 511–526.

D. Gottlieb and C.-W. Shu, On the Gibbs phenomenon III: Recovering exponential accuracy in a sub-interval from a spectral partial sum of a piecewise analytic function. *SIAM Journal on Numerical Analysis*, **33** (1996), 280–290.

D. Gottlieb and C.-W. Shu, On the Gibbs Phenomenon and its Resolution, *SIAM Review*, **39** (1997), 644–668.

D. Gottlieb and C.-W. Shu, A general theory for the resolution of the Gibbs phenomenon. In *Tricomi's Ideas and Contemporary Applied Mathematics, Accademia Nazionale Dei Lincey, National Italian Academy of Science*, **147** (1998), 39–48.

S. Gottlieb and C.-W. Shu, Total variation diminishing Runge–Kutta schemes. *Mathematics of Computation*, **67** (1998), 73–85.

D. Gottlieb and E. Tadmor, Recovering pointwise values of discontinuous data within spectral accuracy. In *Progress and Supercomputing in Computational Fluid Dynamics*, eds. E. M. Murman and S. S. Abarbanel. Boston: Birkhäuser (1985), pp. 357–375.

D. Gottlieb and E. Tadmor, The CFL condition for spectral approximations to hyperbolic initial-boundary value problems. *Mathematics of Computation*, **56** (1991), 565–588.

D. Gottlieb and E. Turkel, On time discretization for spectral methods. *Studies in Applied Mathematics*, **63** (1980), 67–86.

D. Gottlieb and E. Turkel, Topics in spectral methods for time dependent problems. In *Numerical Methods in Fluid Dynamics*, ed. F. Brezzi. New York: Springer-Verlag (1985), pp. 115–155.

D. Gottlieb and E. Turkel, *Spectral Methods for Time Dependent Partial Differential. Equations.* 1983 CIME Session (Italy), *Lecture Notes in Mathematics.* Springer-Verlag, **1127** (1985), 115–155.

D. Gottlieb and C. E. Wasberg, Optimal strategy in domain decomposition spectral methods. The Ninth International Conference on Domain Decomposition Methods for Wave-like Phenomena. *Siam Journal on Scientific Computing*, **22**:2 (2002), 617–632.

D. Gottlieb, L. Lustman, and S. A. Orszag, Spectral calculations of one-dimensional inviscid compressible flows. *SIAM Journal on Scientific Computing*, **2** (1981), 296–310.

D. Gottlieb, S. A. Orszag, and E. Turkel, Stability of pseudospectral and finite difference methods for variable coefficient problems. *Mathematics of Computation*, **37** (1981), 293–305.

D. Gottlieb, M. Y. Hussaini, and S. A. Orszag, Theory and application of spectral methods. In *Spectral Methods for Partial Differential Equations*, eds. R. G. Voigt, D. Gottlieb, and M. Y. Hussaini. Philadelphia: SIAM (1984), pp. 1–54.

D. Gottlieb, L. Lustman, and C. L. Streett, Spectral methods for two-dimensional shocks. In *Spectral Methods for Partial Differential Equations*, eds. R. G. Voigt, D. Gottlieb, and M. Y. Hussaini. Philadelphia: SIAM (1984), pp. 79–96.

D. Gottlieb, L. Lustman, and E. Tadmor, Stability analysis of spectral methods for hyperbolic initial-value problems. *SIAM Journal of Numerical Analysis*, **24** (1987), 241–258.

D. Gottlieb, L. Lustman, and E. Tadmor, Convergence of spectral methods for hyperbolic initial-value problems. *SIAM Journal of Numerical Analysis*, **24** (1987), 532–537.

S. Gottlieb, C.-W. Shu and E. Tadmor, Strong stability preserving high-order time discretization methods, *SIAM Review*, **43** (2001), 89–112.

D. Gottlieb, C.-W. Shu, A. Solomonoff, and H. Vandeven, On the Gibbs phenomenon I: recovering exponential accuracy from the Fourier partial sum of a nonperiodic analytic function. *Journal of Computational and Applied Mathematics*, **43** (1992), 81–98.

S. Gottlieb, D. Gottlieb and C.-W. Shu, Recovering high order accuracy in WENO computations of steady state hyperbolic systems. *Journal of Scientific Computing*, to appear 2006.

D. Gottlieb, J. S. Hesthaven, G. E. Karniadakis, and C.-W. Shu, eds. Proceedings of the 6th International Conference on Spectral and High-Order Methods. *Journal of Scientific Computing*, to appear (2006).

B.-Y. Guo, *Spectral Methods and Their Applications.* Singapore: World Scientific, (1998).

B. Gustafsson, H. O. Kreiss, and J. Oliger, *Partial Differential Equations and Difference Approximations*. John Wiley & Sons, 2001.

J. S. Hesthaven, A Stable Penalty Method for the compressible Navier–Stokes equations: II. One-dimensional domain decomposition schemes. *SIAM Journal on Scientific Computing*, **18** (1997), 658–685.

J. S. Hesthaven, From electrostatics to almost optimal nodal sets for polynomial interpolation in a simplex. *SIAM Journal on Numerical Analysis*, **35** (1998), 655–676.

J. S. Hesthaven, A Stable Penalty Method for the Compressible Navier-Stokes Equations: III. Multidimensional Domain Decomposition Schemes. *SIAM Journal on Scientific Computing*, **20** (1999), 62–93.

J. S. Hesthaven, Spectral Penalty Methods. *Applied Numerical Mathematics*, **33** (2000), 23–41.

J. S. Hesthaven and D. Gottlieb, A stable penalty method for the compressible Navier–Stokes equations. I. Open boundary conditions. *SIAM Journal on Scientific Computing*, **17** (1996), 579–612.

J. S. Hesthaven and D. Gottlieb, Stable spectral methods for conservation laws on triangles with unstructured grids. *Computer Methods in Applied Mechanic Engineering*, **175** (1999), 361–381.

J. S. Hesthaven and C. H. Teng, Stable spectral methods on tetrahedral elements. *SIAM Journal on Scientific Computing*, **21** (2000), 2352–2380.

J. S. Hesthaven and T. Warburton, High-order nodal methods on unstructured grids. I. Time-domain solution of Maxwell's equations. *Journal of Computational Physics*, **181** (2002), 186–221.

J. S. Hesthaven and M. Kirby, *Filering in Legendre Spectral Methods*, Math. Comp. 2006.

J. S. Hesthaven and T. Warburton, Discontinuous Galerkin Methods for the Time-dependent Maxwell's equations: an introduction. *ACES Newsletter*, **19** (2004), 10–29.

J. S. Hesthaven and T. Warburton, High-order nodal discontinuous Galerkin methods for the Maxwell eigenvalue problem. *Royal Society of London Series A*, **362** (2004), 493–524.

J. S. Hesthaven, P. G. Dinesen, and J. P. Lynov, Spectral collocation time-domain modeling of diffractive optical elements. *Journal of Computational Physics*, **155** (1999), 287–306.

I. Higueras, Representations of Runge–Kutta methods and strong stability preserving methods. *SIAM Journal on Numerical Analysis*, **43** (2005), 924–948.

M. Y. Hussaini, D. A. Kopriva, M. D. Salas, and T. A. Zang, Spectral methods for the Euler equations: Part I: Fourier methods and shock-capturing. *AIAA Journal*, **23** (1985), 64–70.

A. Ilin and R. Scott, (eds.), Proceedings of the Third International Conference on Spectral and High Order Methods, Houston, Texas, 5–9 June 1995. *Special Issue of the Houston Journal of Mathematics*, (1996).

K. Ito and R. Teglas, Legendre–Tau approximations for functional differential equations. *SIAM Journal for Control and Optimization*, **24** (1986), 737–759.

K. Ito and R. Teglas, Legendre–Tau approximations for functional differential equations: Part 2: The linear quadratic optimal control problem. *SIAM Journal for Control and Optimization*, **25** (1987), 1379–1408.

A. Karageorghis, A note on the Chebyshev coefficients of the general order derivative of an infinitely differentiable function. *Journal of Computational and Applied Mathematics*, **21** (1988), 129–132.

A. Karageorghis, A note on the satisfaction of the boundary conditions for chebyshev collocation methods in rectangular domains. *Journal of Scientific Computing*, **6** (1991), 21–26.

A. Karageorghis and T. N. Phillips, On the coefficients of differentiated expansions of ultraspherical polynomials. *Applied Numerical Mathematics*, **9** (1992), 133–141.

G. E. Karniadakis and S. J. Sherwin, *Spectral/hp Methods in Computational Fluid Dynamics*. 2nd edn. Oxford: Oxford University Press, (2005).

D. A. Kopriva, *A Practical Assessment of Spectral Accuracy for Hyperbolic Problems with Discontinuities*, J. Sci. Comput. **2**(1987), pp. 249–262.

D. A. Kopriva, T. A. Zang, and M. Y. Hussaini, Spectral methods for the Euler equations: The blunt body revisited. *AIAA Journal*, **29** (1991), 1458–1462.

R. Kosloff and H. Tal-Ezer, A modified Chebyshev pseudospectral method with an $O(1/n)$ time step restriction. *Journal of Computational Physics*, **104** (1993), 457–469.

H. O. Kreiss and J. Oliger, Comparison of accurate methods for the integration of hyperbolic problems. *Tellus*, **24** (1972), 199–215.

H. O. Kreiss and J. Oliger, Stability of the Fourier method, *SIAM Journal on Numerical Analysis*, **16** (1979), 421–433.

P. D. Lax, Accuracy and resolution in the computation of solutions of linear and nonlinear equations. In *Proceedings of Recent Advances in Numerical Analysis*. New York: Academic Press (1978), pp. 107–117.

D. Levy and E. Tadmor, From semi-discrete to fully-discrete: stability of Runge–Kutta schemes by the energy method. *SIAM Review*, **40** (1998), 40–73.

I. Lomtev, C. B. Quillen and G. E. Karniadakis, Spectral/hp methods for viscous compressible flows on unstructured 2D meshes. *Journal of Computational Physics*, **144** (1998), 325–357.

L. Lurati, *Pade-Gegenbauer Suppression of Runge Phenomenon in the Diagonal Limit of Gegenbauer Approximations*, J. Comput. Phys. 2006 – to appear.

L. Lustman, The time evolution of spectral discretizations of hyperbolic systems. *SIAM Journal of Numerical Analysis*, **23** (1986), 1193–1198.

H.-P. Ma, Chebyshev–Legendre spectral viscosity method for nonlinear conservation laws. *SIAM Journal of Numerical Analysis*, **35** (1998), 869–892.

H.-P. Ma, Chebyshev–Legendre super spectral viscosity method for nonlinear conservation laws. *SIAM Journal of Numerical Analysis*, **35** (1998), 893–908.

H.-P. Ma and B.-Y. Guo, The Chebyshev spectral method for Burgers-like equations. *Journal of Computational Mathematics*, **6** (1988), 48–53.

M. G. Macaraeg and C. L. Streett, Improvements in spectral collocation discretization through a multiple domain technique. *Applied Numerical Mathematics*, **2** (1986), 95–108.

Y. Maday and A. Quarteroni, Legendre and Chebyshev spectral approximations of Burgers' equation. *Numerische Mathematik*, **37** (1981), 321–332.

Y. Maday and E. Tadmor, Analysis of the spectral vanishing viscosity method for periodic conservation laws. *SIAM Journal on Numerical Analysis*, **26** (1989), 854–870.

Y. Maday, S. M. Ould Kaber, and E. Tadmor, Legendre pseudospectral viscosity method for nonlinear conservation laws. *SIAM Journal of Numerical Analysis*, **30** (1992), 321–342.

A. Majda, J. McDonough, and S. Osher, The Fourier Method for nonsmooth initial data. *Mathematics of Computation*, **32** (1978), 1041–1081.

C. Mavriplis and J. M. Rosendale, *Triangular spectral elements for incompressible fluid flow.* ICASE Report 93–100, NASA Langley Research Center, Virginia: Hampton, (1993).

B. E. McDonald, Flux-corrected pseudospectral method for scalar hyperbolic conservation laws. *Journal of Computational Physics*, **82** (1989), 413–428.

M. S. Min and D. Gottlieb, On the convergence of the Fourier approximation for eigenvalues and eigenfunctions of discontinuous problems. *SIAM Journal on Numerical Analysis*, **40** (2002), 2254–2269.

S. A. Orszag, Galerkin approximations to flows within slabs, spheres and cylinders. *Physical Review Letters*, **26** (1971), 1100–1103.

S. A. Orszag, Comparison of pseudospectral and spectral approximation. *Studies in Applied Mathematics*, **51** (1972), 253–259.

S. A. Orszag, Spectral methods for problems in complex geometries. *Journal of Computational Physics*, **37** (1980), 70–92.

R. G. M. van Os and T. N. Phillips, Efficient and stable spectral element methods for predicting the flow of an XPP fluid past a cylinder. *Journal of Non-Newtonian Fluid Mechanics*, **129** (2005), 143–162.

S. Osher, Smoothing for spectral methods. In *Spectral Methods for Partial Differential Equations*, eds. R. G. Voigt, David Gottlieb, and M. Y. Hussaini. Philadelphia: SIAM (1984), pp. 209–216.

S. M. Ould Kaber, Filtering non-periodic functions. In *Analysis, Algorithms and Applications of Spectral and High Order Methods for Partial Differential Equations*, eds. Christine Bernardi and Yvon Maday. Amsterdam: North-Holland (1994), pp. 123–130.

S. M. Ould Kaber and H. Vandeven, Reconstruction of a discontinuous function from its Legendre coefficients. *Comptes Rendus Mathématiques*, **317** (1993), 527–532.

J. E. Pasciak, Spectral and Pseudospectral Methods for Advection Equations. *Mathematics of Computation*, **35** (1980), 1081–1092.

R. Peyret and T. D. Taylor, *Computational Methods for Fluid Flow.* New York: Springer-Verlag, (1983).

R. Peyret, *Spectral Methods for Incompressible Viscous Flow.* New York: Springer-Verlag, (2002).

W. D. Pilkey, (ed.) Special issue: selection of papers presented at ICOSAHOM'92. *Finite Elements in Analysis and Design archive*, **16**:3–4. Amsterdam: Elsevier Science, (1994).

A. Quarteroni, Blending Fourier and Chebyshev interpolation. *Journal of Approximation Theory*, **51** (1987), 115–126.

S. C. Reddy and L. N. Trefethen, Lax-stability of fully discrete spectral methods via stability regions and pseudo-eigenvalues. *Computater Methods in Applied Mechanics and Engineering*, **80** (1990), 147–164.

L. G. Reyna, l_2 estimates for Chebyshev collocation. *Journal of Scientific Computing*, **3** (1988), 1–24.

S. J. Ruuth and R. J. Spiteri, Two barriers on strong-stability-preserving time discretization methods. *Journal of Scientific Computing*, **17** (1990), 211–220.

L. Sakell, Pseudospectral solutions of one- and two-dimensional inviscid flows with shock waves. *AIAA Journal*, **22** (1984), 929–934.

S. Schochet, The rate of convergence of spectral-viscosity methods for periodic scalar conservation laws. *SIAM Journal of Numerical Analysis*, **27** (1990), 1142–1159.

C.-W. Shu, Total-variation-diminishing time discretizations. *SIAM Journal on Scientific and Statistical Computing*, **9** (1988), 1073–1084.

C.-W. Shu, A survey of strong stability preserving high order time discretizations. In *Collected Lectures on the Preservation of Stability under Discretization*, eds. D. Estep and S. Tavener. Philadelphia: SIAM (2002), pp. 51–65.

C.-W. Shu and S. Osher, Efficient implementation of essentially non-oscillatory shock-capturing schemes. *Journal of Computational Physics*, **77** (1988), 439–471.

C.-W. Shu and P. Wong, A note on the accuracy of spectral method applied to nonlinear conservation laws. *Journal of Scientific Computing*, **10** (1995), 357–369.

D. Sidilkover and G. Karniadakis. Non-oscillatory spectral element Chebyshev methods for shock wave calculations. *Journal of Computational Physics*, **107** (1993), 10.

A. Solomonoff, A fast algorithm for spectral differentiation. *Journal of Computational Physics*, **98** (1992), 174–177.

J. Strain, Spectral methods for nonlinear parabolic systems. *Journal of Computational Physics*, **122** (1995), 1–12.

E. Süli and A. Ware, A spectral method of characteristics for hyperbolic problems. *SIAM Journal of Numerical Analysis*, **28** (1991), 423–445.

G. Szegö, *Orthogonal Polynomials*. Colloquium Publications **23**, American Mathematical Society, Providence, RI, 1939.

E. Tadmor, The exponential accuracy of Fourier and Chebyshev differencing methods. *SIAM Journal on Numerical Analysis*, **23** (1986), 1–10.

E. Tadmor, Stability analysis of finite-difference, pseudospectral, and Fourier–Galerkin approximations for time-dependent problems. *SIAM Review*, **29** (1987), 525–555.

E. Tadmor, Convergence of spectral methods for nonlinear conservation laws. *SIAM Journal of Numerical Analysis*, **26** (1989), 30–44.

E. Tadmor, Shock capturing by the spectral viscosity method. *Computer Methods in Applied Mechanics and Engineering*, **78** (1990), 197–208.

E. Tadmor, Super viscosity and spectral approximations of nonlinear conservation laws. "Numerical Methods for Fluid Dynamics IV". *Proceedings of the 1992 Conference on Numerical Methods for Fluid Dynamics*, eds. M. J. Baines and K. W. Morton. London: Oxford University Press (1993), pp. 69–82.

E. Tadmor, Approximate solutions of nonlinear conservation laws. In *Advanced Numerical Approximation of Nonlinear Hyperbolic Equations, Lecture Notes from CIME Course Cetraro, Italy, 1997* (A. Quarteroni, ed.), *Lecture Notes in Mathematics*, Springer-Verlag, **1697** (1998), pp. 1–150.

H. Tal-Ezer, Spectral methods in time for hyperbolic equations. *SIAM Journal of Numerical Analysis*, **23** (1986), 11–26.

H. Tal-Ezer, Spectral methods in time for parabolic problems. *SIAM Journal of Numerical Analysis*, **26** (1989), 1–11.

J. Tanner and E. Tadmor, *Adaptive Mollifiers – High Resolution Recover of Piecewise Smooth Data from its Spectral Information*, Found. Comput. Math. **2**(2002), pp. 155–189.

J. Trujillo and G. E. Karniadakis, A penalty method for the vorticity–velocity formulation. *Journal of Computational Physics*, **149** (1999), 32–58.

L. N. Trefethen, Lax-stability vs. eigenvalue stability of spectral methods. In *Numerical Methods for Fluid Dynamics III*, eds. K. W. Morton and M. J. Baines. Oxford: Clarendon Press (1988), pp. 237–253.

L. N. Trefethen, *Spectral Methods in Matlab.* Philadelphia: SIAM, (2000).

L. N. Trefethen and M. R. Trummer, *An instability phenomenon in spectral methods*, SIAM J. Numer. Anal. **24**(1987), pp. 1008–1023.

H. Vandeven, Family of spectral filters for discontinuous problems. *Journal of Scientific Computing*, **8** (1991), 159–192.

J. S. Hesthaven, D. Gottlieb, and E. Turkel, Proceedings of the Fourth International Conference on Spectral and High Order Methods. *Applied Numerical Mathematics*, **33**:1–4, (2000).

Proceedings of the Fifth International Conference on Spectral and High Order Methods (Uppsala, Sweden). *Journal of Scientific Computing*, **17** (2002).

T. Warburton, I. Lomtev, Y. Du, S. Sherwin, and G. E. Karniadakis Galerkin and discontinuous Galerkin spectral/hp methods. *Computer Methods in Applied Mechanics and Engineering*, **175** (1999), 343–359.

J. Andre C. Weideman and L. N. Trefethen, The eigenvalues of second-order spectral differentiation matrices. *SIAM Journal on Numerical Analysis*, **25** (1988), 1279–1298.

B. D. Welfert, On the eigenvalues of second-order pseudospectral differentiation operators. *Computer Methods in Applied Mechanics and Engineering*, **116** (1994), 281–292.

B. D. Welfert, Generation of pseudospectral differentiation matrices, I. *SIAM Journal on Numerical Analysis*, **34** (1997), 1640–1657.

J. H. Williamson, Low-storage Runge–Kutta schemes. *Journal of Computational Physics*, **35** (1980), 48–56.

Index

aliasing error, 30, 39, 41, 42, 54, 114

Boundary conditions, 133–134
Burger's equation, 122, 127
Chebyshev polynomials, 74, 75
  bounds, 74
  recurrence relations, 75
  Rodrigues' formula, 74
  weight function, 74
collocation method, 129–133, 142–145

Derivatives, 6, 15, 17, 30, 31, 80–83
Differentiation matrices, 7, 9
  Chebyshev, 96–98
  Fourier, 32, 33, 57
  Legendre, 93–96
Dirichlet kernel, 156
discrete inner products, 54, 88
discrete norms
  Chebyshev, 89
  Legendre, 89
discrete trigonometric polynomials, 24, 25
  even expansion, 25
  odd expansion, 28
  approximation results, 27, 38, 39
discrete Legendre expansion
  approximation results, 115
discrete Chebyshev expansion
  approximation results, 115, 116
discontinuous Galerkin methods, 246–247

eigenvalue spectrum of differentiation operators
  Fourier–Galerkin approximation, 189
  polynomial approximation, 189–191
electrostatic, 99, 104, 106
elliptic problem, 126–127
energy, 55
even–odd decomposition, 207–210

fast cosine transform, 206
fast Fourier transform, 32, 34, 50, 51, 205
filters, 160–174
  and super-spectral viscosity, 151
filter family, 170
filter function, 169
finite difference, 5, 6, 10, 11
finite precision effects *see* round-off errors
fully discrete stability analysis, 191–192

Galerkin method, 117, 135
Gauss quadrature points, 108
  computation of, 210–213
Gaussian weights
  computation of, 210–213
Gegenbauer reprojection, 178–182
  for linear equations, 182–184
  for nonlinear equations, 185–186
general grids, 236
Gibbs complementary basis, 175–176
  to Fourier basis, 178
  to Legendre basis, 181
  to Gegenbauer basis, 182
Gibbs phenomenon, 154–160
  removal of, 174–182

hyperbolic equations (*see also* wave equation), 6, 55, 119, 130, 137–139, 153

interpolation, 26, 48

Jacobi polynomials, 69–72

Christoffel–Darboux, 71 (Theorem 4.4)
explicit formula, 70
normalization, 70
recurrence relations, 71 (Theorem 4.2), 72 (Theorem 4.5)
recurrence relation in terms of derivatives, 71 (Theorem 4.3)
Rodrigues' formula, 69

Lagrange polynomial interpolation, 91–92, 131
Lebesque constant, 100–102, 105
Lebesque function, 100–102
Legendre polynomials, 72
explicit formula, 73
recurrence relations, 73
rodrigues formula, 72
sturm–Liouville equation, 72

mapping techniques, 225–234
mass matrix, 118
multi-step methods, 193

parabolic equation, 119, 121, 122, 132, 136
penalty methods, 133–134, 238–239
for complex boundary conditions, 243–246
collocation methods, 241–243
Galerkin methods, 239–241
stability of, 145–150
phase error, 9, 10, 12, 13, 18
points per wavelength, 13, 16, 18, 77, 78
polynomial expansions, 79
expansion coefficients, 80
Legendre, 81
Chebyshev, 82

quadrature formulas, 29
Gauss, 85
Gauss–Lobatto, 83
Gauss–Radau, 84
Legendre–Gauss–Lobatto, 86
Legendre–Gauss–Radau, 86
Legendre–Gauss, 87
Chebyshev–Gauss–Lobatto, 87
Chebyshev–Gauss–Radau, 87
Chebyshev–Gauss, 88

round-off errors
in Fourier methods, 214–217
in polynomial methods, 217–225
Runge–Kutta methods, 193–197, 199–201
low storage, 195–196
linear stability regions, 195
strong stability preserving, 197–202

spectral methods, 5
Chebyshev, 117–123
Fourier, 16
Galerkin, 43–48
collocation, 48
tau, 123–129
stability
Chebyshev-collocation method, 142–145
Fourier–Galerkin method, 52, 53, 54
Fourier-collocation
linear hyperbolic problems, 54–62
linear parabolic problems, 62, 63, 64
nonlinear hyperbolic problems, 64, 65
Galerkin method for semi-bounded operators, 135–142
nonlinear equations, 150–152
penalty methods, 145–150
stiffness matrix, 118, 121, 122
strong stability preserving methods, 197–202
Runge–Kutta, 198–201
multi-step, 202
Sturm–Liouville problem, 67, 68
regular, 69
singular, 68
super spectral viscosity, 60, 62, 64, 65
for Chebyshev, 150
for Legendre, 150

tau method, 123–129
trigonometric polynomial expansions, 19, 20, 21, 36, 37
truncation error, 24, 41, 45
ultraspherical polynomials, 76–77
relation to Legendre and Chebyshev polynomials, 76
Rodrigues' formula, 76
Sturm–Liouville problem, 76

wave equation, 6, 9, 17, 133
weight function, 69

For EU product safety concerns, contact us at Calle de José Abascal, 56–1°,
28003 Madrid, Spain or eugpsr@cambridge.org.

www.ingramcontent.com/pod-product-compliance
Ingram Content Group UK Ltd.
Pitfield, Milton Keynes, MK11 3LW, UK
UKHW042338271225
466426UK00005B/85

9780521792110